FLY THE WING

FLY THE WING

SECOND EDITION

JIM WEBB

IOWA STATE UNIVERSITY PRESS / AMES

JIM WEBB, retired Eastern Airlines captain, learned to fly with a barnstormer at the age of thirteen. For fifteen years he worked as an Eastern Air Lines flight instructor and check pilot. During that period he trained almost 500 pilots, with no failures and well above average results. He has also conducted rating rides, proficiency checks, and instruction and checking in simulators. During World War II he was a B-24 pilot in the Central Pacific, flying thirty-three missions and earning two Distinguished Flying Crosses and four Air Medals. After the war Webb flew a variety of charter and corporate planes before joining Eastern. He has logged more than 35,000 hours in the cockpit in aircraft of all types.

Acknowledgment is made to Eastern Air Lines and to Jeppeson and Company for use of their charts and approach plates and to Allen Feldman for all the drawings.

♾ Printed on acid-free paper in the United States of America

First edition, 1971, through five printings, copyright © Iowa State University Press, Ames, Iowa 50010.

Second edition, 1990
Second printing, 1992
Third printing, 1993

Library of Congress Cataloging-in-Publication Data

Webb, Jim
    Fly the wing / Jim Webb.—2nd ed.
       p.   cm.
      ISBN 0-8138-0541-4 (alk. paper)
      1. Airplanes—Piloting. 2. Aeronautics, Commercial. I. Title.
TL710.W37   1990   89-77430
629.132′52—dc20   CIP

CONTENTS

INTRODUCTION

Flying an airplane—any kind of flying: joyriding, aerobatics, military flying, flying for hire, and particularly instrument flight—is a form of human expression. A pilot's skill, the manner in which the flight is planned and executed, and regard for safety are revealing personality traits.

Many of the preemployment procedures used by airlines to test pilot applicants and to screen them thoroughly before putting them into the cockpit may be totally unrelated to flying. Though these procedures are sometimes unfairly criticized by those already in the cockpit, they are designed solely to determine if the applicant has sufficient maturity and stability to become an airline captain. Airlines never hire copilots only—at least not intentionally. They seek captain material. In addition to aviation experience and background they examine ability to learn, to adapt, and to exercise the qualities of judgment and command that the left seat requires.

To fly demands that one adapt to a new three-dimensional environment, learn the vagaries and capricious tendencies of weather and meteorological phenomena, acquire knowledge of the atmosphere, and develop skills and techniques of control manipulation to make the craft respond smoothly and safely. The pilot must also know and thoroughly understand the complex systems and operational limitations of the craft as well as the limitations of personal skills and abilities so that the bounds of either are never exceeded. The pilot must learn the science of navigation and radio aids to navigation; develop self-discipline to accept the responsibility of command and to exercise the degree of judgment the profession requires; and learn to project forward in both time and space to stay ahead of the speeding aircraft in thinking and planning and thereby base decisions and actions upon an extension of the position in flight to a predicted position in time and space to adequately cope with the situation presently at hand.

These technical skills, this required knowledge, and the ability to exercise command judgment are not easily acquired. They come only from experience, good training, and constant practice. Because this is true, they are things that really cannot be taught. Pilots themselves must learn by analyzing every maneuver and phase of flight for desired performance; by striving constantly to improve both their technique and judgment through practice; and by being honestly self-critical in evaluating their mistakes.

I don't believe any flight instructor in the world can actually bring a pilot to a high enough degree of proficiency in these factors to meet the qualifications of an airline captain. The instructor can offer advice, demonstrate various maneuvers, criticize, and impart knowledge and skill; but the acquisition of that knowledge and skill must come from the students. They must learn first how to recognize their own mistakes and shortcomings, then how to correct them quickly and smoothly; therefore, they must teach themselves to a great degree—taking the instructor's advice and help into account—and acquire the necessary skill and ability through their own efforts.

Since flying is a form of self-expression, it may be compared to another form—music. A music teacher can teach the fundamentals, show students how to play the piano, teach them to read music, but cannot teach them how to play the piano. The instructor can only guide; learning to play must come from a student's motivation and desire to acquire the skill, and success comes in direct relation to the amount of effort put forth to achieve it.

As they progress from the fundamentals, pianists develop their own style of playing. No two concert pianists play exactly alike, yet each may very well be recognized as a master of the instrument. And so it is with pilots. No two persons think alike, or react alike, or express themselves in an identical manner; but each—with a skill based on the same theories of flight, aerodynamics, and aeronautical sciences—may well be a master pilot. In every case they must first learn the fundamentals well and then polish and refine their skills as they go on to more difficult maneuvers and advanced phases of flight.

This creates a very frustrating problem for flight

instructors. When a student does an outstanding job in the oral examination and check ride, it is attributed to skill, knowledge, and individual ability. This is as it should be. The student really deserves the credit. But if one does poorly or fails the oral examination or flight check, then it is charged to the instructor for not teaching the fundamentals, not teaching the student the procedures, and not properly preparing the student for the check. And woe unto the instructor who has several below-average students in a row! The FAA soon begins to look critically at that instructor.

Very rarely do airline flight instructors receive any thanks. Their job is to see that the students receive a rating in the aircraft. To do this, an instructor patiently and painstakingly tries to eliminate a student's weak spots and endeavors to train the applicant to a degree of proficiency that will instill the confidence needed to pass the check. The instructor's only satisfaction comes from seeing the student's success.

The instructor gets a share of criticism though. I've seen many instances of students doing a below-average or failing check ride and then criticizing their instructor, the check pilot, the FAA inspector, and the entire training department for their own display of poor airmanship. These severe critics of the training program almost invariably are of two distinct types.

The first type goes to the training center without any preparation whatsoever—hasn't opened the aircraft manual since ground school, hasn't taken time to study the performance section, has taken no interest in reviewing normal operating procedures and emergency procedures, and hasn't even looked at the training section of the manual to learn the profiles of the various maneuvers to be performed. In other words, this type hasn't even tried to learn the memory items that are required and can be learned—before ever flying the airplane—right out of the book. This type seems to think that an instructor can bore a hole in the head, pour in skill and knowledge, and turn out a superior pilot instantly. It can't be done! It requires some effort on the part of the student.

The other type is the lazy copilot, either getting the ATR or upgrading to captain. In some instances, it may even be a captain upgrading to a new piece of equipment. This is the worst type (especially if also unprepared for flight training), who really screams bloody murder when doing a poor job. However, an instructor may be prepared for the worst if time is taken to review the previous flight training records. Most flight instructors, including myself, very rarely look into the old records unless the student is below average and not making normal progress about midway through training. Then the story is an open book. This type has had trouble in every airplane ever checked out in so there's little reason to believe this time will be any different.

All instructors feel a sympathy for this type. It is a regrettable thing, a rut the student has fallen into and a hard one to get out of. It begins with being a lazy copilot, acquiring sloppy flying habits by flying in the "golden chair" with no responsibility. This type hasn't exercised command or judgment or had to make a decision—including what time to eat the crew meal—in years. For a number of years and several thousand flight hours, even when actually flying the airplane, these copilots have leaned heavily on the captain to do all the thinking and planning. They have either lost or never had sufficient ambition and motivation to take advantage of the opportunity to develop their talents under the direction and supervision of the captain. When they progress to the left seat or transfer to new equipment, they are required to think and plan on their own. Fifty percent of the evaluation for grading on the rating ride is based on command ability and judgment. In every training program they are again brought face-to-face with the fact that they have become below-average pilots. No wonder they scream so loud. But seniority alone does not make a good pilot or captain.

This industry was built by the pilots. The slogan of the Air Line Pilot's Association, Schedule with Safety, tells but a small part of the honor and tradition of the airline piloting profession. Those whom we follow built the tradition and bequeathed a noble heritage that the present-day pilot would do well to emulate.

All of us know and have flown with pilots whom we recognize as being masters of the trade. They demand and receive respect by virtue of their skill and ability; they handle emergency situations with a calm assurance; their flight planning and thinking ahead for contingencies are without parallel; they fly the plane as if they were a part of it, and it seems that the controls move but slightly as they maneuver the craft; the instruments seem glued into the proper place and never waver on an instrument approach. They are dedicated, cautious, and love to fly; and flying seems ridiculously easy for them.

I've known, and had the good fortune to fly copilot for many such pilots who would freely impart their skill and knowledge to any copilot showing a willingness to learn. They have a carefully guarded secret to their ability, which I finally learned. That is the reason for this book—to pass that secret on to others. You will find it in the material that follows, in the title, and partially here in the introduction.

Most of those pilots were old-timers whose experience covered all the history of commercial aviation—from the era of the barnstormer and airmail pilot through all the pioneering stages of airline development. These men have long since left the scene. They were replaced by pilots from World War II, whom the old-timers trained. The present age restriction has closed the cockpit to them.

There are many other pilots who also learned the hard way—those who were too young during the barnstorming era and too old for military service and flight training. These were the young men between wars who wanted to fly and needed the necessary experience to get employment during the depression years. They took any job available—flunkied around airports, worked as line boys, did anything at all, preferably around an airport, in exchange for flight time. Now they too, due to the age 60 rule, have left the cockpit.

Airline pilots today are mostly products of the military, with those of the Vietnam era filling the left seat. Even with this background, the overall experience level is dropping. There is also another factor weakening the crew experience level. Since the cockpit force became a three-pilot crew in 1963, the third person has spent too much time as a flight engineer and lost many pilot skills due to lack of practice. The engineer of a ship doesn't learn the skills necessary to meet the qualifications to command or to be first officer of a ship at sea, and neither does the second officer who monitors system operation of a large aircraft practice and learn the skills required to fly it and make decisions. I once had, serving as a second officer in a 727, a former Air Force major, squadron commander, instructor, and check airman in Boeing 707 tankers. Some five years later he upgraded to first officer and, when flying copilot for me, was so slow and had such poor instrument scan that I was virtually flying solo. What a waste! If you were to check the reports of air carrier accidents, you would find that too many of them have occurred when the copilot was flying.

All the good old master pilots, had one trait in common—motivation. They learned the fundamentals well, recognized the value of *any* flight experience, and stored away their experiences. They progressed with the industry to newer and more sophisticated planes, but they also applied the techniques they had learned in barnstorming and aerobatics from the days of needle, ball, and airspeed. They modified and improvised upon these lessons from the past to adapt to present-day concepts. They knew that any airplane—a J-3 Cub, a Pitcairn Mailwing, or a DC-8—is just a powered wing! Procedures may change, aircraft may become more sophisticated and complex, techniques may vary slightly due to individual flight characteristics of more modern aircraft, but the theory of flight never changes. The three controls used to fly the J-3 and Mailwing—elevator, rudder, and aileron—are still used to fly the DC-8.

They also believed in themselves as pilots. All good pilots know their own ability and take pride in their flying. They have to have self-confidence or they will be unsure of themselves, plagued by doubts, and never become outstanding pilots.

It has been said that pilots who have been in aviation for 10 years, having reached the age of 32, are the type of pilot they will be for the rest of their lives. If excellent, average, or below average at that stage, they will be exactly the same at age 60, if health still permits them to fly. What they learn during this period—the skills of flight proficiency and the habits they develop in their youth and during early stages of their flying careers—will never be changed.

I believe this to be true, and that's another reason I'm writing this. It's too late perhaps to help those who, through their own laziness, have allowed the opportunity to learn and develop proficiency pass them by. But those who are just getting started, new commercial and corporate pilots, and new-hire airline pilots have a great opportunity to make aviation a very rewarding career. If they can learn even one thing—one item of knowledge, technique, or anything at all that will prove useful in pursuing the left seat—then the time I have spent writing this will not have been in vain.

This material is not intended to take the place of any airline's training manual or the procedures set forth in any training program. It is not a book of instruction as such nor a primer on how to fly (though I will describe various maneuvers and their performance in some detail) but is intended as a guide to supplement training material. It is specifically intended for those who already have a commercial license and instrument rating. It should also prove helpful even to the beginner, particularly in basic instrument flight. I hope it will be interesting reading and provide a subject for discussion even among those who may disagree with my rules of thumb, philosophy, and suggestions relating to technique.

There will probably be many who do not agree with all I say. This is as it should be. I don't profess to be an expert nor to present myself as being the world's greatest authority on aviation. Far from it! These are merely methods that have proved useful to me (which I have learned from experience and from others) and helpful to my students. I realize full well that there are probably better methods and easier ways, and I'm eager to learn them. Those who can improve any phase of flying and flight instruction have an obligation to pass it on so that others might benefit from it. If the information is made available, I will be among the first to receive it with an open mind and give it a try.

While still in the introduction, I'd like to give the first suggestion I have to those just entering commercial aviation. Never think like a copilot! Don't get trapped in the rut of the lazy copilot, unless of course your intent is to be a career copilot and you have no ambition to ever occupy the left seat. If you are in the industry as a pilot, you'll probably move to the left seat more quickly than you think. The industry, and its demand for pilots, is moving faster than ever before. If you don't aspire to the captain's chair from the beginning, you'd be better off to seek another way to make a living *now.*

But if you ever intend to fly the left seat, take advantage of your on-the-job training: plan each flight; review the weather; select a route and altitude; figure your fuel requirements and time en route; select an alternate; know your route and destination and alternate airports; have your charts available and up to date; solve your problems of runway requirements for takeoff and landing; work closely with the other members of your crew. In other words, accomplish all the functions just as you would if you were the captain. Be familiar also with the captain's plan for the flight. And always have a plan in mind during the entire flight, since the second in command may suddenly become the pilot in command if the captain is incapacitated. Don't be caught short and not know what to do.

Work as a teammate with each captain with whom you fly. Compare your plan for the flight (with the captain's permission, since you in no way should ever try to usurp the captain's authority) to the captain's and see which is better. See if the altitude or route you would have selected would have been better. Who is more accurate in figuring time and fuel requirements? Why? Find out what factors the captain may have considered that never occurred to you. Take every advantage to learn from the captain. Compare each captain's methods; retain what seems best to you, and file information in your mind that might help you develop your own methods, which may be a combination of many things you have learned from many people.

When it's your leg to fly (and you never "own" a leg but always fly at the captain's pleasure), fly as if you were in command and accept the captain's criticism as being constructive. Express yourself as positively and tactfully as you can, and fly to the best of your ability at all times. Make climbs and descents, approaches and landings, without the use of autopilot as often as possible to develop and maintain your proficiency. Never forget that the captain is actually in command, is responsible for all your actions and mistakes, and will not be bashful about correcting you if you infringe too greatly upon pilot prerogatives and responsibility.

Be your own greatest critic. Accept each flight as a challenge. There is great satisfaction to be found in flying a trip—planning the flight, flying the weather en route, and finally making a good approach and landing. Work to your maximum ability; endeavor at all times to fly clearances exactly; stay right on your heading and course and altitude; try to fly so smoothly that the passengers will never know when you've made a mistake. You will know when you've flown a good flight, and your self-satisfaction will surpass any compliment that may be given you.

Second Edition

Since this book was first published in late 1971, changes have taken place in the industry. Simulators are used to a much greater extent in both training and checks. But this, in reality, is a plus factor; much more comprehensive training is possible in the simulator and it is both less expensive and much safer. No one is injured in a simulated crash; you simply push a reset button, positioning the simulator in takeoff position, and take off and try again. This is particularly true in learning how to cope with wind shear.

Ground school has also made positive changes. There is a greater use of cockpit trainers; better mechanical devices and visual aids provide the pilot an even greater knowledge of the aircraft and its systems than ever before.

For this edition, I find that some material must be updated, but not much. There have been significant changes in ground school, flight training, and check and rating rides. New chapters have been added on flying in thunderstorms and low-level wind shear.

The aircraft I use as examples (such as the Convair 440, Lockheed Electra, Sabreliner, and DC-9), though some are out of date and obsolete (however, many are still flying), are typical of aircraft types. The Convair 440, for example, you may consider typical of a multiengine propeller-driven aircraft; the Electra, of a turbo-prop of any type; the DC-9, of any jet T-category aircraft; and the Sabreliner, of any small jet. Simply use the performance data of the aircraft you are flying or using for training and apply the techniques, rules of thumb, and procedures herein. You will find they work right on the money.

FLY THE WING

1

Ground School and Study Habits

■ The "new-hire" pilot's first exposure to ground school with an airline will begin with several days of learning company procedures. Ground instructors will guide students patiently through the multitudinous intricacies of the Flight Operations Manual, the various forms to be used as an operating crew member (the pay form is considered by most captains as being the first officer's greatest responsibility), and the various subjects required by the FAA for pilot indoctrination and recurrent training. These will include jet upset, high-altitude meteorology, takeoff and landing minimums, weight and balance, takeoff and landing data cards, company operation specifications, etc. These subjects are fairly standard for every airline. Some I consider the province of the ground instructor and have omitted them from this book. Others, related more directly to actual flight, we'll get into a bit later. I know of no pilot who has not successfully completed this portion of training.

After completing indoctrination, the new-hire will be assigned a category in specific equipment and given a base assignment. Insofar as possible this assignment will be in accordance with the pilot's stated preference, but the actual assignment will be predicted on the airline's need at the moment. If, for example, the most pressing need at the time is for Boeing 727 second officers in New York—flight engineers in 727 equipment—the new hire will enter second officer training for the Boeing 727 and be based in New York upon completion of training. Future assignments and training will be on a bid basis of stated preference and seniority. With an equipment assignment the trainee will begin the type of ground school training to be repeated many times as progress is made (rewarded by higher pay) to higher paying equipment and categories demanding more skill and training in specific types of aircraft.

Ground Instructors

The ground instructors will be highly trained educators with a wide diversity of backgrounds. Some will be maintenance specialists with years of experience on the particular aircraft; others will be retired military or airline pilots with thousands of hours of flight experience. They will teach the mechanics as well as the operation of the many systems of the complex transport aircraft. They will also recommend the trainee for the oral examination, which is given by the FAA for type ratings and certificates and/or by a check pilot for first officers.

It is vitally important that the pilot know the aircraft and its systems thoroughly. Therefore, every modern educational aid is used. The airlines have millions of dollars invested in training aids—films, slides, videotapes, system mock-ups (electronic boards duplicating various systems), and very realistic cockpit procedure trainers. Every new and proven technique and aid to education is used to make sure that the pilot learns everything needed. It is desirable that the size of the class be restricted to eight during the first week, which is spent with the airplane manual and visual aids; then there should be one instructor for two students during the operational study system operation using the mock-ups and cockpit trainer.

When this book was first published in 1971, there was little tolerance for failure at this stage of training—70% was no longer a passing grade. The written examinations given at intervals throughout the course and a final exam upon completion required a minimum grade of 85%. It is to be hoped that the standards remain the same. That sounds tough! But actually the class average is usually in the high 90s, and the minimum grade is rare.

After completing the classroom work, the ground

instructor takes two students at a time into the cockpit procedural trainer, in most cases a full-sized cockpit environment capable of simulating and accurately displaying system operation and having some limited flight capability. It is, in fact, a flight simulator without full flight capability. Here normal, abnormal, and emergency procedures are taught, with the student in the normal operating position. Each procedure, beginning with cockpit preflight, may be drilled until recognition of desired responses as well as malfunction becomes second nature.

The ground instructor also teaches aircraft performance, usually along with cockpit procedures. Pay close attention and spend some time on your own review of the performance charts for your aircraft. They will probably have sample problems drawn on them, but you should actually work a few examples yourself. Use different values from the sample problem and work out at least one (preferably more) problem on each chart. I would suggest that you actually draw it in and make some notes on the margin of the page. This may prove helpful later when you take your oral.

The state of the art of ground school training has reached the point where the student takes the oral exam upon the completion of ground school and should be well prepared to pass it easily. This allows the flight instructor to dwell more upon system operation and flight maneuvers in the simulator. Actual flight training in the aircraft has been drastically reduced because of operating cost. The enhanced safety allows more depth of actual practice in critical operations that would be extremely hazardous in the aircraft, and the student (who already knows how to fly) learns far more from hands-on in-flight practice. The simulators of today are far more realistic, perform with precise results, may be frozen in any stage of flight to thoroughly discuss a maneuver or procedure, and create an uncanny illusion of actual flight. The flight instructor, to turn out students as well-qualified pilots, now has more time to go even deeper into systems and flight techniques than ever before and has time to evaluate students, observe their progress, and cover each system in even greater detail. The instructor can quickly spot a weakness in a particular system, procedure, or flight technique and then has time to concentrate on what the student needs most.

Aircraft Systems

It seems that every individual has a weakness in some particular system that causes the most trouble, even after completing ground school and passing the oral. It is variable but normal. Usually the problem lies in not knowing how to study aircraft systems. I've tried many instructional methods, with varying degrees of success, and have evolved a method of studying an aircraft that is both simple and effective.

Many pilots, particularly those with little or no experience in large aircraft, become bogged down in the many systems and their complexities. They store away in their minds a huge hodgepodge of information about a particular system and then have difficulty relating that knowledge to actual system operation. They try to memorize the manner in which the system is constructed rather than to understand system operation. While it is true that there are some minimum and maximum numbers relating to quantities, pressures, and temperatures that must be memorized, everything else about a system should be considered from a strictly operational standpoint. When learning these numbers, therefore, learn also where they are sensed, where they came from.

There is also a commonality of systems. A hydraulic system, for example, is a hydraulic system. Its construction and operation may vary from airplane to airplane, but many of its features are common to all hydraulic systems. However, many experienced pilots, who have been to many ground schools and studied many comparable systems, seem to have difficulty in learning the same systems in a new aircraft. There is an easy way to learn those systems. Consider what you will need to know to safely fly the airplane and pass your oral and flight check. You will need to know limitations verbatim; normal, abnormal, and emergency operating procedures, systems operation, both normal and abnormal; and emergency equipment location. Let's take these one at a time and discuss how to study them, though not in the order indicated. Let's take the pure reading and memory work first.

Limitations are to be memorized and may be broken down into four sections: (1) weights and speeds, (2) pressures and temperature, (3) operational limitations, and (4) equipment operational limits. By breaking limitations down into these categories, you may take a whole limitations section, reduce it to two pages of notes, and memorize it easily.

Emergency equipment location is most easily learned in a similar manner. Most aircraft manuals give you a sketch of the floor plan of the aircraft, showing the emergency equipment location. Break these down into groups of equipment according to type rather than trying to learn the locations of various equipment from front to rear. The emergency equipment required in transport aircraft is the same in all of them. The only variation is in the number required and the location. For example, you might make a list of the type of (1) fire extinguishers (both water and chemical agent) and location, (2) cockpit emergency equipment, (3) oxygen bottles (both ship's system and walk-around), (4) number and location of emergency exits, (5) life rafts and evacuation slides, (6) escape ropes or tapes, (7) first aid kits, (8) spare and supplemental oxygen masks, and (9) floor access to lower compartments. It's much easier to learn the equipment by classification, number, and location than it is from a chart.

Let me suggest at this point that you will need to do your homework by manual review and study to prepare for the next day's ground school subjects. I also suggest that you use the training aids available, including the cockpit procedural trainer, after classroom hours to increase your knowledge and proficiency. You will need some recreation and adequate rest, but also set aside some study time.

Next, look at the various systems. You'll need to know them to fly the airplane and perform the operating procedures. Your study of the manual and ground school knowledge can best be used by trying to understand the operation of the system by how it is put together. It is apparent that certain items of quantity (pressures and temperatures, revolutions per minute, voltage and frequency, etc.) must be learned. These are usually maximum and minimum values, which you may have encountered in limitations, but you should also learn the normal values.

Now take every system diagram in your aircraft manual (or the system sketches you may have been furnished in ground school) and the pictures of the system control in your manual. Take every control in the system and visualize in the system diagram just what happens as you operate that control. Trace out the action for each control and learn where to look for positive reaction, i.e., the indication that the control worked. Train yourself to make this looking for a response become second nature. Practice this technique and procedure in the cockpit procedural trainer, the simulator, and the airplane. It's a good habit that may prevent an accident.

Using system diagrams, locate all pressure and temperature sensing locations. Relate them to the system limitations and indications of abnormal functions. Try to visualize what is going on within the system that may be associated with the warning and caution lights of the annunciator panel.

Take each item of the normal checklist and see what it is doing and how it is indicated in system operation until you know what is going on in any system during any phase of flight or operation and with any system-control use. You want to know what happens when you activate a control and where to look to verify that the desired action took place. I'll guarantee that if you have spent a few hours on your own with this type of study of a system, coupled with what you have learned in ground school and cockpit procedure trainer use, you will know the system. Then you are ready to turn the checklist over and get into abnormal and emergency operating procedures.

Emergency Procedures

Abnormal procedures are not meant to be memorized verbatim. They are related to situations or occurrences considered less hazardous than emergency procedures and are usually handled by use of the flight manual (some operators provide a book of color coded systems) as a work list. Each flight crew member, however, is expected to have close familiarity with actions to be taken when abnormalities occur. There are some abnormalities, however, that do not afford the time to consult the flight manual for corrective action. The items that must be memorized are usually marked by hash-mark enclosures preceding them in the outline of procedures in the abnormal section of the flight manual. In these cases each crew member should be able to perform the appropriate corrective action and then consult the manual to confirm it.

Upon observing an abnormal condition, any crew member should bring it to the captain's attention and/or the attention of the full crew. At that point the captain must become a "crew manager," delegating responsibilities, calmly seeing that all required procedures are carried out, and seeking the input of other crew members based on their knowledge and experience, especially if the situation is or becomes more abnormal than expected.

It is strongly recommended that the captain, upon being advised that an abnormal condition exists, consider the aircraft's altitude, speed, and configuration. If airborne, a safe speed and altitude should be maintained or achieved before beginning corrective action.

The captain may delegate the actual flying of the aircraft to the first officer, specifying the altitude, speed, heading(s), or profile to be flown. Emphasis is on someone flying the aircraft, handling radio communications and navigation, and not being too deeply involved in the problem.

The wise pilot will read through the abnormal procedures and will commit to memory those that require action before using the manual as a work list, just as if they were emergency procedures.

Emergency operating procedures break down into two parts: (1) immediate action items (these you must memorize) and (2) clean-up items that may be done from the checklist after the immediate action is completed. The immediate action items are just that, those that require immediate action to cope with the emergency situation. The less important clean-up items are not to be done from memory and may be left to be handled later after the immediate action has taken care of the emergency. But don't try to handle the memory items too fast. Immediate means *first* items and is not to imply great speed. Being too fast often leads to mistakes that could be serious, like shutting down the wrong engine.

Several things are classified as emergency procedures: engine failure, fire, or separation; electrical fire; loss of pressurization; etc. Do not let the term "emergency procedure" scare you. An emergency is only an unforeseen condition requiring immediate action. When such an event occurs, even under adverse conditions, you will still be in control of the airplane and the proper corrective action will handle the situation. Learn the immediate action items of your checklist for such conditions and perform them by command and with deliberation. As you go over the checklist, learn what happens—the whys and wherefores of every step—in the procedure. That way you will know what you are doing, remember the procedures more easily, and have more confidence in yourself and your equipment to handle these emergencies.

In general, the correct manner to handle emergencies is as follows: Aural warnings should be silenced as soon as the crew recognizes the emergency; then, when the emergency is recognized, the proper corrective actions should be accomplished.

The captain's primary responsibility is the safety of the aircraft, crew, and passengers. At low altitudes the captain may elect to fly the aircraft and may give commands such as "Fight the fire!" or whatever. On takeoff with an engine failure and/or fire, I recommend "Silence the bell!" and continuing to fly the aircraft with no further action until in stabilized flight and when minimum obstruction altitude has been reached. Then, step by step, the pilot flying should call out the "immediate" action items, pausing between commands until they are in fact accomplished and verified.

At a safer altitude I recommend that, if you are the captain and if the particular procedure will not cause you to be the only one left with flight instrument and communication and navigation capability, you should consider delegating the flying of the aircraft to the first officer so that you may supervise completion of the appropriate corrective actions. It is then the second officer's responsibility, with a crew of three, to read the checklist or abnormal procedure aloud, including the response as the controls are placed in the appropriate position. With a crew of two, the pilot *not* flying and accomplishing the procedure should call out the items and responses as they are accomplished.

The procedures outlined for the more serious and quickly developing "emergencies" are divided into "immediate action items" and "secondary action items." It is expected that *all* crew members will commit to memory the *entire* immediate action items of the checklist. Secondary action items should be accomplished as soon after completion of immediate action items as is practicable under the circumstances, using the checklist as a work list, since it is not expected that crew members memorize these items.

Let me illustrate how you might handle an engine fire in a Boeing 727-225. I want to emphasize the importance of knowing what happens step by step in the immediate action items and to stress the deliberate method of executing them. Believe me, this will not be easy. I haven't flown a 727 for over 10 years and turned in my manual when I checked out in the 1011. But I used my methods in my own study and have a clear memory of every aircraft I've flown.

In the center and above the instrument panel just under the glareshield is the fire protection panel. In the center of the panel are three engine-fire switches; beneath them and guarded when the fire switches are in are extinguisher discharge buttons; between the number 1 and 2 fire switches is the bottle transfer switch; between the number 2 and 3 fire switches and to the lower left of the number 1 fire switch are the bottle discharge lights; above the lower left bottle discharge light is the wheel well warning light; to the right of the number 3 fire switch is the fire test switch, to the right of it the bell cutout button, and to its right the detector test button. The engine-fire switches are powered by the essential AC bus, all other switches are powered by the battery transfer bus.

Now we have a fire warning, the fire bell alarm goes off, and engine-fire switch number 1 shows a red light indicating a high-temperature condition (fire) in the related engine area. I immediately call out "silence the bell!" to the first officer, who then pushes the bell cutout button. In this quieter environment I call out, "Essential power," to the the second officer who verifies that the essential power selector is not positioned to the engine on fire by replying, "Operating generator." I either close or command the number 1 throttle to idle and it is verified verbally. This drastically reduces the fuel flow to the engine, and I will pause for a second to see if the fire warning goes out or is still present. If it is still present, my next command is "Start lever," and the first officer places it to cutoff and verifies verbally; this shuts off fuel to the engine at the fuel control unit. If the fire persists, the next command is, "Engine-fire switch," and the first officer deliberately pulls number 1 and verifies, "Pulled." When this action is taken, the following occurs: engine-fuel shutoff valve, closed; engine fire extinguisher button, armed; generator field relay, tripped after a 5- to 10-second delay; hydraulic supply shutoff valves (number 1 and 2 engines only), closed; and hydraulic low-pressure light circuits disarmed on 1 and 2. Next, "Bottle discharge switch," response, "Pushed," held 2 seconds. The last immediate action item is, "Bottle discharge light," and the response is, "On," verifying that the bottle has fired.

The throttle, start lever, engine-fire switch, and bottle discharge switch actions may cause the fire indication to go away and the warning light to extinguish in the fire switch. When the fire indication no longer exists, I recommend (though it's not on the checklist) that the detector test button be pushed. This performs a continuity test of the firewall, engine, wheel well fire detectors, and the detector circuit ground lights. If it tests normally, you can be reasonably certain that the fire no longer exists. If not? . . .

That little exercise was for the purpose of showing you, by example, the type of knowledge you should acquire about every system and procedure.

Apply the same philosophy to all the operating items on the checklist. Use the system diagrams to trace what is going on as you run through the checklist and accomplish the items. Check pressure and temperature sensing, control and actuating power, action in the system, and the indications with each control function. You'll begin to see the reason for the sequence in the checklist and it will be impressed more firmly in your mind.

That's it. That's the way to learn systems on the airplane as well as the philosophy of coping with abnormal and emergency procedures. This is the easiest method I've ever found, and I know from experience that it works. It takes a bit of homework, but it's worth it. I would also point out that in electrical fire procedures, loss of all generators, and in some electrical system problems, the captain will be the only pilot with flight instrument and navigation and communication capability. My recommendation is for the captain to fly the airplane in the presence of *all* electrical problems.

As rules of thumb, teach your eyes to look for actuation in the proper place every time you move a control, learn what happens in the system as you move the control, learn where pressures and temperatures are taken, learn control and actuating power for each control, and remember that something that can be turned on can also be turned off. Check operation when turning something off just as you do when turning it on.

Summary

There are several subjects you will study in ground school "jet indoctrination" that relate to flying and jet operation and must be well understood. I think they are worth going into (in the three next chapters) from a pilot's and an instructor's viewpoint. They can be confusing to some students in ground school, highly technical, and more on an engineering level than a pilot's operational level, so I have tried to reduce them to a more operational form. One subject, high-altitude meteorology, I'm omitting because it is so complicated and wide in scope that it would take another book to cover it adequately. However, I would recommend that a pilot who intends to fly in the jet levels learn something of the weather phenomena peculiar to the upper atmosphere.

Just one further suggestion before we move on. Some of the best students I've had and some of the most senior and best-qualified pilots I know have reduced their notes to a card file. They make notes about the aircraft systems, maneuver profiles, procedures, etc., on 3" × 5" index cards and use them for a refresher. One side of the card will pose a question about a system or procedure, and the other side will give the answer. They carry these cards around with them and use them in spare moments to refresh and retain their knowledge. This system seems to be particularly effective for people who learn best by writing and making notes.

I've also noticed that the best pilots carry different sections of the aircraft manuals with them on every trip and spend at least an hour on layover in study of the airplane. They are conscientious to the degree that they could pass an oral or instrument and/or proficiency check on any given day without any prior preparation or study of any sort. I'd say they have a highly professional approach and attitude.

2

Basic Aerodynamics

To really fly an airplane, to be a really good pilot, and to utilize the theory of "fly the wing" and the instrument scan techniques necessary in instrument flight, a pilot must of necessity have some understanding of the principles and natural laws involved in the phenomenon of flight—basic aerodynamics. This is not an engineering type of knowledge or deeply scientific but an operational type of understanding about how and why the airplane flies and what the pilot is doing to vary these aerodynamic forces acting upon the craft in flight.

Aerodynamic Forces

Since I mentioned the forces acting upon the aircraft in flight, we may as well dispose of them right now. Most pilots will tell you that they fully understand aerodynamics, that they learned this in ground school before first solo and see no reason to review aerodynamics any further. They will tell you that they know the forces acting upon an aircraft in flight—gravity, lift, thrust, and drag—and know the theory of Newton's laws of motion and Bernoulli's theory of venturis. They may be able to state that the greater the pressure, the slower the speed; and the greater the speed, the less the pressure. They go on to say that a body continues in its state of rest or of uniform motion in a straight line except as it is compelled by force to change that state; that a change of motion takes place in the direction of the straight line in which an applied force acts and is directly proportional to the amount of that force; that every action has an equal and contrary reaction; and that an airplane flies as a direct result of these laws of nature. From these laws are derived the four forces acting upon an aircraft in flight—*lift*, which

must overcome *gravity*, and *thrust*, which must overcome *drag*.

All of the above is true, and I have no desire to enter into theoretical argument with those for whom this explanation of aerodynamics will suffice, but it never seemed to me to be enough. It doesn't explain or adequately define the forces acting upon an aircraft in flight—aerodynamics. I've never completely bought the theory that aerodynamics boils down to lift overcoming gravity or weight of the aircraft and thrust overcoming drag. I prefer to think of aerodynamics as the science of, or relating to, the effects produced by air in motion—more particularly the *force* produced by air in motion. Since we can't cause the wind to blow or cause the air to be in motion, we cause the airplane to move by applied thrust, and from then on *all* the forces acting upon the plane in flight are controlled variables. And these forces *are* changed, varied to best utilize the desirable features of each as the pilot causes the plane to maneuver, and are directly proportional to speed and relative motion through the air. Lift must overcome more than the weight of the plane if it is to climb and less if it is to descend. Thrust must overcome more than the inherent drag if the plane is to accelerate in speed, and drag must be greater than thrust if speed is to decrease. To really fly the wing, aerodynamics must be considered in a light different from the classic definitions.

The wing—the airfoil or lift surface—is the whole secret of flight. When an airfoil is moved through the air, a stream of air flows under, over, and around it. If the airfoil has been well designed (by the engineers who need all the theoretical knowledge), the flow will be smooth and will conform to the shape of the moving airfoil. If, in addition, the airfoil is set at the proper angle and is made to move fast enough, the airflow will support the weight of it. And since the airfoil or wing is attached to the airplane, it will support the entire weight of of the aircraft. This then is the whole story—the nature of the action that enables wings to sustain, to furnish enough lift, and to support heavier-than-air craft in flight.

Airfoils are usually curved surfaces to take advantage of Bernoulli's theory. But if you hold a piece of paper in front of you, you will find that you can cause the paper to rise simply by blowing over the top of it. All you have done is to create a low-pressure area over the paper. The pressure beneath being greater does the rest. If, in addition, you hold the paper at a slight angle, causing an airstream to strike the bottom of the paper while it is being held at an angle to the wind, the resultant dynamic force will contribute to the force lifting the paper. Consequently—and this must be apparent—the force exerted on a surface held at an angle to the relative airstream around it is a result of pressure difference created above and below the surface.

The importance of the speed and angle of the lifting surface on flight performance is easily seen by sticking your hand out the window of a speeding automobile. If you hold your hand flat and level, no sensation is felt other than that of the air flowing over, under, and around your hand. If you raise the forward part of your hand, causing it to move through the airstream at an angle, you will immediately encounter a force than tends to lift your hand. You have just induced lift. You may also tilt your hand downward and cause this lift to disappear and your hand to react accordingly. It must be obvious that these forces, which you feel applied to your hand, either increase or decrease as the speed of the car—and therefore the speed of the relative airstream—increases or decreases. And it should be pointed out that the forces exerted on your hand are not due solely to the dynamic pressure of the airstream your hand is encountering. Rather, this lifting force in either direction is a result of the difference of pressure created above and below your hand.

If you will now compare these little analogies to what you are doing when you raise and lower the nose of a plane with the elevators and thereby change the angle of the wing as it moves through the air, you will see at once how lift may be varied by a change in pitch attitude. You can also see the relationship of pitch attitude to speed; you must position the wing at a greater angle at low speeds to create enough lift to sustain flight. The "angle of attack" is the angle of the wing relative to its motion through the surrounding air.

Believe it or not, we've already completely covered almost the entire essence of 20 hours of classroom instruction on aerodynamics—but in a much more operational manner. We've proved, with a piece of paper and our hand, that a flat surface can be made to support weight or create lift by virtue of its forward

motion and angle of attack through the air. But many years ago someone—I think it was Leonardo da Vinci—discovered that a slightly arched or cambered airfoil surface is much more efficient. So most of the wings with which we are familiar—the airfoil shape that gives the most lift with the least drag (at least in flight below the speed of sound)—are those with a rounded nose at the leading edge, a smooth cambered upper surface, and a sharp tail or trailing edge. Figure 2.1 compares this angle of attack and lift versus drag in a relative airflow.

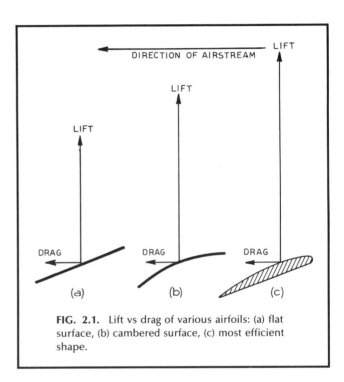

FIG. 2.1. Lift vs drag of various airfoils: (a) flat surface, (b) cambered surface, (c) most efficient shape.

Lift and Drag

The two most important concepts associated with flight are lift and drag. Lift is usually thought of as the force that acts to overcome the weight of the aircraft. But a little thought will prove this statement to be completely inaccurate. The weight of the aircraft is a function of gravity only. But think what happens when gust loads (G loads) or centrifugal forces in steep turns are applied to your airplane. They increase your wing loading, in effect increasing the weight your wing must support. In anything but level flight (normal constant altitude flight) the lift is never always exactly equal to the weight of the aircraft. Lift, accurately defined, is always perpendicular to the airfoil and is the force that acts perpendicularly to the drag of the aircraft. This is quite different from comparing lift to gravity when you also consider its relationship to drag. Look again at Figure 2.1, note the angular relationship of lift versus drag, and keep this relationship in mind as we get a little deeper into this subject.

Drag, on the other hand, is the force that acts in opposition to the plane's forward motion. Lift is a desirable quality. Drag is undesirable and must be overcome by lift and *compensated* for by the thrust of the engines.

Now that we've begun to explore some of the mysteries of aerodynamics and talk about lift and drag as if we know what they are, maybe we should more adequately and properly define them. To do so, we must get a bit technical (I don't like to be too technical), but it is not really necessary for you to understand the mathematical formulas. They may appeal to and please some of the more educated minds who read this but are not really essential to fully understand lift and drag.

Lift is the most important. As we have already illustrated, lift depends greatly on the shape of the wing or the form of its airfoil. Lift also depends on the size and position of the wing (angle of attack), the density of the air through which it moves, and the speed with which it moves through the air. The manner in which lift varies with speed may be seen in Figure 2.2.

Lift may be accurately and exactly defined by the relationship:

$$\text{Lift} = \tfrac{1}{2}p\,V^2 S C_L; \text{ or lift} = qSC_L$$
where p = air density
V = speed of the aircraft
S = plan area of the wing
C_L = lift coefficient
$q = \tfrac{1}{2}p\,V^2$

Diagrams and formulas of this type are usually passed over quickly by most pilots—perhaps from a lack of understanding them—but don't brush over them too quickly. Look at them with a critical and open mind and try to find the message they are trying to present. From this illustration and lift equation, you can see that for a given angle of attack the lift of a wing will vary with the density of the air and the speed of the aircraft. While the angle of attack (α) does not appear in the lift equation, it is there—hidden in C_L, the lift coefficient. The lift coefficient is determined by the design of the wing and varies with the angle of attack. Thus, by changing the angle of attack of a wing—a function of pitch attitude (all other conditions remaining the same)—we vary the lift coefficient in the equation. Consequently, the lift coefficient really gives a true picture of the lifting quality of the wing.

Incidentally, as a result of this it is customary to define the lifting performance of a wing in terms of lift coefficient and rarely if ever in terms of the actual lift. For aircraft speeds below Mach 0.3, lift coefficient depends only upon angle of attack and can be considered as being completely independent of the velocity. Above Mach 0.3 the speed of the aircraft exerts some influence, but the main factor that controls lift coefficient is still the angle of attack.

This may be seen in Figure 2.3, which depicts the working range of a wing in terms of lift coefficient and

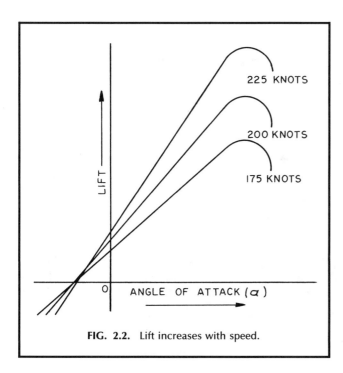

FIG. 2.2. Lift increases with speed.

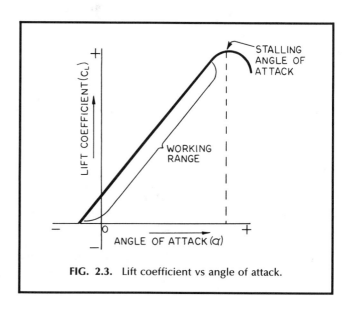

FIG. 2.3. Lift coefficient vs angle of attack.

angle of attack. Notice that the lift coefficient is zero or less at a negative angle of attack, which you induce by pushing the nose over to descend or dive, and rises in almost a straight line until a certain angle of attack is reached. At this point, the curve usually flattens out and begins to fall, usually steeply.

The straight-line portion of the curve is called the working range of the wing. In this range the wing works well and provides the necessary lift, and the lift varies directly in proportion to the angle of attack. The working range of a wing will, of course, vary with design, but a typical wing will have an angle of attack between −3 and +12°. The upper value of the angle of attack (this highest point in the curve) is called the stalling angle of attack. When the angle of attack exceeds this value, the lift begins to decrease and the wing is in a stalled condition.

The aerodynamics of stalls is covered fully in Chapter 12 pertaining to the stall as a required flight maneuver, so it is not gone into here. But an understanding of the relationship of angle of attack to a stall, recognizing that a stall is not solely a function of speed, will make the chapter on stalls more easily understood. Drag also comes into the picture in a stall, and the comments that follow about drag should also be remembered in the performance of flight maneuvers.

Drag is defined by a relationship similar to that for lift:

Drag = ½p V^2SC_D
where p = air density
         V = speed of the aircraft
         S = plan area of the wing
         C_D = drag coefficient

A plot of drag coefficient versus angle of attack is shown in Figure 2.4.

The numerical value of the drag coefficient of a practical wing is naturally less than that for the lift coefficient in the working range of the wing. This must be true, since the wing must certainly produce more lift than it does drag if the plane is to fly at all. And the variation of the drag coefficient with the angle of attack is quite different, for the drag coefficient values do not follow a straight line. At the stalling angle of attack, the drag coefficient is increasing rapidly. The lift:drag ratio usually reaches a maximum at small angles of attack. After the stalling angle of attack has been reached and passed, the lift:drag ratio falls off very rapidly.

Air Pressure

Earlier we talked about pressures and pressure differences causing lift. Pressure distribution around

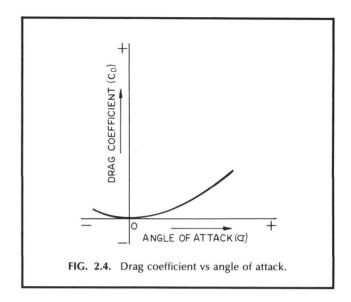

FIG. 2.4. Drag coefficient vs angle of attack.

an airfoil is the whole key to flight. One condition, and one condition only, must be present for the lifting action of a wing to occur: Air pressure above the wing must be less than the pressure below the wing. This has been determined by experiments connecting manometers to holes made in a test wing at pertinent points and taking pressure readings when the wing is subjected to an airstream in a wind tunnel. Figure 2.5 shows the static pressure distribution readings taken at the middle section of a wing at various angles of attack.

The total lift force exerted on a wing is proportional to the area between the curves of the top and bottom of the wing. That is, the greater the pressure differential, the greater the lifting capability of the wing. As shown in Figure 2.5, the major portion of the lift is caused by the negative pressure on the top side of the wing and occurs near the leading edge of the wing. You can easily imagine that lift will be obtained only as long as the high-pressure airstream below the wing has a solid surface to push against. The sudden appearance of holes in the wing would present the high-pressure air with a convenient shortcut to the upper surface, tending to equalize the pressure differential, and a noticeable decrease in lift would occur.

The lower air pressure above the wing of an aircraft in flight results from the fact that the speed of airflow over the top of the wing is greater than the speed of airflow below the wing. This is in accordance with Bernoulli's theorem as applied to airflow: As air increases in velocity it decreases in pressure and vice versa. If you have always associated high winds with high pressure, this may sound confusing. The thing to remember is that when you stick your hand into a rapidly moving airstream, you feel the pressure, not of

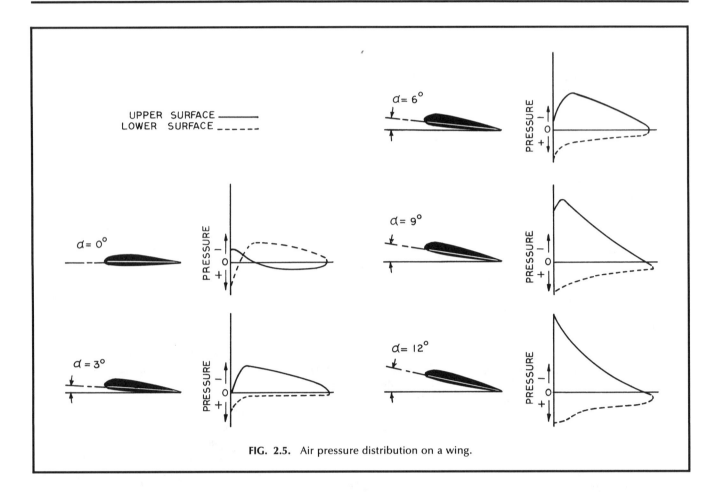

FIG. 2.5. Air pressure distribution on a wing.

the airstream, but of the air that has piled up as the hand stopped its flow. The instant you remove your hand, the air is free to flow again, and the pressure drops as the air "spreads out" and picks up speed. In other words, the pressure you feel is a result of your decreasing the velocity of the moving air by an obstacle.

So the difference in pressure of the air above and below a moving wing is caused by the greater relative speed of the airflow over the wing. This leaves us with the logical question: Why is the airflow above the wing faster than the airflow below the wing?

The answer lies in the typical profile shape of a wing used for aircraft at speeds in the subsonic range. The leading edge is rounded, the upper surface is curved, and the trailing edge is sharp. Each part serves to assist in the creation of an airflow that will produce lift. Figure 2.6 shows a simple profile of an airfoil shape and illustrates Bernoulli's theorem simply.

During flight, part of the airstream is forced to flow up over the leading edge of the wing while the remainder flows below the wing. The round leading

edge is designed so that the approaching airstream is split near the bottom of the leading edge to achieve this desired characteristic. Here the air is completely—though momentarily—stopped, and a stagnation point or high-pressure area is formed. As you might imagine, it is important that the leading edge be designed in such a manner as to cause this stagnation

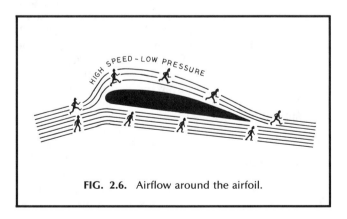

FIG. 2.6. Airflow around the airfoil.

point to be created at the bottom of the leading edge and contribute to lift, rather than to have the stagnation point be created at the center of the leading edge and thereby merely constitute an increase in drag. It is by utilization of this fact, by increasing this stagnation point, that leading edge devices such as slats are so effective in producing high lift at low speeds.

These phenomena, the major forces on a wing due to airflow, are shown in Figure 2.7. Below the wing, the airstream striking the flat surface piles up, thereby creating a high-pressure area. The air flowing over the wing, on the other hand, is forced to follow the contour of the wing. The air, being forced to travel a longer path, picks up speed and forms a low-pressure area.

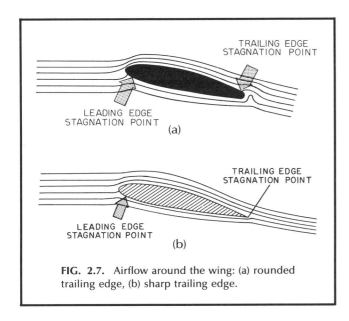

FIG. 2.7. Airflow around the wing: (a) rounded trailing edge, (b) sharp trailing edge.

The net result of the difference in pressure is an upward lifting force acting on the wing. This increase in pressure differential with an increase in the angle of attack results from the fact that at high angles of attack the air flowing over the wing has farther to travel and therefore travels faster. Similarly, at high angles of attack, the air slamming into the lower surface of the wing strikes it at an increased angle and produces an increased high-pressure area. In other words, varying pitch attitude (angle of attack) has an effect on the airflow about a wing and changes its pressure differential.

Now that we've taken care of the leading edge of the wing, let's take a look at the trailing edge. Here the air coming over the wing is about to meet the high-pressure airstream from the lower surface. The two streams will again intermingle into a common mass

with equal pressure and speed, and there will be an area of turbulence created as this occurs. This creates a vortex, which you should avoid by not following too closely behind large aircraft in flight.

If the trailing edge was rounded (Figure 2.7a), the lower stream of high-pressure air would tend to curve around to the upper surface in an attempt to fill the low-pressure area of the upper wing, creating turbulence on the trailing edge of the wing. This would result in a second high-pressure stagnation point on the upper surface of the wing, which would oppose lift. As a result of making the trailing edge sharp (Fig. 2.7b), the airstream does not make the turn to the upper surface and no undesirable forces are produced on the after part of the wing.

The greatest pressure difference occurs near the leading edge of the wing. Since this is the pressure difference that produces the lift, the greatest portion of a wing's lift is therefore produced near the leading edge. The after part, or trailing edge, of the wing produces very little lift and serves primarily to prevent formation of any forces that would oppose the lift forces produced near the leading edge. This is true even with trailing edge flaps. Some types of flaps, such as the Fowler flap, serve to increase the plan area of the wing and change the airfoil curve to increase the airflow over the wing at low speeds; other types, not increasing the wing area, merely change the airfoil shape; but all types are designed so as not to oppose but rather to increase the lift forces at the leading edge of the wing.

This creation of pressure differential by controlling the airflow about a wing is the primary lift-producing factor in the subsonic region. It should be pointed out, however, that for airflow above Mach 0.3 another factor enters the picture, i.e., changes in the nature of the air due to its ability to be compressed. At any speed, the air being encountered by the wing is capable of being compressed, and this changes the nature of the air as far as aerodynamic considerations are concerned. At low speeds these changes are so small as to be insignificant and can be ignored. Thus for airflow below Mach 0.3 we treat the air as if it was incompressible. As the speed increases above Mach 0.3, these changes in the nature of the air increase and can no longer be ignored. Then the air must be treated as a compressible fluid, and the wing is said to be in compressible flow. But this begins to get into high-speed aerodynamics (see Chapter 3) and is only touched on briefly here.

Up to this point, we have been content to state that lift is produced as a result of a pressure differential below and above a wing's surface. I have attempted to reduce aerodynamics to its barest essentials, but you should realize that this pressure differential is not free but has many strings tied to it. An aeronautical engineer and aircraft designer need to be familiar with this, but a pilot does not; therefore, discussion has been omitted.

In addition to induced drag, you should readily realize that the surface of the earth must somehow ultimately support the weight of the aircraft in flight. It takes but little imagination to realize that this is true for all heavier-than-air craft. Since the earth supports the air, and the air supports the plane, the weight of the aircraft is transferred to the earth's surface.

The process by which the earth supports the weight of the aircraft in flight is really a simple one. The pressure directly below the plane (or more specifically, at right angles to the bottom of the lift surface), which is created in the production of lift, is transmitted to succeeding layers of a column of air until it reaches the earth. By the time this happens, the force has been spread over such a large area that the force is barely perceptible unless the aircraft is flying within a few feet of the ground. (This is the ground effect cushion with which all pilots are familiar.) Thus, the air merely transmits the weight of the aircraft to the ground, distributing it over a large area in the process.

Aircraft Control and Stability

Any discussion on basic aerodynamics must include aircraft control and stability. Aircraft control and stability in the transonic and supersonic speed ranges differ from those encountered in subsonic flight. In order to have a foundation as a starting point in talking about stability and control problems of high speed, we must review the more basic facts of subsonic control.

An airplane must have an inherent stability. In addition to providing means of exerting control forces on the airplane's direction and attitude, the design of the various control surfaces and wings is important in imparting this desired stability to the airplane. The plane must not only be capable of holding a normal flying position when little or no control force is exerted, but it must also have the inherent stability to exhibit what might be called "return characteristics." Should a sudden gust lift the nose and change pitch attitude, it is desirable that the tail surface of the horizontal stabilizer act to dampen this effect and tend to cause a return to a level position.

This should also be true in the design of a wing and in the case of an airplane being rotated about its roll axis when one wing is lifted by a gust of wind. In such cases, the wing that dips encounters increased lift; the opposite occurs to the raised wing and as a result the roll is checked very rapidly. In some cases the plane may be left in a list, while still in a bank, which can result in a sideslip. To counteract this, some planes are designed with a lateral dihedral. The dihedral, together with the sideslip velocity, effectively produces a greater angle of attack on the low wing, and a rolling motion is induced that returns the plane to its normal position of wings level.

Such return characteristics should be present in yawing and pitching as well as in rolling, and the degree to which this stability should be present will depend upon the use for which the aircraft is intended. Stability is highly desirable in air carrier operation; generally, in airline aircraft additional stability devices are used for pitch and yaw at the extreme ranges of their operating speeds. The yaw return characteristic, at all speeds but specifically low speeds, is assisted by a "yaw damper" as part of rudder control; a Mach, or "pitch trim compensator," is incorporated to apply a back pressure on the elevator pulling the nose up when the "tuck" characteristic begins at high speed.

An airplane moves is three dimensions; therefore, we must describe the movement of the airplane about three separate axes. Basically, the airplane pitches about one axis, rolls about another, and yaws about another. The basic and essential flight controls are designed to be capable of producing motion about these three separate axes, and their functions never change.

The rudder always causes or controls yaw; it enables the aircraft to yaw right or left. It does this by all the principles we've discussed pertaining to an airfoil. It changes the airflow around the vertical stabilizer and rudder and creates a pressure differential as it is deflected to one side or another and "flies" the tail from one side to another.

The elevator does the same thing on the horizontal stabilizer; it varies the load on the tail, thereby serving as the pitch control. It raises or lowers the tail, giving the illusion of raising or lowering the nose, and changes the angle of attack as it moves the plane about the longitudinal axis.

The ailerons move in opposition to each other, change the lift coefficient of the wings by changing the airflow and pressure differences, and vary the load on the wings, which causes the plane to roll about its lateral axis.

There is one other control—a vitally important flight control—that must be considered. To provide the necessary airflow for the wings and controls to become aerodynamic, to create lift and flight control capability, the aircraft must be in motion. Therefore, the throttle (or power levers), which is used to vary engine thrust and affect the speed of the aircraft, must also be considered as a flight control.

3

High-Speed Aerodynamics

Today, flights in the transonic and supersonic ranges are common occurrences. Most of our present-day airline craft cruise in a speed range of Mach 0.78–0.85+, speeds that were never dreamed possible not too many years ago. With these high speeds have come new aerodynamic problems. It has taken years of work and flight research, beginning in the early 1930s, for the basic problems associated with the behavior and characteristics of aircraft at such high speeds to be solved. This chapter is devoted to seeing how aircraft behave while operating in a transonic speed range—the range at which modern aircraft operate.

Speed Range

It is the practice nowadays to refer to nearly all flight, particularly jet operation at high altitude and in the regions of high-speed flight, in terms of the speed of sound. The terms used are as follows:

1. *Subsonic flight*—flight at which the airflow over the surfaces of the aircraft does not reach the speed of sound. The subsonic region starts at Mach 0.00 and extends to approximately Mach 0.75. However, this upper limit will vary with the design of the aircraft.

2. *Transonic flight*—flight in which the airflow over the surfaces of the aircraft is *mixed* supersonic and subsonic. The limits of this region vary for different aircraft depending on the design of the aircraft, but in general, the transonic range is considered to be between Mach 0.75 and 1.2.

3. *Supersonic flight*—flight in which *all* the airflow over the surfaces of the aircraft is supersonic, or faster than the speed of sound. This region extends from the upper limit of the transonic range, Mach 1.2, to infinity.

Much has been written about aircraft in the transonic range. We've all read stories and seen movies about the first attempts to penetrate the sound barrier and fly faster than the speed of sound. Most of it was pure nonsense, fanciful flights of imagination, filled with such fearful sounding and eye-catching phrases as "penetrating the sonic barrier." The mysterious effects, sometimes catastrophic, were depicted in buffet, loss of control, and disintegrating aircraft when the plane reached the speed of sound. But it wasn't all nonsense. Any pilot who has inadvertently flown into the transonic region, well below the speed of sound for the aircraft, and accidentally exceeded the critical Mach number will tell you that.

Research, flight tests, and experience have found answers to all the strange effects that have been encountered in the transonic region. The steps taken in design to alleviate them and the pilot knowledge necessary to avoid them are the subjects of this chapter. In addition to explaining the theory of "fly the wing," the preceding chapter helped to provide a sound foundation for discussion of the phenomena peculiar to high-speed flight.

The effects of transonic flight are varied, depending mainly on the type of aircraft involved. With aircraft designed specifically for flight in the transonic regions, such as all of the present jet air transports, the effects are usually not too objectionable, since they have been taken into account in design. (But the effects are still present, can cause trouble if not understood, and may be inadvertently encountered in flight. Therefore, we'll discuss them thoroughly in Chapter 4 and learn to recognize them by making them occur, within acceptable limits, during a training flight.)

With aircraft not designed for flight in the transonic region, however, a venture into that region is apt to have some disconcerting and frightening effects. The nose may "tuck under," and the pilot may be dismayed to find that it can't be pulled up. The pilot's strength may be exhausted by pulling on the elevator to no avail. Some pilots have been fortunate and have been able to reduce their speed and pull out of such a predicament; many have not. In some airplanes the wings may begin to buffet and the controls may start to vibrate or "buzz" when entering the transonic speed range. The controls may actually reverse; you might start a turn to the right and be astonished when your airplane turns to the left. There are many weird effects that may be experienced, depending on the design of your plane; all are disconcerting to say the least. But one fact is generally true; when the limiting Mach number of *any* aircraft is exceeded, the majority of effects experienced are detrimental to both pilot and aircraft.

In Chapter 2, which related to subsonic flight, we said that air changes in nature due to its ability to become compressed. It is disregarded in subsonic flight—fortunately, compressibility is not just a detriment below Mach 0.3; it is almost the entire story at transonic flight speeds.

As a pilot, you need to gather only one important fact: Compressibility is not something that hits you suddenly with a hard jolt the second you reach your critical Mach number and transonic speed. Rather, it builds up slowly, with increasing speed; it is only after you have passed into the transonic speed region that is is capable of producing its weird effects, commonly called compressibility effects.

To understand the nature of compressibility, you must realize that when an object moves through the air, it continuously creates small pressure waves or disturbances in the airstream as it collides with the various air particles in its path. Each disturbance— actually a small change in the pressure of the air—is transmitted in the form of a weak pressure wave from the point where it was created. This expanding pressure wave travels at a constant rate—the speed of sound. This is logical, since sound waves themselves are nothing but very small pressure waves. (Remember, I am talking about weak pressure waves only. Large pressure waves, like those created in an explosion, often travel faster than the speed of sound.)

Although each pressure wave expands equally in all directions, the important direction to consider is that in which the object is traveling. This is because the pressure wave effectively serves as an advance warning to the air particles in the path of the moving object, informing them that the object will soon be coming along.

As long as the object is traveling at low subsonic speeds, below Mach 0.3, the pressure wave travels up-

stream well ahead of the object that created it and causes the air particles to change direction and conform to the shape of the moving object before it arrives on the scene. This, of course, is an advantage in subsonic flight and helps smooth out the airflow around an airfoil.

Now let's see what happens as the speed of the object approaches the speed of sound. Figure 3.1 shows the pressure wave formation ahead of an object moving at various speeds in relation to the speed of sound, or Mach numbers, and the resultant airflow distribution around the object. As the object's speed increases, the object comes closer and closer to the air particles (shown at point *B* in the sketches) before they are warned by the advancing pressure wave and can start to change direction. This means that the greater the object's speed, the fewer the number of air particles that will be able to move out of its path. As a result, the air begins to pile up in front of the object and the air density increases.

When the object's speed has reached the speed of sound (Fig. 3.1d), the pressure waves can no longer warn the air particles ahead of the object because the object keeps right up with the wave. Therefore, the air particles in the path of the approaching object are not aware of its presence until it collides with those particles that are piled up in front of it. As a result of these collisions, the speed of the airstream directly before the object slows down very rapidly, while at the same time its density and pressure increase accordingly.

As the object's speed is increased beyond the speed of sound, the pressure and density of the air ahead of it are increased correspondingly just ahead of it. The region of compressed air extends some distance ahead of the object, the actual distance depending on the speed and size of the object and the temperature of the air. Thus we have a situation in which at one point in an airstream the air particles are completely undisturbed, having had no advance warning of the approach of a fast-moving object, and then in the next instant are compelled to undergo drastic changes in temperature, velocity, pressure, and density. Because of the sudden nature of these changes, the boundary line between the undisturbed air and the region of compressed air is called a "shock wave" (Fig. 3.2).

Shock Waves

In summary then, we can say that the true nature of compressibility is this: The greater the speed of a blunt object moving through the air, the greater the air density and air pressure directly in front of it, and the less smooth the flow of air around it. In an airstream flowing at a speed less than about three-tenths the speed of sound, Mach 0.3, the density changes of

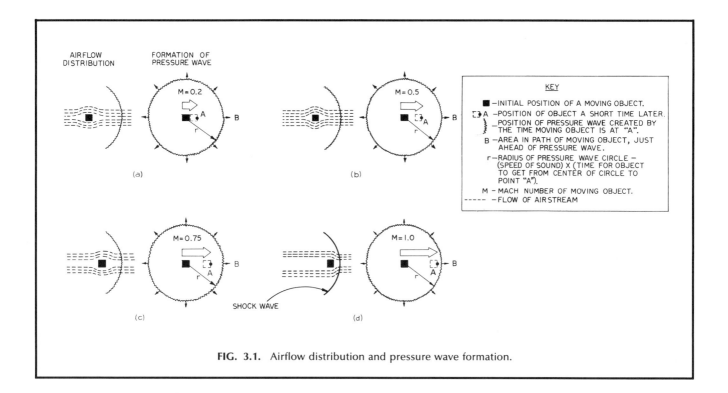

FIG. 3.1. Airflow distribution and pressure wave formation.

the air due to its compressibility can be ignored. However, above this speed, the density changes are large enough to affect the performance of the airplane and must be considered. In general, this effect is an aid to lift until the airstream on the wing's surface exceeds

the speed of sound. When this happens, compressibility effects begin to appear due to the creation of shock waves.

It should be apparent that I am talking about the speed of an airstream and not the speed of an airplane. You must remember that the speed of the air flowing over the wings of an airplane is greater than the speed of the plane itself. In fact, the speed of the air over the wings may be more than twice the speed of the plane at high angles of attack. For this reason, the air flowing over the wings may reach the speed of sound and form shock waves, even though the plane is flying at a speed well below the speed of sound. This is the reason that the limiting Mach number of today's aircraft is below the speed of sound, Mach 1.0.

The limiting Mach number is the maximum operating speed in relation to the speed of sound and is expressed as a percentage of the speed of sound. This limiting Mach, depending on your particular airplane, may be as low as Mach 0.75 and as high as Mach 0.90. Just remember, operation above that speed will subject you and your plane to all the bad features that may be expected from compressibility effects.

Shock waves contribute greatly to compressibility effects, so a jet pilot should know something about them. There are several different types of shock waves—normal, oblique, and expansion—and each has its own characteristics. But there are three characteristics common to all shock waves:

FIG. 3.2. Air passing through a strong shock wave.

1. The density and pressure of air flowing through a shock wave increases suddenly.

2. The velocity of the air measured perpendicular to the shock wave decreases suddenly from supersonic to subsonic speed.

3. The temperature of the air flowing through the shock wave increases suddenly.

Shock waves, because of the first characteristic common to all of them, are often called "compression waves." The normal shock wave is so named because it lies perpendicular, or normal, to the direction of airflow. It is *always* a strong shock wave, since the changes that occur across it are great. A normal shock wave can be formed by putting a blunt object in a supersonic stream of air, as shown in Figure 3.3. Looking at the illustration, we can see that at *A* the air is traveling at supersonic speed, completely unaware of the approaching object. The air at *B* has piled up and is traveling at subsonic speed, simply trying to slip around the front of the object and merge with the airsteam. The supersonic air from *A* slows up immediately, increasing in pressure and density as it does so. As pointed out, a rise in temperature also occurs. The center part of the shock wave, lying perpendicular or normal to the direction of the airstream, is the strong normal shock wave. Notice that above and below this portion of the wave, the airstream strikes the shock wave in an oblique direction, giving rise to an oblique shock wave.

Like the normal shock wave, the oblique shock wave in this region is strong, since the air passing through it is reduced to subsonic speed. The primary difference is that the airstream passing through the oblique shock wave changes its direction; the airstream passing through the normal shock wave does not.

The dotted lines in the illustration outline the area

of subsonic flow created behind the strong shock wave. Particles passing through the wave at *C*, outside this area, do not slow up to subsonic speed. They decrease somewhat in speed and emerge at a slower but still supersonic velocity. At *C* the shock wave is a weak oblique shock wave. Farther out from the point at which the shock wave is normal to the flow, the effects of the shock wave decrease until the air is able to pass the object without being affected. Thus the effects of the shock wave disappear, and the line cannot be properly called a shock wave at all; it is called a "Mach line."

Normal shock waves occur whenever air slows down from supersonic to subsonic speeds without a change in direction. Consequently, a normal shock wave can be created without an interfering body in the airstream. A good example of this occurs in an airflow through a constricted tube or venturi. Subsonic air enters the tube at a high speed, but less than the speed of sound, and increases in speed as the area of the tube decreases, until the airstream reaches the speed of sound. When the area of the tube increases again, the airstream decreases slightly in speed for a short distance and then suddenly drops back to subsonic speed by passing through a strong shock wave.

Here we see Bernoulli's theorem in action again. We've already agreed that it has a great deal to do with creating the pressure differential over a wing that is needed to produce lift. Now it's about to get us into trouble at high speed. Because of the difference of speed in the air over a wing to create a pressure difference, normal shock waves can occur over the wing surfaces of a subsonic aircraft that exceeds its critical Mach number. This is illustrated in a very simplified manner in Figure 3.4. The high-speed (but still subsonic) airstream flows up over the leading edge of the wing, increasing in velocity as it does so, and passes the speed of sound. The air flowing over the wing increases in speed for a short distance and then passes through a normal shock wave, decreasing from supersonic to subsonic speed in the process. Notice that the shock wave was formed *only* during the instantaneous decrease in speed from supersonic flow to subsonic flow and not during the gradual increase in speed over the leading edge of the wing.

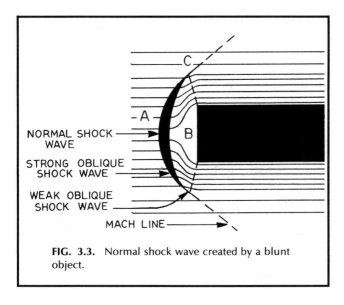

FIG. 3.3. Normal shock wave created by a blunt object.

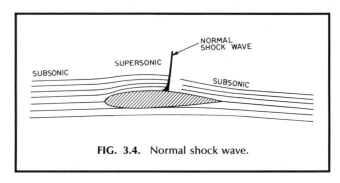

FIG. 3.4. Normal shock wave.

This clearly depicts a rule that always holds true: The transition of air from subsonic to supersonic is always smooth and unaccompanied by shock waves; but the change from supersonic to subsonic flow is always sudden and accompanied by rapid and large changes in pressure, temperature, and density across the shock wave that is formed.

Therefore, when a plane flies fast enough, the air flowing over the leading edge and top of the wing may increase to supersonic speed. This air then decreases in speed as it flows over the wing, and a shock wave is formed. This wave is usually near the center of lift when it first forms and is scarcely noticeable to the pilot at this point. But let's see what happens as we go faster.

In Figure 3.5, we can see that as the speed of the air over a typical wing section increases, the area of supersonic flow increases, and the shock wave begins to move back toward the trailing edge of the wing. The air passing under the wing also forms into a supersonic flow, and another shock wave is formed beneath the wing as this airflow reduces to subsonic speed. Finally, when the wing itself has reached supersonic speed, the shock waves on both the bottom and top of the wing have moved all the way back to the trailing edge. At the same time, a new shock wave forms in front of the leading edge due to the air piling up at that point.

This new shock wave ahead of the wing, or formed on the leading edge of the wing, is separated from the wing by a small area of subsonic air, like the shock waves created in front of blunt objects that we men-

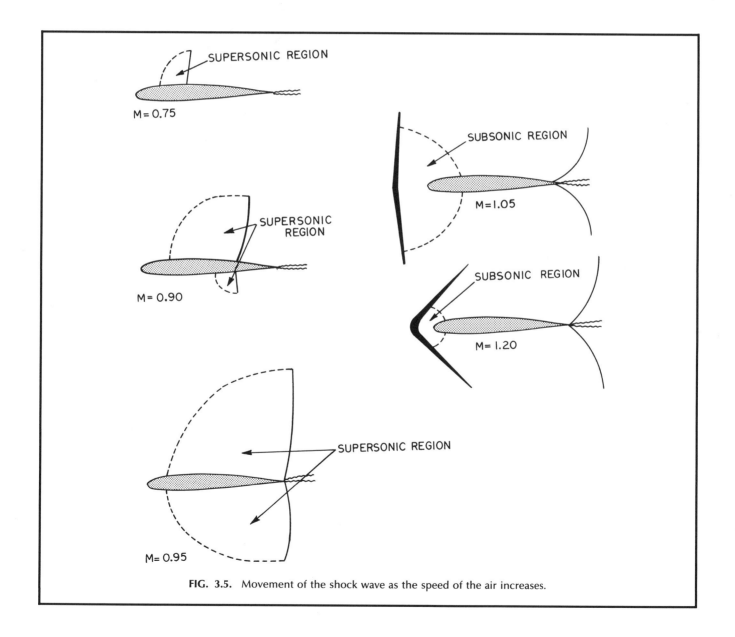

FIG. 3.5. Movement of the shock wave as the speed of the air increases.

tioned earlier. This is exactly what it is—a shock wave formed by a blunt object. When a shock wave is formed and separated from an object in this manner, it is a "detached shock wave."

When it is first formed, a detached shock wave lies in a plane perpendicular to the airflow. As the speed of the airstream is increased, the shock wave bends toward the object, forming what is known as a "Mach cone." The sharpness or shape of this Mach cone created by an object is an indication of both the speed of the airstream and the shape of the object. By either increasing the speed of the object or making the leading edge sharper, or both, we can create a sharper Mach cone. We're not going to get into supersonic flight here, but this is the reason for wings being formed and shaped differently on supersonic aircraft. An attached shock wave forms *on* an object, and its formation is dependent upon the airfoil shape of the wing, the shape of the leading edge of the wing, and the speed of the airflow.

In Figure 3.6 I have used an artillery shell as an illustration, but it shows the attached and detached shock waves just as they would be formed by a supersonic aircraft. In (a), the formation of a detached wave is shown. When the speed of the artillery shell increases, the Mach cone becomes sharper and the attached shock wave is formed (b). The same situation can be achieved by increasing the sharpness of the object, as in (c).

COMPRESSIBILITY EFFECTS

If you thoroughly understand the foregoing material, you now have a pretty good idea of the nature of compressibility, understand the nature and origin of shock waves, and are ready to move on into the compressibility effects that are associated with them. If you're going to fly high-speed jet aircraft at high altitude, a good working knowledge of compressibility and shock waves may someday save you a lot of trouble.

When the normal shock wave first appears on the wings of an airplane, the air passes through the shock wave in the same direction for a short distance. As the plane's speed increases further, the shock wave begins to move toward the trailing edge of the wing, and separation of the airstream occurs immediately behind the shock wave. That is, the air cannot continue to follow the normal flow pattern created by the wing surface and, as a result, tumbles in a random turbulent motion very much like a stall.

When this turbulence starts to grow, it makes itself known to the pilot by certain annoying and disconcerting compressibility effects. Jet pilots call these the "Machs," probably because planes are not affected by them until they are flying at high, and yet subsonic, Mach numbers. This turbulence occurs when the limiting Mach speed is reached and serves as a

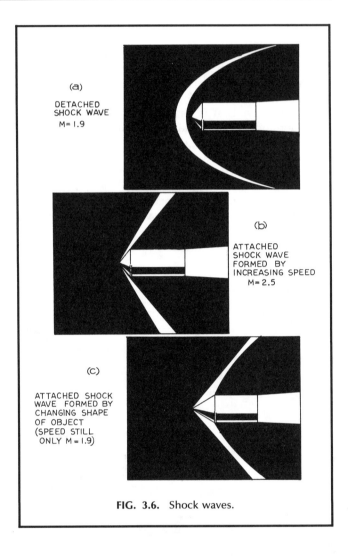

FIG. 3.6. Shock waves.

warning that there is much worse yet to come unless the pilot takes prompt action to get the plane's speed below the limiting Mach. Here are a few of the first "Machs":

1. The airplane may start an unaccountable roll. That is to say that lateral, or aileron trim, changes may be noticeable to say the least. This is because any tendency to roll that has been previously corrected by aileron trim adjustment may reappear as the trim tab begins to lose its effectiveness in the separated airflow induced by the high speed. Also, small differences in construction between the left and right wings (like more dents on one side than the other), which would be completely unnoticed in low-speed flight and noncompressible airflow, will suddenly show up and induce a rolling motion.

2. The ailerons may begin to vibrate rapidly, and this noticeable "buzz" will be felt in the control column.

3. The wings may begin to twist. This tends to upset the passengers, if not the pilot.

These are general symptoms. Different types of airplanes will naturally exhibit slightly different effects; sometimes these effects occur in one part of the plane and sometimes in another. Whatever gets there "firstest with the mostest" in supersonic airflow produces the shock wave and separated airflow over its surface and begins to suffer compressibility effects first. Due to its curved airfoil shape, the wing is usually where the first shock wave forms.

In any event, when you first notice these compressibility effects, the one sure cure is to reduce speed. You might throttle back, reduce your rate of descent slightly, and use speed brakes if necessary. But here's a word of warning. It is imperative that you be familiar with the characteristics of your speed brake at different speeds. There are some types (spoilers on the upper wing surface) that may cause even heavier buffeting at high speeds; there are others that may cause a considerable change in pitch attitude and resultant angle of attack. A "tuck-under" would certainly be undesirable right at this point.

At any rate, if you don't get your speed reduced, you'll soon regret it. The separation of the airflow over the wings will begin to produce severe turbulence, and this will cause even more trouble.

Buffeting is the next compressibility effect that is almost sure to show up. The extremely turbulent air begins to bang against the horizontal stabilizer, in a manner that may best be described as "irregular," and disturbs the flow pattern around it, causing it to buffet. The rudder and vertical fin may also come in for a share of this turbulence, and the buffet will not only be apparent through the seat of your pants but may also be transmitted through the elevator control and rudder pedals. The ailerons will probably buzz about now, and vibrating ailerons combined with a buffeting yoke and rudder pedals have a detrimental effect even on the coolest-nerved pilot. This type of "Mach" has also been known, if allowed to continue, to cause the tail surface to part company with the aircraft; then it really gets hard to fly!

If you haven't slowed down and escaped compressibility effect by now, your reflexes, knowledge of high-speed aerodynamics, or both are totally inadequate. There is another compressibility effect that will show up about now—loss of longitudinal stability. Your airplane may either "tuck up" or "tuck under," depending on its design. The tuck-under, of course, is the most dangerous, since it tends to increase the speed of the plane even more.

Tuck-under is caused by two things: (1) The airplane may begin to descend because it goes into a "shock stall," similar to an accelerated stall that may occur in subsonic flight, or even a normal stall. Like any stall, this is caused by the large separation of airflow over the wing. (2) The resultant loss of lift is only partially responsible for tuck-under; the primary cause is loss of downwash on the horizontal stabilizer.

When the flow of air over the wings is normal and smooth, the airflow over the tail has a downward direction; when separation occurs, downwash is lost and the normal forces pushing down on the tail disappear. The result? The tail goes up, the nose goes down, and down you go!

The important thing to remember is this: The first sign of buffeting or tuck-under calls for immediate pilot action. The whole trick is to get the power off to reduce airspeed and ease the nose up. The power reduction has to be fast, for when the tuck starts and the plane starts into a dive, the situation is going to get worse rapidly. As the plane picks up speed in the dive, the separation will become more pronounced and the severity of the buffet greater, and it will become harder and harder to pull out of the dive. (There have been some who haven't.) The greater the turbulence of the airstream over the tail, the greater will be the elevator angle and stick force required to pull out (assuming, of course, that the tail is still attached).

The graph in Figure 3.7, constructed from actual flight test data, shows the stick forces required to trim a subsonic airplane in *level flight* as the Mach number increases. Starting at Mach 0.6, an increasing forward or pushing force is required for trim. It reaches a maximum of about 15 lb. at about Mach 0.72. Compressibility effects begin to occur at about the same speed, and the plane starts to tuck under; and increasing pull force on the elevators is required to prevent an immediate entry into a dive. By Mach 0.83 it takes a full 70 lb. of back pressure to hold the nose up. (The flight test was terminated at this point by reducing power and speed.) The rapid rate at which the stick force curve moves upward between Mach 0.72 and 0.83 indicates clearly that fast corrective action is needed when these effects first make themselves known.

It is important for you to bear in mind that I am discussing general characteristics and that these situations vary from one to plane to another. Some planes may tuck under (as in Fig. 3.7), which is most com-

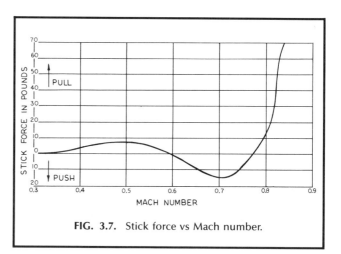

FIG. 3.7. Stick force vs Mach number.

mon; others may develop tuck-up characteristics. The stick forces required when the limiting Mach number is exceeded depend entirely upon the design of the plane.

You should note one very important fact in connection with tuck-under. While it is perfectly possible to get into tuck, buffet, etc., by exceeding your limiting Mach number in level flight or in shallow, straight dives, the effects are even more serious and pronounced if you exceed the limiting Mach number while maneuvering, especially at high altitude. This stands to reason because (1) during maneuvers the angle of attack is exaggerated (increased by back pressure used in a turn as an example), producing greater separation effects; and (2) in the thin air at high altitude you can pick up speed very, very rapidly. So as a rule of thumb: *Watch your attitude at altitude!*

I'm sure you've gathered by now that tuck-under is a pretty dangerous thing. Under certain conditions tuck-up can be equally dangerous. In an airplane with tuck-up characteristics, the nose may come up even with a full-down flight control position. This may also occur as a secondary characteristic after tuck-under. There you are, in a fast, steep dive, speed still increasing as you pull mightily back on the useless elevator, and you may reach a point where the turbulent airflow over the tail suddenly produces tuck-up! You will suddenly be in a violent pull-out that the aircraft may not be able to withstand. It is at this point that the wings separate from the aircraft and you are only seconds from eternity.

All this sounds frightening, but don't let it scare you too much. There are solutions to all the problems of compressibility effects, and no one but the uninformed gets into any trouble. The most obvious solution to the problem of compressibility effects, from a cockpit viewpoint, is to keep your plane below the speed at which these effects occur. The designers have given some thought to these problems too, and maybe we'd better determine why jets are designed with certain features not found in slower planes. But before we get into design features, let's define critical Mach and limiting Mach.

Critical Mach, or Mach crit, is the Mach speed at which a shock wave first appears on the aircraft. This is usually on the wing and is far enough forward, near the center of pressure, that separation has not yet occurred and compressibility effects have not begun. This shock wave is sometimes visible on the wing; it looks like a dancing pencil line running on the wing and is most easily seen in conditions that will form contrails or when moisture is present, as in cirrus clouds. It is not yet dangerous, and the normal cruise Mach is usually faster than Mach crit.

Limiting Mach is generally considered to be the speed at which the compressibility effects produce a noticeable lack of control of the plane. The "barber pole" on your airspeed indicator shows the speed corresponding to your limiting Mach number at all times. If you fly *beyond* this limit, you will encounter all the compressibility effects we've mentioned unless you reduce speed at once.

With each new jet produced, we find this limiting Mach becoming faster and faster. The fact that the limiting Mach number becomes higher on each new type of plane is a real tribute to our aircraft designers and manufacturers. They are continually making planes both faster and safer by designing features to lessen compressibility effects and raise the limiting Mach numbers.

Of the various reasons for tuck-under, the effects of turbulence and lack of downwash from the airflow over the wing onto the tail seem to be the most powerful forces acting on the tail and causing tuck. The designers have solved this by positioning the horizontal stabilizer as far as is practical from the turbulence off the wings. In today's modern jet aircraft this solution has been put into effect by placing the horizontal stabilizer, the tail assembly, considerably above the wings. The T-tail jet didn't get that way by accident.

Placing the horizontal stabilizer above the level of the wing alleviated the turbulence effects on the tail, reducing the tuck-under characteristic and the pitch buffet. But in so doing the designers created a new compressibility effect. Separation of the airflow over the horizontal stabilizer now showed up and took place at the junction of the horizontal stabilizer and vertical fin. Result: rudder buffeting! In addition to giving the rudder and vertical fin a good pounding, this can cause a yaw effect. Thus vortex generators— little airfoils to straighten out airflow—have been put in various places on the wings and vertical fin of high-speed jet aircraft to reduce this tendency.

This new characteristic is a good example of the problems transonic aircraft designers have learned about the hard way: when you begin to chase the Machs out of one part of the plane, they are very apt to crop up in another. It's back to the old drawing board, and then it takes a lot of flight testing, sweat, and tears to get all the Machs balanced out to an acceptable level. But it has been done. Other innovations to reduce rudder buffeting, in addition to vortex generators, are thin tail surfaces and smoother aerodynamic fairing of intersecting bodies.

Strangely enough, the bad effects of compressibility can be further eliminated by increasing the limiting Mach number. After solving the problems of compressibility effects, it is necessary to get around them—lessen their danger to faster flying. The accepted methods of doing this are also present and easily seen in most jet aircraft—thinner wing sections and sweptback wings with low aspect ratios.

4

High-Altitude Machs

Chapter 3 on high-altitude, or high-speed, aerodynamics—with its emphasis on the frightening things that happen when you exceed your airplane's limiting Mach number—sounds a bit scary. And it was so intended. "Jet upset," or loss of control at high speed and high altitude, can be a very dangerous thing. The whole idea of the high-speed aerodynamics chapter was to give you a background on which to base your operating technique and flight planning in high-speed sweptwing jet aircraft. This knowledge is essential for flight in turbulence, selecting the proper altitude according to your gross weight and buffet onset boundary, etc. But I don't want to talk you out of flying high and fast completely. Now that you know what causes Machs—what they are and how to avoid them—let's go "upstairs" and actually cause these undesirable things to happen. We'll kind of stick our toe in the water to familiarize ourselves with the particular Mach characteristics of our airplane, make the shock wave and Mach buffet appear and the airplane tuck, and then fly just beyond our Mach limitation to cause the overspeed warning to sound. We'll see how the airplane flies and feels, and then pull our toe out before we get our whole foot in.

Dutch Roll

Before we go to high altitude, however, let's look into "Dutch roll." This isn't a pastry from the Netherlands but a characteristic of sweptwing aircraft. Like nearly everything in aviation it is variable, more critical in some aircraft than others, but it is something you should be familiar with if you're going to fly jets. It is caused by one wing getting ahead of the other—a yaw—which may be brought about by poor rudder control, may be the result of a sideslip caused by turbulence, or may be pilot induced by overcontrolling. This can happen at any speed but, since the swept wing is less efficient at low speeds, usually occurs as a result of the pilot having a heavy hand and foot while maneuvering at slow speeds.

Most high-performance sweptwing aircraft (like the DC-9 we'll use for our flight) and other T-category airline jet aircraft have a yaw damper that is very effective in preventing and stopping Dutch roll. This functions through the rudder autopilot servo, which is on all the time, and a control switch to turn the yaw damper on and off. All the "black boxes" are working through the yaw damper, and we'll soon see its effectiveness.

Figure 4.1 shows what happens to a swept wing in a slip. Imagine what would happen if you were to cause one wing to get ahead of the other by yawing the aircraft with too much rudder. The wing going ahead would produce more lift and climb. Eventually the aircraft would slip, the low wing would produce more lift, and the aircraft would roll back in the opposite direction. This is a Dutch roll. Since it is usually pilot induced, you could, with sufficient altitude, just turn it loose and let the aircraft take care of itself. It will wallow around a bit, and then the inherent stability of the aircraft will return it to straight flight. But I've seen this happen close to the ground, right on flare for landing or in V_2 climb after cutting an engine at V_1 on takeoff. The landing Dutch roll is usually caused by the pilot overcontrolling the aileron; the V_2 climb problem is induced by improper use of the rudder for directional control in climb-out. In either case, close to the ground it is risky, so let's do a few Dutch rolls and recover at a safe altitude just for practice.

At a speed not to exceed the maximum at which we can use full control deflection (220 knots in the DC-9), I'll turn off the yaw damper and get a Dutch roll

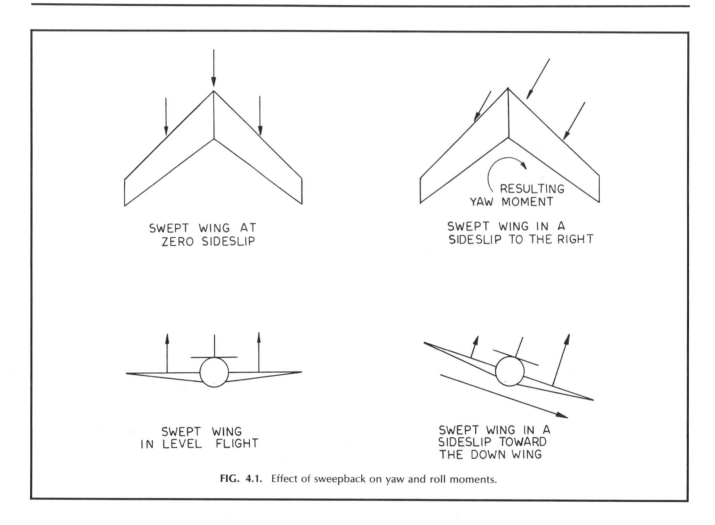

SWEPT WING AT
ZERO SIDESLIP

RESULTING
YAW MOMENT

SWEPT WING IN A
SIDESLIP TO THE RIGHT

SWEPT WING
IN LEVEL FLIGHT

SWEPT WING IN A
SIDESLIP TOWARD
THE DOWN WING

FIG. 4.1. Effect of sweepback on yaw and roll moments.

going. When it is yawing and rolling about 20° to either side, I'll give the airplane to you and you correct it.

Well, you finally got it stopped. I deliberately didn't tell you how because I thought you might learn faster if I didn't. You did the thing that appeared to be the best procedure, and it stopped the roll after going about three times to either side. Knowing I caused it with rudder, you were trying to catch it with rudder and aileron together, coordinating. You may also have felt the rudder trying to move under your feet. You can see what happens. The whole tail is swinging from side to side (putting a lot of side stress on the vertical fin), and the rudder is moving back and forth from side air loads.

I'll start the Dutch roll again, and this time recover by locking the rudder firmly in the neutral position and try to catch the roll with opposite aileron as the airplane comes through the level position. Say the left wing is down and the plane begins to roll to the right. As the wings are level, quickly apply left aileron to try to stop it from going farther right. Just lock the

rudder and try to hold the wings level with aileron. If it starts back left, quickly apply right aileron. You'll stop the roll much faster. If you try to use rudder, as the yaw damper does, you'll always be behind and may even aggravate the situation.

But the yaw damper can do a good job. Let's start the plane yawing and rolling again, and this time turn the yaw damper back on and don't even touch the controls. The yaw damper may stop it in a bank from which you can easily recover, but it will stop the Dutch roll smoothly and quickly. The whole story is to use the yaw damper by having it on at all times, don't overcontrol, and don't fight the rudder.

Now let's go up to 31,000′ or more. Anywhere from 31,000 to 35,000′ will be high enough in the DC-9; 35,000–40,000′ may be required in other aircraft, depending on wing design, critical Mach number, etc. We're going to make the tuck characteristic appear, demonstrate the use of the Mach trim compensator, fly into a high-speed Mach buffet, and then cause the overspeed warning to sound as we exceed the Mach limits of the aircraft momentarily. There is no cause for any undue concern. None of the weird effects described in the high-speed aerodynamic chapter will occur or, if they do, be serious enough to cause any problems. These things are dangerous only when the Mach limitations are exceeded, and we'll exceed them only very briefly and under controlled conditions. The purpose of this flight is to show the Mach effects you might expect to encounter as warning signs if you continue to accelerate or operate under conditions that might cause a Mach buffet.

First we'll cause the airplane to tuck, after a brief review of the Mach numbers with which we'll be working. The DC-9 cruises at Mach 0.78 (some operators use Mach 0.80), but the critical Mach—the speed at which the first shock wave appears—is Mach 0.75. This is a very mild shock wave, even at Mach 0.78, and has no noticeable effects on the aircraft. Tuck-under will begin to occur between Mach 0.80 and 0.82. Here again this is a mild effect, is easily counteracted, and is not the dangerous tuck you might experience after exceeding the Mach limits. The limiting Mach is 0.84. At this point the "clacker" (overspeed warning) will sound and we'll slow down to a normal speed. Except for Mach buffet, we will now have experienced high-altitude "Machs" in their first stages.

Since we have a Mach trim compensator (sometimes called a pitch trim compensator)—a mechanical device that receives information from the black brain boxes in the aircraft and applies up elevator to compensate for tuck—we'll have to turn it off to begin the tuck. You'll see that there is plenty of warning as the tuck begins and that it may still be stopped and controlled with elevator and stabilizer trim; then we'll turn the Mach trim compensator back on to show its effectiveness.

Before we cause the tuck, let's talk for a moment about what is going to happen and why. You know, of course, that uncontrollable tuck is caused by a loss of downwash on the horizontal stabilizer from the shock wave turbulence and separation of the air flowing over the wings. But the first tuck tendency comes as a result of force divergence. When the actual speed of the aircraft is greater than the critical Mach number, the shock wave (and the beginning of separation) will cause variations in the aerodynamic coefficients. There will be an increase in the drag coefficient for a given lift coefficient; there will also be a decrease in lift coefficient for a given angle of attack; and a change in the pitching moment coefficient will occur.

This may be shown by a graph plotting drag coefficient against Mach number for a constant lift coefficient. The Mach number at which a sharp change in the drag coefficient is produced is called "force divergence," "drag rise," or "drag divergence." For most airfoils this force divergence usually occurs at 5–10% over the critical Mach number. In the DC-9 it is about 9–10% over a Mach crit of 0.75. Therefore,

when this force divergence first occurs, the tuck may be expected at about Mach 0.82. Since we have a swept wing that is thicker at the root where the shock occurs first, we will tuck under rather than up. The wing center of pressure contributes to the trim change and pitching moment. The root shock first moves the wing center of pressure aft and causes the first diving tendency. The graph in Figure 4.2 shows the force divergence plotted for a critical Mach of 0.75 and is almost exactly the picture of what we'll do in flight.

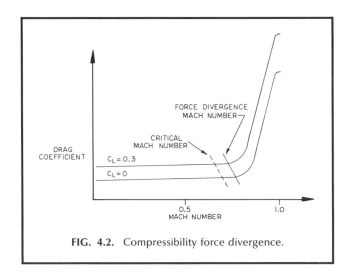

FIG. 4.2. Compressibility force divergence.

With the Mach trim compensator off, we'll accelerate to Mach 0.80 at 33,000′ and trim for level flight at that speed and altitude. Without any further trimming, let's continue to accelerate to Mach 0.82. You'll notice a slight tendency to climb at first, since the wing, by going faster, is creating more lift. But as we reach Mach 0.82, the force divergence begins to occur and the nose begins to drop into a shallow dive, which would continue to get worse unless we decelerated but which may be controlled, since we are still below the Mach limit of 0.84 for the aircraft. By placing the Mach trim compensator in the *on* position, it will, after about 5 seconds needed for it to synchronize with the black boxes, correct the tuck. Or we could place the Mach trim compensator switch in the *test* position, causing it to actuate until the Mach trim position is correct as indicated by the Mach trim inoperative light going out on the annunciator panel, and correct the tuck. You will actually see the Mach trim compensator apply back pressure to the elevator, smoothly correcting the tuck, and will see the indications of its actuation.

That's all there is to that. In the early stages, below the limiting Mach, we have a mild tuck characteristic that is easily controllable. Now let's get into Mach buffet.

Mach Buffet

Shortly after the force divergence begins and we experience the trim and stability changes, we may expect buffet to begin and control effectiveness to decrease. This is caused by the shock wave becoming a strong one, moving aft on the wing, the boundary layer no longer having sufficient kinetic energy to withstand the large adverse pressure gradient, and separation beginning to occur. At speeds only slightly above the critical Mach, the shock wave formed is not strong enough to cause separation or any noticeable change in aerodynamic force coefficients. The critical Mach of 0.75 didn't give us any indication at all. But as we accelerated to Mach 0.82, we experienced the first change in aerodynamic force coefficients and began to tuck slightly. We are now in a speed range where much further increase above critical Mach will form a stronger shock wave that can cause boundary layer separation and produce much more sudden changes in the aerodynamic force coefficients. If we were to accelerate to a speed well above Mach crit, we could, by inducing G loads or abrupt changes of angle of attack, get ourselves into serious trouble. The shock wave separation could subject the controls to a high-frequency buffet, causing them to "buzz," and the change in hinge moments might produce undesirable control forces. Going much beyond Mach 0.84 would be a violation of operating limitations, and the buffet would be severe; if prolonged, structural damage and failure would almost certainly occur.

We're going to cause a high-speed Mach buffet, but at a speed below our Mach limits, to demonstrate how a pilot can get into Mach buffet by flying too high for the plane's gross weight and encountering turbulence or by overcontrolling the aircraft in maneuvering. It will feel like an accelerated stall, which in effect it is, and the student pilot should have no fear or concern. The purpose is to demonstrate the first warning of Mach buffet and proper corrective measures. Before we execute the maneuver, study Figures 4.3 and 4.4 for a better understanding of what will happen to the airflow over the wing as we cause the buffet.

To cause a Mach buffet to occur, accelerate to 0.02 under your Mach limit at an altitude no lower than 4,000′ below the maximum certified for your aircraft (we're using a DC-9 as an example at 33,000′ and Mach 0.82) and make a 30° banked turn. Do not lose any altitude. Use sufficient elevator to keep the altitude constant. You may enter the turn smoothly, but the back pressure required to hold altitude in the turn will increase the angle of attack enough to cause a light to moderate high-speed Mach buffet. The airplane will shake and shudder just as if it was in an accelerated stall. To recover from a banked high-speed buffet below Mach limits, just sacrifice altitude, release the back pressure, roll the wings level, and re-

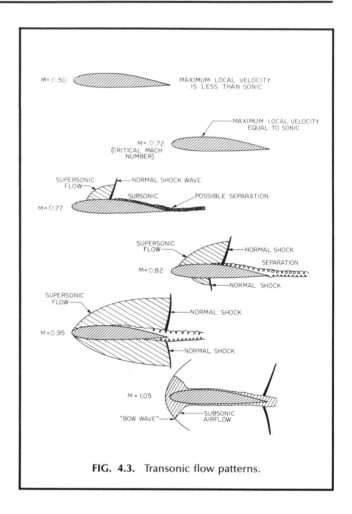

FIG. 4.3. Transonic flow patterns.

duce speed. If you continue to accelerate, the buffet will get worse. It's not really dangerous at this point, so we'll keep a speed of Mach 0.82 and make the buffet occur several times by causing the airflow to speed up over the wing by increasing the angle of attack in a turn until you feel comfortably sure of recognizing the buffet and preventing it from progressing.

The point I want to emphasize is that the buffet is caused by the airflow passing over the wing and creating a shock wave, and this increase in the speed of airflow may be a result of either aircraft speed or increased lift with a higher angle of attack at speeds just below those at which buffet will be felt in level flight. The lift that a wing must produce is, of course, dependent upon the weight it must support, and later I will illustrate a method of avoiding Mach buffet in normal operation.

Now that we've experienced tuck and Mach buffet, let's get a clearance to descend to an altitude below positive control and accelerate in descent until the Mach overspeed warning sounds (Mach 0.84 in the DC-9). As you start your descent, you may experience a very mild buffet just below your Mach limit. This of course is a warning that normally we'd recognize as a signal to slow down, but we want to hear the "clacker" of the overspeed warning so we'll let the plane continue to accelerate until it occurs.

The overspeed warning is usually a very accurate device and goes off right at Mach limit. But always be aware that, like any mechanical device, it can fail and

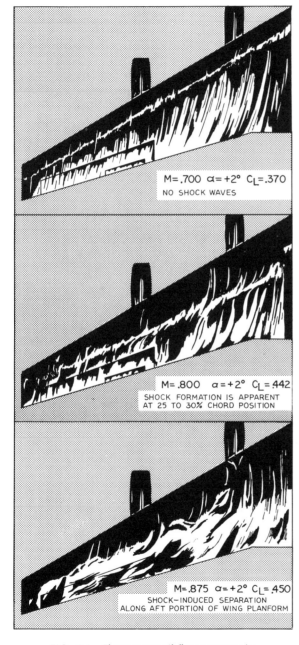

M=.700 α=+2° C_L=.370
NO SHOCK WAVES

M=.800 α=+2° C_L=.442
SHOCK FORMATION IS APPARENT
AT 25 TO 30% CHORD POSITION

M=.875 α=+2° C_L=.450
SHOCK–INDUCED SEPARATION
ALONG AFT PORTION OF WING PLANFORM

FIG. 4.4. Fluorescent oil flow patterns in transonic flow.

that it has a performance tolerance of Mach 0.02. If it was inaccurate on the low side and sounded off at Mach 0.82, it wouldn't be of much concern. But if it was on the high side, Mach 0.86, it would cause a bit of trouble. The moral is: Watch your speed, recognize the other warnings of tuck and buffet, and don't rely entirely on the overspeed warning to keep you out of trouble. When the clacker sounds, you'll be in a slight buffet, and the correct procedure is to reduce the power and slow down. If you hold your altitude, nothing will happen as the plane slows down.

Now you've heard the "clacker" and felt the slight buffet in level flight at the Mach limit, so start reducing to Mach 0.80 and continue the descent. Mach 0.80 is the recommended high-speed descent number until the Mach crosses V_{mo} minus 10 knots for an air-speed limitation in normal operation. We'll still be at a relatively high altitude (25,000–30,000') when I'll tell you to make a 30° banked turn to some particular heading in descent; the speed will be at Mach 0.80 when I'll suddenly tell you to level off immediately! The combination of bank and suddenly induced angle of attack as you stop descent will put you into Mach buffet again. It's a demonstration of high speed, gust loads induced by heavy control handling, and weight loads imposed by angle of bank. The moral of the story is smooth control usage and judgment in relation to speed.

Thus far we've caused the Mach buffet to occur by increasing our angle of attack, imposing a lift demand greater than actual gross weight on the wing, at high speed. The speed of the aircraft in level flight and descent in our demonstration has been well above the critical Mach, though still below the limiting Mach, and we've simply made the airflow above the top surface of the wing speed up to a point where the shock wave begins separation and a random tumbling airflow very similar to a stall condition. We also felt a slight high-speed Mach buffet beginning to occur as we demonstrated the overspeed warning in straight flight descent. Obviously it would have gotten much worse had we continued to accelerate beyond our Mach limit, but the key to the Mach buffet demonstration is not high speed alone. We demonstrated the Mach buffet at high speed because that's where it is commonly encountered, and because the buffet characteristics are more pronounced at high speed and easier to induce and detect, but the point is that we caused them to occur by increasing the angle of attack.

The Mach buffet is a function of the airflow over the wing, not necessarily the actual speed of the aircraft, and will also occur at a speed far below the normal cruise speed of the plane. This is known as "low-speed Mach buffet." If a plane is flown too fast at any weight, or if too great a lift demand is made on the wing near limiting Mach, then a high-speed buffet will occur. However, if a plane is flown too slowly for its weight at high altitude, the angle of attack necessary for it to maintain altitude will cause the airflow over the wing to speed up to the point where the buffet will occur again. In the earlier jets, the range of speed between high-speed buffet and low-speed buffet was very narrow at high weights and high altitude and was called a "corner" or "box." The speed range is much wider in modern jets, and the box is not so critical, but the low-speed buffet may be encountered in some cases.

The lift created by a wing is in direct proportion to the weight of the aircraft. The equivalent gross weight—the additional lift demand above actual gross weight that will cause the angle of attack to speed up to the point of Mach buffet—may then be plotted for weight, altitude, and speed (Figs. 4.5 and 4.6). When this equivalent gross weight is exceeded by any means—turbulence gust loads, artificial gust loads in-duced by too rapid control movement, or speed changes—then Mach buffet will occur.

The manner in which Mach buffet may be encountered by the unwary pilot may best be described in a hypothetical flight. Our pilot has a long-range flight to make in a DC-9 and elects to fly at maximum certified altitude of 35,000′. The air is clear, there are no clouds over the entire route, and our pilot wants to make good time with a tailwind he'll be picking up at this altitude. He knows the recommended cruise speed is Mach 0.78–0.80 but figures if he cruises at 0.82, just under a Mach limit of 0.84, he'll pick up about an additional 25 knots true airspeed and really make good time. And why not? He's still not above the Mach limit, and the Mach trim compensator is taking care of the little bit of tuck he's experiencing. He's at a gross weight of 95,000 lb. (well under max gross), and there's no turbulence forecast, so he's sure he'll have a nice smooth ride. Then he encounters clear air turbu-

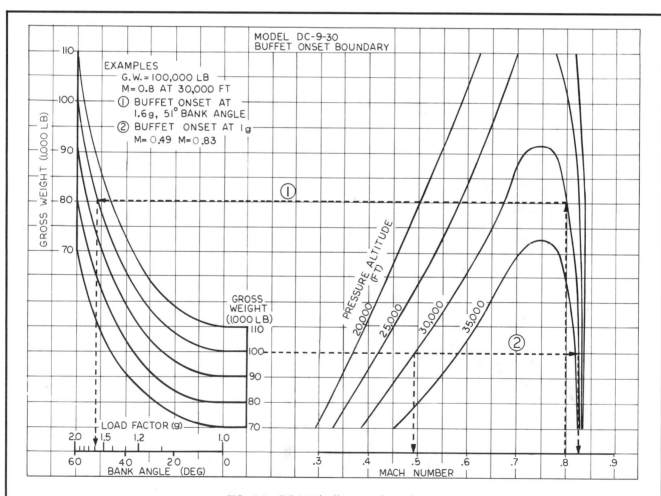

FIG. 4.5. DC-9-30 buffet onset boundaries.

G-LOAD 1.25
WEIGHT & ALTITUDE LOW (LO) & HIGH (HI) SPEED BUFFET MARGINS

GROSS WT.		70,000	75,000	80,000	85,000	90,000	95,000	100,000	105,000
FLT ALT		LO--HI	LO--HI	LO--HI	LO--HI	LO--HI	LO--HI	LO--HI	LO--HI
350	M	.525-.826	.550-.824	.575-.821	.596-.818	.617-.815	.639-.815	.660-.803	.680-.794
	IAS	172--280	181--280	190--280	197--278	204--277	212--274	220--272	227--268
330	M	.496-.829	.519-.826	.542-.824	.564-.821	.585-.819	.605-.814	.625-.810	.645-.805
	IAS	170--294	178--294	187--294	195--292	202--290	210--289	217--287	224--286
310	M	.460-.831	.484-.829	.506-.827	.525-.825	.545-.823	.566-.820	.586-.817	.606-.813
	IAS	165--309	174--308	181--307	188--307	196--306	204--304	212--303	219--302
290	M	.435-.832	.456-.831	.477-.830	.496-.828	.515-.826	.535-.825	.555-.822	.575-.820
	IAS	162--323	170--323	178--322	186--321	194--320	201--320	209--319	217--318
280	M	.425-.833	.445-.832	.465-.831	.483-.829	.500-.827	.520-.825	.540-.824	.560-.823
	IAS	162--330	170--330	178--330	185--329	192--327	200--327	208--326	216--326
260	M	.398-.834	.418-.833	.438-.832	.454-.831	.470-.830	.490-.829	.509-.828	.527-.826
	IAS	158--345	166--344	174--344	181--343	188--342	195--342	203--341	210--340
240	M	.372-.835	.390-.834	.408-.833	.424-.832	.440-.832	.458-.831	.476-.831	.495-.830
	IAS	154--359	162--359	169--358	176--358	183--358	191--358	198--358	206--356
220	M	.354-.835	.370-.834	.386-.834	.401-.834	.416-.834	.432-.833	.449-.833	.465-.832
	IAS	153--374	160--374	167--374	174--374	180--375	187--374	194--374	202--373
200	M	.335-.836	.350-.835	.365-.835	.378-.835	.392-.835	.407-.834	.423-.834	.439-.833
	IAS	151--390	158--390	165--390	171--390	177--390	184--389	192--389	198--388

G-LOAD 1.50

FLT ALT		70,000	75,000	80,000	85,000	90,000	95,000	100,000	105,000
350	M	.597-.818	.623-.812	.649-.806	.670-.796	.691-.787	.706-.775	-----	-----
	IAS	197--277	206--275	215--273	223--269	231--266	236--261	-----	-----
330	M	.563-.822	.589-.818	.614-.814	.634-.808	.655-.802	.673-.793	.692-.784	.709-.771
	IAS	194--293	203--291	213--289	220--287	228--285	235--281	242--277	248--273
310	M	.526-.825	.550-.822	.575-.819	.595-.815	.616-.811	.636-.806	.656-.802	.675-.796
	IAS	189--307	198--305	208--304	215--302	223--301	231--299	239--297	246--294
290	M	.497-.829	.521-.827	.544-.825	.564-.822	.585-.819	.605-.815	.625-.811	.640-.806
	IAS	186--321	196--320	205--320	213--319	221--317	229--316	237--315	243--312
280	M	.484-.830	.507-.828	.530-.826	.550-.824	.570-.821	.589-.818	.609-.815	.624-.811
	IAS	185--329	194--328	204--327	212--326	220--325	227--324	235--323	242--321
260	M	.455-.831	.477-.830	.500-.829	.519-.827	.538-.825	.556-.822	.575-.820	.591-.817
	IAS	181--343	190--342	200--341	208--340	216--340	223--339	231--338	238--336
240	M	.425-.832	.445-.832	.466-.832	.484-.831	.504-.829	.523-.827	.542-.825	.555-.823
	IAS	177--358	185--358	194--358	202--357	211--356	219--355	227--355	233--354
220	M	.401-.833	.421-.833	.440-.833	.457-.832	.475-.831	.492-.830	.510-.828	.525-.826
	IAS	174--374	183--374	191--374	199--373	207--373	215--372	223--370	229--370
200	M	.379-.835	.397-.835	.415-.835	.431-.834	.447-.832	.463-.831	.480-.830	.495-.828
	IAS	171--389	179--389	188--389	195--389	202--388	210--387	218--386	225--385

FIG. 4.6. DC-9-31 low-speed and high-speed buffet margins.

lence and a real good chop and experiences gust loads of 1.5 Gs. He's in real trouble! He has exceeded the maximum equivalent gross weight for his altitude and is experiencing a very heavy Mach buffet in addition to turbulence.

Or let's say he doesn't encounter any turbulence at all. It's perfectly smooth and he's cruising along at the same gross weight and a speed of Mach 0.82. He has just crossed an omni station and has a fairly large change of course to make, so he rolls into a 30° banked turn. Wham! Mach buffet again! Just the change in angle of attack, brought about by the slight back pressure necessary to hold altitude and the resultant slight increase in wing loading, has increased the speed of the airflow over the wing to bring about a high-speed buffet.

But let's take another case. Still at the same altitude and weight he is given an en route delay, so he slows to 220 knots indicated to fly a holding pattern. His Mach number is about 0.64, his true airspeed is 375 knots, and once again he encounters moderate turbulence and experiences Mach buffet. It's a low-speed buffet now. His speed is well below that of the limiting Mach, but the angle of attack required to hold altitude at that weight, speed, and altitude is such that he will experience Mach buffet in level flight if his speed becomes less than Mach 0.596, or about 197 knots indicated airspeed. At his holding speed, gust loads can cause an angle of attack producing an airflow over the wing with sufficient transonic speed to cause Mach buffet.

If, with flight demonstration and hypothetical flight, I've been able to convince you that you can bring about a Mach buffet with a heavy hand on the elevator and with bank angles, visualize what can happen to you in turbulence. The gust loads imposed by turbulence, combined with the control loads as you fight the effect the turbulence has upon your plane, can cause you an exciting few moments. You can prevent this to a large degree by preflight planning. Select an altitude according to gross weight and forecast turbulence that will give you some latitude in relation to your buffet onset boundary, widening the range between your high-speed and low-speed buffet.

Figure 4.5 is a chart that every jet aircraft should have. If your aircraft flight manual doesn't contain one, ask the manufacturer to supply it. If there is turbulence of any type forecast or expected on your route of flight, use this chart to select an altitude where an equivalent gross weight of 1.5 times your gross weight will not put you into Mach buffet.

I'm using as an illustration the buffet onset boundary chart for a DC-9-30. It's a very simple graph, easily understood and easy to use. A similar chart for your aircraft, properly used, can make life easier in turbulence and may prevent "jet upset."

You'll notice that the chart shows all the things I've talked about and demonstrated in flight—buffet onset at high and low speeds, load factor (equivalent gross weight, which may be from turbulence or control use), angle of bank, and the Mach numbers where the buffet will occur in relation to gross weight and altitude. To use it for flight altitude planning, draw a line from the 1.5 load factor up through the gross weight and extend it across to where your normal cruise Mach crosses. This will show you the buffet margins for an equivalent gross weight of 1.5 times actual weight, or a gust load of 1.5. If you don't fly at an altitude higher than the buffet boundaries you've established for your weight, you'll stay out of most exposure to Mach buffet from turbulence.

However, the graph is rather hard to use in flight, so most airline operators make up a numerical buffet margin chart similar to Figure 4.6, which is easier to read but takes a lot of work to make up. Anytime there might be known turbulence ahead, such a chart would be a big help in making a good altitude decision.

Note

Just a word of caution about turbulence avoidance. Some people seem to think that climbing is always the answer. It isn't! Even a jet will not always be able to get above all the weather. And remember the clear air turbulence that may be indicated by cirrus clouds, the altitude of the tropopause, the position of the jet stream, etc. Unless you absolutely know that a higher altitude will be smooth and free of turbulence, use the buffet onset boundary chart or a numerical buffet margin chart and select the best altitude for turbulence penetration considering your weight. With the use of radar, some knowledge of high-altitude meteorology, proper turbulence penetration speed and technique, and the correct altitude for your weight, you'll have a safe flight.

5

Approach Speed Control and Target Landings

One of the most difficult things in converting to jets is flying a good landing approach at the proper speed, angle, and rate of descent. The techniques and habits you might have developed in prop aircraft can quickly get you into trouble in shooting an approach to a landing in a jet. You might touch down short of the runway or, just as undesirable, overshoot and touch down too far down the runway. Most landing accidents are caused by poor approach technique.

In Chapter 14 we'll go into the flight characteristics of jet aircraft in slow flight. I'll recommend that you practice acceleration and deceleration, familiarize yourself with the effects of gear and flap extension and retraction, and practice maneuvering at the approach speed of your airplane. Every one of the maneuvers is put into practical application in flying the traffic pattern and making the final approach to land. There are several treatises on the subject, written by engineers, that make it appear very complicated and difficult, but it isn't. It's still just flying the wing, knowing what's going on, using the proper control and speed techniques, and flying your plane right down the groove. Let's take the whole thing apart, consider all the factors involved from a cockpit viewpoint, and then put it back together in a manner we can use.

First of all, just what are target landings and approach speed control? They are, very simply, a landing aimed at an exact touchdown or flare point from an approach at the correct speed, angle, and rate of descent. They may be broken down into three parts:

1. Aiming at a point on the runway 1,000′ from the approach end and maintaining a close control over airspeed, approach angle, and rate of descent during the final approach.

2. Making a slight but definite flare to reduce rate of descent just prior to touchdown.

3. Getting the main gear wheels onto the runway immediately after the flare, even if the forward speed is in excess of the desired touchdown speed.

By consistently following these procedures in approach and landing, there will be less chance of undershooting or overshooting the runway. If runway conditions are less than ideal (wet or slippery), using up available runway while "floating" will make it difficult to stop the airplane within the remaining runway. Floating (riding the ground-effect cushion too long) and letting the pavement slip behind you can put you in a position where you are too low for the ailerons and too high for the brakes. An airplane will stop better every time when it is firmly on the ground and the brakes are applied.

Think of an approach as aiming for a keyhole in the sky. The ultimate goal, even in an instrument approach, is to get the airplane into a "slot" at the proper speed, alignment, attitude, and rate of descent from which a landing may be made visually at a selected point. And never forget that so far every landing is made *visually*; thus, at a certain point in the approach, it becomes a matter of judgment and feel. Good landings, even in the biggest jets, are still due to the seat-of-the-pants sensitivity and eyeball judgment of the pilot.

Eye Reference

A pilot's eye reference, the plane of vision from the cockpit position, is perhaps the most critical factor in making good landings. When a pilot's eye is lower than the correct position (the normal position), the aircraft appears to be too high and the landing is misjudged, sometimes resulting in a hard or short landing. The pilot has a strong tendency to lower the nose in order to acquire additional ground visibility, especially in low-visibility conditions.

Back in 1963 when Category II approaches began, the International Air Transport Association (IATA) held an all-weather conference in Lucerne, Switzerland, and recommended certain standards for cockpit visibility during approaches. This all-weather criteria required that, at a 100′ decision height with visibility restricted to 1,200′ runway visual range (RVR), an external field of view that provides at least 3 seconds observance of an object on the ground before it disappears under the nose of the aircraft (glareshield cutoff) be designed into jet transports.

The field of vision from the cockpit is controlled by the up-and-down and fore-and-aft seat position and the seat must be adjusted correctly for the eye to be in the optimum position. Some aircraft have an eye-position locator, a small metal ball either suspended from the cockpit ceiling and retractable or fixed in the center of the windshield. In either case, simply adjust your seat so that indicator is exactly at eye level without having to raise or lower your eyes. In the DC-9, I taught the use of the glareshield (see Fig. 5.1). Geometrical analysis would show that sitting too low costs 1.4° for every centimeter.

In the L-1011 when the pilot is correctly seated for the optimum eye position, the windshield cutoff angle is 21° below the horizon. During a low-visibility approach (1,200 RVR) at 130 knots, the aircraft is moving at 220′ per second. At the 100′ decision height, a reference point on the runway becomes visible. In 3 seconds, this point will have moved 660′ closer to the aircraft. Even with a pitch angle of 6° on approach, the L-1011 exceeds the IATA recommendations, since the referenced point would be visible over the nose for more than 3 seconds. However, pilots who sit too low or too far back will experience reduced over-the-nose visibility. Eye position is not optimum. They are handicapping themselves because visual cues will be hidden under the nose of the aircraft, whereas these cues would be available if the optimum eye position was selected.

Sitting 1 inch below the optimum eye position in the DC-9, will reduce the cutoff angle by 3.25°. This moves the nearest point that can be seen over the nose 260′ further down the runway. At optimum eye position, a point on the runway 600′ ahead of the aircraft (half the 1,200 RVR) would be visible at the 100′ decision height. By sitting 1 inch too low, the pilot would have sacrificed the best 43% of the visible runway; the last 340′ of the 1,200 RVR is all that could be seen. In this condition, the tendency to drop the nose is overwhelming (see Fig. 5.2).

Figure 5.2 shows values that vary from those above. It is a DC-10 illustration prepared by Douglas, but I'll guarantee you that my figures (based on operational experience) are nearer what will actually be seen.

I have found that most instructors do not go any further than merely telling their students "eye position is important" and not explaining why. Seat and eye position *are* important, but I have found that teaching *why* is equally so. The pilot who *knows* why certain things are taught has a greater tendency to make correct seat and eye position an ingrained habit.

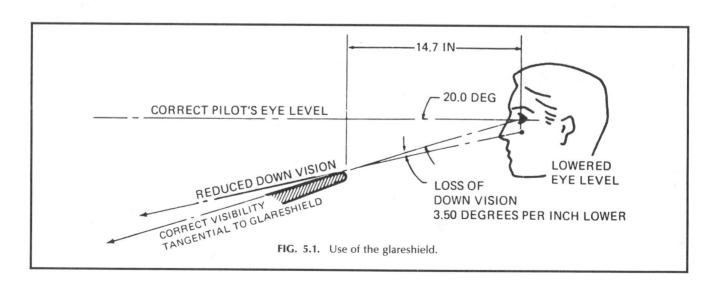

FIG. 5.1. Use of the glareshield.

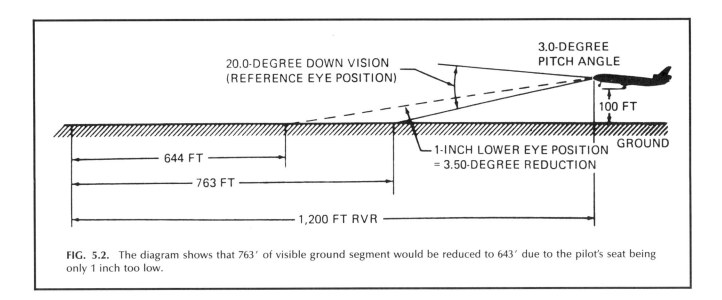

FIG. 5.2. The diagram shows that 763′ of visible ground segment would be reduced to 643′ due to the pilot's seat being only 1 inch too low.

Pilots want to take advantage of everything that will help them make good landings, especially in low-visibility approaches. Why not start with the correct seat adjustment for the optimum eye position?

Angle and Rate of Descent

Landing characteristics of the various transport jet aircraft vary widely, but a pilot with a really sensitive posterior and a good pair of eyes can land any of them just like a J-3 Cub. Under ideal conditions it is possible to flare, feel and ride the ground-effect cushion, and hold the aircraft off smoothly until it touches gently. Landings executed in this manner are so smooth that it is virtually impossible to tell when the main gear touches. But the ideal conditions don't always (very rarely in fact) exist, so this landing technique is not always the best way to land a jet. There's a much better technique, more consistent under every condition and sometimes with equally smooth results, which should be used every time to practice for consistent performance—simply a "wheel" landing. It's not a bit different from the wheel landings made in DC-3s, Twin Beeches, or any other conventional tailwheel aircraft. You break your descent slightly, ease your plane into a level position just off the ground in the ground-effect cushion, ease the power off as you gently reduce your back pressure on the wheel to counteract the ballooning tendency of the ground effect, and "roll" the wheels on. In aircraft like the Boeing 727, DC-9, BAC-1-11, and virtually every large T-tail airplane, it has the effect of actually slowing the sink rate of the landing gear and can result in a very smooth landing.

Let's see how this works. Aircraft with rear-mounted engines have the wings farther aft to support the weight of the engines. In relation to the cockpit, where the eyeball judgments are made, this also places the wheels far aft of the pilot, and they are very close to the ground at flare. The pilot is riding ground effect, and when power is reduced the airplane will want to come down. By relaxing back pressure, raising the tail, and decreasing the angle of attack, the pilot will get the airplane on the ground a bit quicker than by holding it off. At the same time, easing forward on the wheel to raise the tail will rotate the airplane slightly about its center of gravity axis, and the landing gear, being well aft of the rearmost center of gravity, will be reduced in descent just prior to touchdown.

Every airplane has slightly different characteristics due to design, but the basic principle is the same. This landing technique can be used in any airplane but is particularly effective in T-tail jets with aft-mounted engines. It still takes skill, feel, and judgment but is the easiest and safest way to obtain the best landing performance and roll to a stop with maximum control most consistently.

So much for landings. Now let's back up and consider the approach. Good landings are at least 90% the result of a good approach—in jets it may even be 95%. Since the approach breaks down into angle of descent, rate of descent, and approach speed, we'll take them in that order.

It has been found, through experience and accident evaluation over the years, that flying a jet—or any airplane for that matter—at a flatter angle than the 2.5–3° recommended makes judgment of height above the ground and glide path aim point extremely difficult, particularly at night. If the approach angle is too flat and conditions are less than optimal (gusty wind, precipitation, an unlighted approach area over the water or open fields), the result could be unexpected touchdown short of the runway. Such landings are usually spectacular, noisy, and quickly lighted by the flames of a burning aircraft. I strongly recommend, particularly in jets, that you practice the 3° glide path approach, even in VFR conditions and when not using the ILS.

This angle of approach is closely tied to speed control and will result in a rate of descent, at approach speed, of 500–700′ per minute. Since approach speed is related to stall speed in approach configuration for various weights, the only variable affecting approach speed will be wind and gust components; the resultant ground speed will affect rate of descent.

Now we've covered two components of the approach—angle and rate of descent. These two boil down to a 3° angle (just like an ILS glide slope) and a descent rate of 500–700′ per minute. When done visually, not using glide slope information, fix your eyes on the target and adjust your speed, angle of approach, and rate of descent to get the desired angle and rate of descent.

Approach Speed

These then are the targets of desired performance—the touchdown point and the angle and rate of descent with which we approach it. Now comes the hard part—the speed control of the aircraft while flying this approach groove. The proper approach speed in jets is absolutely essential. During the final approach to flare and landing, a pilot must maintain strict control of the speed and rate of descent. A speed that is too high when the wheels touch the ground takes more runway to stop because the kinetic energy of the airplane that must be dissipated during the rollout to stop varies roughly as the square of the airplane's speed. On the other hand, a speed that is too low on approach may cause the airplane to touch short of the runway.

The best recommended approach speed is 130% or 1.3 over stall. This is called V_{ref}, or reference speed, and varies according to gross weight and configuration. Normal landing approaches are made with full flaps and a V_{ref} computed for the aircraft's weight, and this becomes the approach speed to which wind and gust components are added for desired speed to the flare point. However, landing performance is predicted on the speed being no more than 1.3 over stall at the 50′ threshold point in every case; 1.5 over stall is the minimum recommended maneuver speed in the clean and takeoff flap configuration and should be the minimum speed while flying the traffic pattern and maneuvering for the approach. This may be reduced to

1.42 over stall in the approach flap configuration—an intermediate stage between takeoff flaps and full flaps—and may be used for maneuvering. However, when on approach in any configuration and maneuvering for alignment is limited to a bank of 15° or less, then 1.3 becomes the desired approach speed.

There have actually been short, or hard, landings where the flight recorder showed the aircraft to be right on the proper approach speed and rate of descent and then suddenly be too slow and hit like a ton of bricks. This is caused by a falloff in wind velocity as the airplane gets closer to the ground and is a complicating factor to approach speed control. It is not too apparent in props but has a profound and sudden effect on jets. I will have more to say about this phenomenon, but first let's consider basic approach speed control.

It is dangerous to believe that rate of climb or descent is strictly and solely a function of power, that an airplane goes up or down as a result of applying power or pulling it off. A very basic understanding of aerodynamics proves that a wing in flight climbs or descends as a result of angle of attack and lift variations at various angles of attack. Power or thrust merely pushes or pulls the plane through the air fast enough for the wing to become aerodynamic, to furnish lift. From then on, climb and descent are functions of pitch attitude (lift coefficient) and are controlled with the elevators. The power controls speed—period! This is the whole essence of flight, particularly jet flight.

Figure 5.3 is from flight test data showing the difference in stall speeds and rate of climb in relation to takeoff speed between prop and jet aircraft of similar

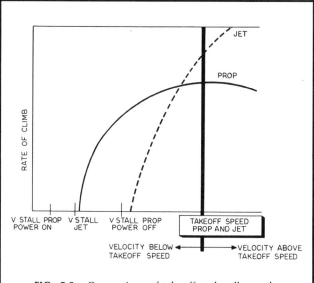

FIG. 5.3. Comparison of takeoff and stall speeds of prop and jet aircraft of comparable gross weights.

weights and takeoff configuration, e.g., a Douglas DC-6 and Douglas DC-9. It graphically illustrates the difference in stall speed of propeller-driven aircraft with power off and power on and also shows that jet aircraft exhibit essentially the same stall pattern with or without power.

First, let's consider that prop airplanes have two stalling speeds—one with power off and a lower one with power on. This difference in stalling speeds is due to the slipstream effect from the props around the wings. (We're talking about multiengine aircraft, of course.) When power is on, the propellers generate a flow of air around the wings that does not occur when the power is off. This increased flow of air generates additional lift and allows a prop plane to fly more slowly than if the airflow around the wings comes only from the plane's forward motion.

Jet airplanes, on the other hand, show a much smaller difference between power-on and power-off operations at low speeds. Airflow around the wing varies only when the forward speed of the airplane varies. A jet wing is just a board—a lift surface; it goes where you point it, and its speed varies according to the power or thrust applied to it.

This difference in lift characteristics between prop and jet aircraft with and without power requires a different technique in handling approaches. In a prop aircraft you may safely use the power for rate of descent control, but in a jet you really fly the wing.

For example, if you're flying a prop airplane and getting a little low, apparently settling a little short of the intended touchdown point on the runway, you simply ease on a little power. Immediately, as the engines and props increase power, the additional airflow around the wings generates more lift. With more lift, the airplane will reduce its rate of descent and stretch out the approach to reach the runway and touchdown point. Since this power application is usually slight, and any increase in lift is also a change in angle of attack, which also produces more drag, the actual speed of the aircraft will only vary a knot or two in forward speed, and very little if any elevator control is necessary.

But this isn't the way it works in a jet because of its lack of propeller slipstream. When a jet appears to be low, settling in short of the desired touchdown point, the nose should be raised with elevator control to create a higher angle of attack and increase lift to establish the new desired approach path. *At the same time* throttles should be advanced to provide the additional thrust needed to counteract the added drag resulting from the increased angle of attack and to maintain the desired approach speed.

In other words, without props to generate additional airflow, the only way to quickly generate additional lift to change the glide path is to increase the angle of attack of the wing. With this type of correction, the desired approach speed should be maintained, rather than increased, with engine thrust as the approach angle of attack is changed by elevator control. A jet's glide path can be readily adjusted if enough elevator control and thrust are applied quickly enough. It works both up and down; pull the nose up to go up and apply a little power to maintain speed that may be lost from increased drag, and drop the nose to descend if you're high. It's the old story of pitch attitude and speed being irrevocably tied together, a change in one affecting the other. However, it takes time for a speed change to affect pitch attitude in a jet; therefore, changing pitch attitude first is most effective in controlling descent and climb. This principle is even built into the flight director systems in use today, giving you a "fly-up" and "fly-down" indication to fly the glide slope in an ILS approach.

But pushing the nose down and reducing power brings up another factor that can be dangerous in jet operation. If a jet engine is pulled all the way to flight idle, there will be a delay in getting it back up to power. This delay varies with different engines. In a single-spool compressor engine such as the small business jets are powered with, it may be only 3 seconds; in large two-stage compressor section engines used in large airline jets, it may take as long as 15 seconds. This delay may be the difference between life and death at low altitude and slow speed.

This characteristic comes from a bleed valve (also called bleed strap and surge valve), which is open at lower engine speeds to relieve the load on the compressor section during start. The engine is referred to as being "unspooled" in the range of revolutions per minute below which this valve opens. Air is being bled off the compressor through the open valve, and the engine is producing virtually no power. When power is applied, it takes a few seconds for the compressor section to come back up to an RPM where the engine can sustain itself under a loaded compressor and for this valve to close.

The opening and closing of this bleed valve will occur at a percentage of RPM, usually N1 or first section of the compressor stages, and it is a good idea to learn the RPMs associated with bleed valve operation. You should know the RPM indications showing that the bleed valve has closed during power application (such as at takeoff). Operation of the engine at or very near the RPM where the bleed valve will begin to open will result in engine vibration and should be avoided. Operating at the RPM where the bleed valve will be on the verge of closing and the engine idling will allow you to return the engine power without the long delay you would expect from a full unspooled condition. For example, the JT8D engine used in the Douglas DC-9 and Boeing 727 may be set at 40% N1 RPM while at idle. The engine then is just on the verge of closing the bleed valve, and power may be returned in 1–1.5 seconds rather than the 8–12 seconds required from a full flight-idle condition.

Briefly, then, the "wheel landing" technique in a jet involves using an approach speed of 30% over stall, flying a glide path angle of approximately 3°, controlling your glide path angle and rate of descent with pitch attitude and elevator, and flying your speed with power. The only factor left to consider in approach is wind and its effect on an airplane.

Wind

The very earliest pilots knew that wind had a profound effect on approaches and landings. Uneducated pilots realized that if they had an approach speed of 90 in an absolutely calm wind, their ground speed would also be 90 at flare; and that if they were landing in a 20-mile wind and used an approach speed of 100, ground speed at flare would be 100 minus 20, or 80, and actual landing performance would be better than it was at 90 with zero headwind. From this basic knowledge, pilots began to apply a little extra to their approach speed for better control close to the ground and to correct for the lift variations they experienced in gusty wind.

They found that winds tend to decrease in velocity near the ground and that the greatest change in wind gradients occurred between 300′ and the earth's surface. They strongly suspected, based on experience that many years later was confirmed scientifically, that wind gradients are affected by terrain contours, wind velocities at altitude, temperature lapse rate, and convective influences caused by heating and cooling of the earth's surface. They discovered that while wind velocities above the ground are hard to predict, wind direction remains fairly constant below 300′ above the ground.

The best and most experienced early pilots, as a rule of thumb, added one-half the headwind component and all the gust to their approach speeds. Then along came the jets, and the engineers came up with the same application to approach speed. The same rule of thumb—one-half the headwind and all the gust (not to exceed 20 knots except under extreme conditions)—is modified to adding 5 knots to V_{ref} as a minimum for unknown wind in calm wind conditions. This is necessary because of the nonuniform height at which wind direction and velocity are measured at airports. All wind factors in landing performance are based on a wind measured at a 50′ height, but wind-measuring heights vary from 5′ to 100′ at different airports. Unless you happen to know where the anemometer is located, there is no way to know where the wind is measured, hence the 5-knot factor for unknown wind in calm wind conditions.

These wind gradient effects produce an instantaneous change of airspeed in a jet, which can be read immediately on the airspeed indicator. Their effect on lift is also apparent at once. Normally, wind gradients assist an airplane during takeoff because as the airplane climbs into an increasing wind velocity the indicated airspeed increases faster than the aircraft actually accelerates. But just the opposite occurs on landing. A high-level headwind that decreases as you approach the ground causes a decrease of indicated airspeed that could cause you to touch the wheels to the earth before the pavement is under you. This could be disconcerting, particularly if coming in over trees, buildings, or other obstructions.

To avoid this, you should be in landing configuration and on V_{ref} speed plus at least one-half the headwind reported when passing through 300′. You now have a built-in cushion for the wind changes that will be readily apparent in descent. You may expect some bleed-off of airspeed, and you should be mentally prepared to add considerable thrust if necessary to accelerate the aircraft if this bleed-off is more than expected as a result of the wind gradient being more than expected. Sometimes this wind gradient may be considerable, as Figure 5.4 illustrates.

This chart shows that a 30-knot wind at 300′ may vary from 0.5, or 15 knots, to 0.9, or 27 knots, at an average height of 50′. If you've done much jet flying, you've had experience with this factor, and the factors applied here are temperature and lapse rate only. The wind may be reported as 3–5 knots on the ground, and you are all stabilized on approach at V_{ref} plus 5, when suddenly the airplane starts to fall out of the sky at a low altitude and you have to apply power to fly it up to the airport. This usually occurs at 300′ above the ground or less, just beneath the base of the clouds on an instrument approach with a 300–500′ ceiling or on a clear day without a cloud in the sky when there is a wind aloft and it is calm on the ground. If you know the lapse rate, you can compensate for this; otherwise, simply add a normal wind component.

In considering the wind factor, it is necessary to consider only the headwind component, which will vary with the angle of crosswind. But even in cases of no headwind component, it is good practice to add 5 knots for unknown wind.

Gustiness must also be considered. In theory at least, gust effects should be added above any wind gradient correction. However, because you can (partially at least) compensate with elevators and power, and because some margin of safety is already built into the 1.3 V_s reference speed recommended for approach, the total speed correction added to the V_{ref} need not exceed 20 knots in most aircraft—a maximum of combined wind gradient and gust factor.

Theoretically, headwind component becomes almost 0 at 90° of crosswind, and the wind gradient correction would also become 0. But the gust factor should still be applied regardless of wind direction because of the possible shift of direction of the gusts.

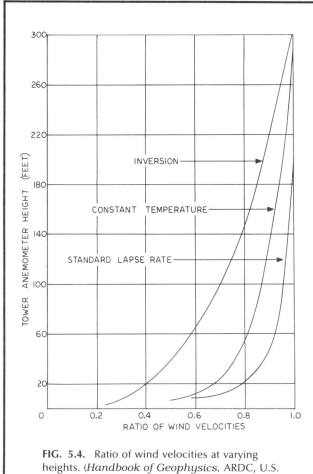

FIG. 5.4. Ratio of wind velocities at varying heights. (*Handbook of Geophysics*, ARDC, U.S. Air Force)

This should be added on top of the minimum of 5 knots added for unknown wind.

The Douglas DC-9 *Flight Crew Operating Manual* states it thus: "The full value of the gusts should be added to V_{ref} in addition to the allowance for the wind gradient effect, except that the total velocity increment for both gusts and wind gradient should not exceed 20 knots. In cases of crosswinds, the component of wind down the runway only need be considered for gradient allowance; however, the full gust allowance would still apply regardless of wind direction."

How would you apply these corrections to V_{ref}? Let's suppose the wind is reported right down the runway at 20 knots with gusts to 30. This could be considered as the "classic" example because of its resultant gust and wind gradient correction factor—20 knots added to V_{ref}. But let's see how the 20 knots is obtained. The wind is 20 knots, and one-half the wind is 10 knots. This gives you a wind gradient factor of 10 knots added to V_{ref}. The gust factor is the difference between the wind, 20 knots, and the velocity of the peak gust, 30 knots. Here again you have a correction of 10 knots, but this time it is a gust factor—the difference between 30 and 20—and this 10 knots is added in addition to the 10-knot wind gradient factor for a total of 20 knots added to the basic V_{ref}.

In another example, you might have a wind of 10 knots with gusts to 35 from an approaching thunderstorm. You take one-half the wind, 5 knots, and all the gust above the wind, 25 knots; the two added together gives you 30 knots. This is an example of where you would stop at the maximum of 20 knots added to V_{ref}.

Just how important are these factors in flying proper approach speed? Let me show you what could happen if a headwind should fall off only 10 knots below 300'. Figure 5.5 shows the wind gradient effect of a normal glide slope (profile A), which could change the flight profile as indicated unless the pilot is alert to the changes required to continue to the desired touchdown point. The second portion of the diagram (profile B) shows that a flat glide slope requires greater correction to prevent landing short when wind velocity falls off close to the ground.

Both diagrams show the flight paths followed by airplanes that hold the same airspeed as the wind velocity falls off by dropping the nose and keeping a constant power. The short approaches indicated would occur only if a pilot made no correction to airplane attitude to control the glide path and any power changes required for maintaining correct approach speed. The diagrams also point out two most pertinent facts: (1) As a pilot you must be alert to changes in the glide path and make appropriate corrections to enable your glide path to fly you to a touchdown point 1,000' down the runway; when an allowance for headwind has been added to the reference or approach speed of 1.3 V_s, you should allow the airspeed to bleed off close to the ground rather than attempt to hold the approach speed plus one-half wind allowance. (2) Corrections are more difficult during a flat approach with a low rate of descent than during a normal approach of 3° or a normal ILS glide slope angle.

In the final analysis, the pilot's judgment should remain the controlling factor during any landing, since the actual wind pattern and behavior on every landing approach can vary considerably. Any rule or statement regarding wind characteristics near the ground can at best be only a crude generalization, but wind gradients in relation to lapse rates are usually quite reliable.

If you're wondering about that 1,000' aiming point, take another look at Figure 5.5 and note the aircraft altitude when crossing the end of the runway. The normal glide path shows the aircraft to be about 50' high at runway end, and the shallow glide path shows the aircraft about 35' high at the same point. If you've ever watched a jet approach and land, you've

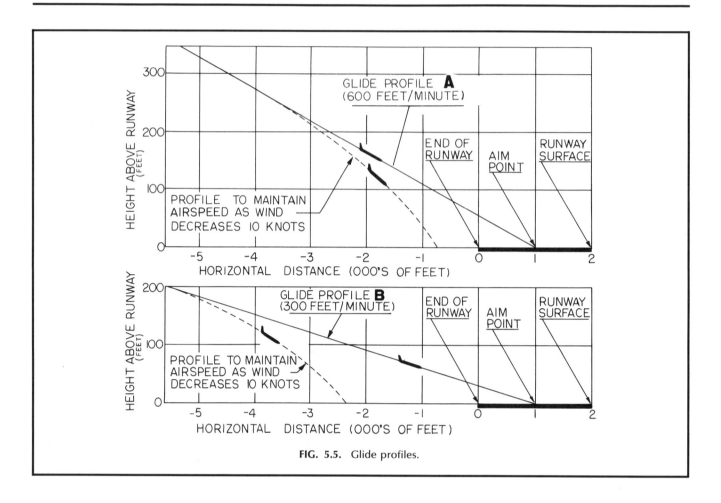

FIG. 5.5. Glide profiles.

no doubt noticed the nose-high attitude. This is to compensate for the size of the plane and distance aft from the cockpit that the main gear is located. If you aim any shorter than 1,000′ down the runway, you'll very likely stick the wheels into whatever happens to precede the runway pavement. You're aiming the cockpit, and you're 50′ high, *but*, depending on the airplane, the gear may cross the end of the runway only 5′ high. The longer the fuselage, the farther aft the gear is from the cockpit, so the closer to the ground the wheels are when the cockpit is 50′ high in approach. You are landing a set of wheels that may be 50–100′ behind you, and a short aiming point will put them on the ground before you want them there.

Preparation for Flight Training

After completion of ground school and while engaged in "home study" of your airplane manual while getting ready for your oral, you should also be preparing yourself for flight training.

Training programs have not changed a great deal in general aviation. However, the major airlines, flying the large jet aircraft, have made substantial changes, both in ground school and flight training. As mentioned in Chapter 1, the excellent devices and training aids now available in ground school are also virtually a phase of flight training, with the extensive use of very sophisticated cockpit procedure trainers that are actually better than the first simulators used for the DC-7 and L-188 (Lockheed Electra) in the 1960s. As a result, it is now the ground school instructor who provides the first cockpit experience and recommends the student for the oral. However, it is still the flight instructor who recommends the flight check.

The major reason for this change is that it is too expensive to use large aircraft extensively in flight training. The direct operating cost is considerable, but the revenue loss of taking the aircraft out of scheduled service is tremendous. All of the flight training and the rating and/or check ride is conducted in the simulator. There are two exceptions: a second officer upgrading to first officer or a first officer on the first upgrade to captain. Then the aircraft is used for three takeoffs and landings to meet minimum recent experience requirements.

This brings about an unusual circumstance. I have been told by captains upgrading that their first takeoff and landing in the aircraft was during line operation with a load of passengers! This is not an unusual or hazardous happening. The upgrading captain is under the supervision of a checker in the right seat until completing the initial line qualification in the aircraft, and the simulator training has been more than adequate for the applicant to be thoroughly competent to fly the airplane.

The simulator is the best training device the industry has ever known. Its performance capability is identical with that of the aircraft, and it provides the capability of training in situations that would be impossible to meet in the aircraft with safety. Use of data derived from accidents may be programmed into the simulator and flown in a practical real world environment. Emergency procedures may be used that would be extremely hazardous in the aircraft. The simulator may be slowed down, frozen, backed up, etc., at critical points so that the pilot can analyze mistakes in each phase, and the learning potential is tremendous. In my opinion pilots that receive all their training in the simulator are far better trained than ever before. When you see the space shuttle landing on television, for example, that pilot was trained in the simulator and may be making his first landing ever in the aircraft . . . and with no power!

The ground school format is the first step in preparation for flight training. But the professional pilot will also be preparing for flight training. Every well-written manual for airline aircraft will have a section devoted to flight training, explaining in some detail every maneuver in the flight training curriculum. This material is added to the manual by the flight training department of the particular airline, usually as a part of the normal operating procedures, though it may be a separate chapter; in some instances it may be written as a separate Flight Standards Manual for the particular aircraft. In either case you should review each maneuver and procedure thoroughly, paying particular attention to entry speeds, expected performance, stall entry and recovery, profiles of traffic patterns, instrument and ILS approaches, engine failure on takeoff and V_2 climb-out, and engine-out landings. Knowing what is expected of you and how each maneuver and procedure is to be performed makes flight training much easier for both you and your flight instructor.

Some instructors furnish outlines of training flights so that the student knows in advance just what to expect on each flight, assuming normal progress. These outlines, coupled with study of the training section or Flight Standards Manual, make training easy for the pilot who is really a professional. If, in addition to such an outline, the pilot is furnished with a complete description of each maneuver and procedure, the check-out is simply a matter of becoming familiar with the cockpit and airplane; the systems; normal, abnormal, and emergency procedures; and practicing to a degree of proficiency.

You may expect at least eight training periods in the simulator; followed by a "pre-rate" check, then the rating ride itself. With a good instructor and diligent preparation on your part, you should have no difficulty and be well trained and prepared to fly the aircraft upon completion of training.

The newer aircraft such as the Boeing 767 and 757, the MDA-80, etc., are a new world for the pilot. They have navigation systems and autopilots comparable to the space shuttle, which make the pilots virtually computer programmers and monitors. It is important to remember that "garbage in equals garbage out." Be particularly deliberate in making entries, and they should be rechecked by both pilots before using the data entered for flight operation. The performance capabilities of these newer aircraft, instruments, and navigation systems offer a precision that is incredible and make them a delight to fly. You will enjoy every minute in the air, but always remember that computers do not think and make decisions.

Preflight

Prior to flight, either in the simulator or aircraft, a thorough cockpit and/or airplane familiarization will be given each captain and first officer. This will include the location and use of emergency equipment.

Adequate preflight training will be accomplished using visual aids, perhaps a preflight film, and a static aircraft. An instructor will conduct you through a thorough walkaround and an internal preflight inspection of the aircraft. If the aircraft is used in training, you will be expected to accomplish a preflight.

The inspection should be conducted in accordance with the preflight procedure outlined in the aircraft manual. The purpose of this portion of training is to provide a systematic program with which the pilot can determine the airworthiness of the airplane with efficiency and certainty.

Then the pilot will be expected to conduct normal cockpit procedures from the operating position and to ascertain that all necessary flight instruments and navigation equipment are functioning properly.

Briefing

A thorough briefing will be conducted by your flight instructor before each flight period and will include discussions of all maneuvers and emergency procedures to be accomplished during the period. The first briefing period will include crew duties, checklist use, engine starts, and taxiing. Each flight period will, of course, be followed by a critique.

General

All normal, abnormal, and emergency operating procedures will be accomplished in accordance with the flight manual. All initial, transition, and upgrading flight training will be conducted in accordance with the maneuvers and procedures set forth in the manual. Recurrent flight training will be accomplished in accordance with FAR 121, Appendix E or F.

Engine Start

The first part of flight training will no doubt be starting the engines and be accomplished in accordance with procedures outlined in the flight manual. Jet engines are very expensive; certain malfunctions while they are at relatively slow speeds in starting, such as hung starts and hot starts, can cause a lot of damage. You may expect training, therefore, in both normal and abnormal starting procedures. The trainee should review engine starting procedures thoroughly. Since these malfunctions will not damage the simulator, your instructor will give you enough problems to ensure your immediate recognition and appropriate corrective action.

Potential hot starts can be detected and avoided by comparison of engine RPM and exhaust gas temperature (EGT) rise. In the 757, when fuel is fed, ignition occurs, and combustion is indicated by EGT, a comparison of N3 RPM and EGT rise is necessary. If EGT rise is more rapid than N3, over 300° vs 30% N3, 400° vs. 40%, etc., a hot start is almost assured and the start should be aborted.

Taxiing

Jet engines produce exhaust that reaches high velocities. When maximum taxi power is used, jet wake velocities may be as high as 50 MPH at 150′ behind the engines, reducing to 25 MPH at 300'. To preclude damage to fixed physical facilities or ground support equipment due to jet exhaust blast, throttles should be moved slowly forward (maximum of 50% N1 RPM with this engine) after receiving the "clear to taxi" signal, and this value should not be exceeded. Know the maximum value for the engine you are operating.

When the aircraft was used for flight training, proper taxi technique was easily taught. It can also be taught in the simulator but should be discussed, especially for those making the first transition into jets. From the preceding paragraph it is apparent that good taxi technique requires an awareness of the proximity of obstacles, the effects of excessive noise, and the force of jet exhaust. Therefore, the following are important considerations when taxiing:

1. Do not taxi with engines in reverse except for power back from the gate or ramp. During power back the ground crew should watch the wing tips and engine nacelles carefully to assure that they will clear equipment on the ramp, remembering that the wing tips effectively move outboard in a turn.

2. Differential braking is not normally used to taxi and should not be used in conjunction with nosewheel steering. This can result in excessive side loads on the nose strut. Differential braking *may* be used if the nosewheel steering becomes inoperative.

3. Make all turns with as large a radius as possible, considering width of ramp areas, taxiways, and runways.

4. Rudder pedals provide nosewheel steering that is usually sufficient for takeoff, landing, high-speed turnoffs, and minor turns while taxiing. In the rare case where rudder pedal steering is inoperative for takeoff, use nosewheel steering (wheel or tiller) until rudder effective speed is attained.

5. Make all turns slowly and smoothly, returning the steering to center and allowing the airplane to roll forward, thereby relieving nosewheel and main tire twisting stresses.

6. Never exceed maximum taxi power and occasionally look out the side window to judge taxi speed.

7. Relatively high idle thrust may cause acceleration to a higher taxi speed than desired. It is not desirable to continually use the brakes to prevent the aircraft from accelerating, since there is very little brake cooling when the brakes are in continuous contact. Allow the airplane to accelerate, then brake to a slow taxi speed, release brakes, and repeat the sequence. Intermittent brake usage in this manner provides a cooling period between brake application.

8. Use of an engine (with aircraft with engines mounted outboard on the wings) will assist in minimum radius turns. Use the least thrust possible and do not allow the airplane to stop while turning unless necessary.

Summary

Flight training is an important part of a pilot's career. Reports of each training period, grading performance, and progress will become part of pilot's training files kept by the airline throughout their entire careers. This is also true of rating rides and proficiency checks. The FAA also keeps a record. Pilots, therefore, should thoroughly prepare for each training period and check ride and strive for above average performance.

7

Basic Instrument Scan

Instrument flying, or any type of flying for that matter, is merely flying the wing. You position the wing into the desired attitude and control its speed for the desired performance to execute any maneuver you can think of. It may be compared to a mathematical formula: Attitude + speed = desired performance. The maneuver may be climb, descent, roll, turn, or a combination of these; if you correctly control the wing's attitude and speed throughout the maneuver, you'll execute it perfectly.

Performance (the desired maneuver) remains a constant on its side of the equation, but the attitude and speed are variables. Being on the same side of the equation, a change in the value of one will affect the other. A change in pitch attitude, up or down as if to climb or dive, will affect the speed and either increase it or reduce it; conversely, a change in speed will cause a change in pitch attitude.

That's the whole secret of flight in any airplane and for any maneuver. It doesn't matter if the plane is large or small; just fly the wing, make it work for you at all times, and you can make the plane do anything its limitations will allow.

Instrument flight is no longer a matter of needle, ball, and airspeed. The modern flight director systems are designed for and fully capable of flight by attitude reference during any maneuver or aircraft attitude.

The horizon, or approach horizon, is the basic instrument of the attitude group, but much of the technique learned in needle, ball, and airspeed has been carried over and incorporated into attitude instrument flight. Controlling an airplane's attitude by instruments is very similar to flying by visual reference. The instruments merely replace the natural horizon and in large aircraft supplement visual reference to a great degree even in VFR operation. When used properly, flight instruments produce a more precise control than is possible by visual reference alone. To be proficient in their use, you must be familiar with the flight instruments used to control attitude, be able to interpret their indications, and be able to use instrument indications to control the plane in the desired attitude.

Instrument scan (reading, interpreting, cross-checking, and using flight instrument indications to control the attitude of the plane) is the key to instrument flight. A pilot who uses the proper scan for the performance of various maneuvers and phases of flight invariably performs much better and adjusts to new equipment more easily and rapidly than the pilot with poor scan. In transition the pilot may also be learning the use of new types of instruments, such as the various integrated flight systems, but the basic scan technique is still the same.

One of the biggest difficulties experienced by new-hire pilots in initial flight training (and occasionally by more experienced pilots) is a lack of this elusive thing called instrument scan. Even though they have an instrument rating, they still apparently do not understand the instruments and therefore are unable to use them most effectively. This chapter is written especially for pilots with limited experience, but a basic discussion of instrument flight may prove useful to veteran pilots as well.

The British are credited with being the first to use the full panel and to teach attitude instrument flight rather than the 1-2-3 system of the needle, ball, and airspeed. In the United States, even as late as 1954, military and private aviation was being taught with needle, ball, and airspeed. Aircraft were equipped with gyros (artificial horizons and gyrocompasses), which were subject to precession and tumbling in extreme attitudes. It was necessary that a pilot be able to fly and recover from unusual attitudes by use of the turn and bank indicator in conjunction with airspeed.

The airlines, followed later by the military, were the first in this country to use and develop attitude flying. Eastern Air Transport, when it was a part of North American Aviation, along with Sperry and other aviation corporations of the General Motors complex, pioneered the use of attitude instruments. In 1930 the first automatic horizon was developed by Sperry for greater safety at night, and instrument flight was installed in a Pitcairn Mailwing. It wasn't long until this instrument was installed in all Eastern aircraft and adopted by other airlines and the military. But it wasn't until 1956, when reliable instruments were developed, that the military began to teach attitude instrument flight as a primary method. The lessons learned over the years were incorporated into a new primary instrument system, which the FAA and civil aviation (other than airlines) began to fully utilize in the late 1960s. I believe it is the best, simplest, easiest, safest, and most reliable system of instrument flight in the world.

This "primary system of attitude instrument flight and scan" is based on the theory that attitude breaks down into three parts: pitch, bank, and speed control. The instrument or instruments that are to remain constant are the primary instruments, the starting point for scan in relation to the three phases of attitude, for any maneuver or phase of maneuver. Since these primary instruments vary for different maneuvers, there is no set pattern of instrument scan. You vary your scan in relation to attitude control, always beginning your scan for pitch, bank, or speed with the primary or constant instrument for that particular attitude.

The three phases of attitude (pitch, bank, and speed control) are combined in flight but are learned most easily by considering them separately. When understood separately, they are easily combined using the primary system of scan. In this chapter we will discuss the three parts of attitude and the instruments related to each; the combined scan technique will be dealt with in the chapters on flight maneuvers.

Pitch Control

HORIZON

The horizon is used to maintain proper attitude to hold altitude in level flight or to establish the proper pitch attitude in maneuvering. Since it pictorially displays the aircraft attitude in relation to the actual horizon, it is considered to be the basic instrument of attitude flight. But it isn't necessarily the primary instrument for pitch attitude; it is primary only when you wish to establish a constant body angle in climb or descent or when you want to hold the pitch attitude fixed for some specific purpose. In every other case it is secondary. You discover your pitch error in deviations from the desired altitude, rate of climb, or airspeed and use the horizon to change the pitch angle to correct these errors.

The horizon also has an acceleration and deceleration error in its indications. With a change of speed, the horizon, if followed, will result in a gain or loss of altitude. If you decrease your speed, holding the same pitch attitude, you will descend. Conversely, by increasing speed, holding the same pitch attitude, you will climb. From this comes the impression that an aircraft climbs and descends solely as a result of power. This is a fallacy, because if you hold the nose high enough to get on the back side of the power curve, the aircraft will stall and descend even with full power. If you hold the nose down in a dive, the aircraft will descend with power on or power off. Climb and descent are functions of pitch attitude, establishing the proper angle of attack; and the *rate* of climb or descent is in direct relation to power.

Thus it is important to recognize the function of the horizon (and its errors) in pitch control; to remember that altitude, rate of climb, and airspeed are related to pitch, each affecting the other when changes are made in any one of them, and that they are irrevocably tied together; and to realize that the plane's pitch relation to the artificial horizon is *always* controlled by the elevators and that smooth pressures are required to prevent overcontrolling.

ALTIMETER

Level flight may be defined, with a constant power setting, as constant altitude and attitude. The altimeter is the instrument that should remain constant in level flight and is the *primary* instrument for pitch while maintaining a specific altitude.

I'll repeat that, phrasing it a little differently. Maintaining a constant or specific altitude is a must in most maneuvers. The parameters for check flight purposes are plus or minus 100'. Anything outside those parameters is unsatisfactory: 70–100', below average; 40–70', average; 0–40', above average. Good altitude control may be obtained by recognizing and using the altimeter (since it must remain constant or be controlled within these parameters) as the *primary* instrument for pitch control when a constant altitude is to be maintained.

If it is apparent that altitude is being gained, the nose is too high; if altitude is being lost, the nose is too low. If altitude is being gained or lost rapidly, there is also a large change in attitude that may be read on the horizon. By interpolating the rate of movement of the altimeter (after you have become familiar with the particular aircraft in which you are practicing and its control pressures at various speeds), you should be able to visualize the approximate change in pitch attitude.

There is a lag in the altimeter. If you make an abrupt change in pitch attitude (overcontrol in making a pitch correction), there will be a momentary lag, but then the altimeter will catch up and show the resultant change in altitude, which may be considerable. If pitch corrections are made in small increments and smoothly, this lag is not apparent, and the altimeter can be considered as giving immediate indication of a change in pitch attitude.

Using the altimeter as the primary instrument when flying an assigned altitude, and beginning your pitch scan with the altimeter, a continuous cross-check of the horizon must be maintained and all corrections made smoothly and small. Errors of 50' of altitude, well within the average parameters, may be corrected by an attitude change of only ½ bar width on the old-type horizon, or about 1° or less change in pitch attitude.

RATE OF CLIMB OR VERTICAL SPEED INDICATOR

The rate of climb or IVSI (instant vertical speed indicator) gives an immediate indication of a change in pitch attitude, though a short time elapses before it settles down and gives an actual rate of vertical speed. This time lapse in normal operation is negligible with the IVSI. It is an extremely accurate and useful instrument in normal operation, but it is affected considerably by G forces and may give indications induced by these G forces rather than indicate actual vertical speed. This characteristic must be taken into account under such conditions. Steep turns, as an example, induce G forces of a magnitude sufficient to affect the IVSI indications.

The relationship between the rate of climb and the horizon is dependent on airspeed, the horizon (pitch attitude) changing but little in relation to vertical speed at high speeds and by a greater amount at low speeds. Therefore, it may be said that the lag in this instrument is proportional to airspeed and magnitude of pitch change. Even though it gives an immediate indication of pitch change, due to this lag it should not be chased and considered as a trend instrument in level flight.

At speeds between 200 and 150 knots (the range of many maneuvers) 200' per minute on the rate of climb is seen as a very small movement of the horizon. Therefore, it is obvious that this instrument shows a change in pitch attitude faster than the altimeter does and is easier to detect than movement of the horizon.

To maintain altitude in level flight, a vertical rate of no more than 200' per minute is sufficient to correct altitude changes and errors of 50', and the rate of climb becomes a primary instrument during the few seconds required to correct altitude. The rate of climb must be included in instrument scan and properly related to the horizon and altimeter indications, using small control pressure for corrections, to maintain level flight and a constant altitude.

When an altitude error is discovered in your pitch scan, check your horizon reference, change pitch attitude to a desired rate of vertical speed to correct the error, and go back to altimeter reference as primary to cease the correction.

The rate of climb is the primary instrument for pitch in climb or descent at a specified rate and in correcting altitude errors in level flight. In all other cases it is a trend instrument.

AIRSPEED INDICATOR

With a constant power setting in level flight, maintaining a constant altitude, airspeed will remain constant. But small pitch changes will produce slow and slight changes in airspeed, and large changes will produce faster changes. There is no appreciable lag in this instrument, but there is an apparent lag due to the time required for the aircraft to accelerate or decelerate. So the airspeed indicator must also be included in the pitch instrument scan.

You will be required to maintain constant airspeed in some maneuvers, and a proper interpretation of all pitch instruments is required to do this well. The more quickly a deviation is noticed, the smaller the amount of control pressure necessary to counteract the deviation, and the smoother and more accurate the resultant performance.

Remembering that speed is affected by pitch attitude changes and that airspeed also comes under the third phase of attitude control as speed control and is therefore primary for that purpose, we'll discuss it here only in the light of pitch control.

Airspeed is primary for pitch for climbs and descents with a constant power setting and when airspeed alone is the predominant factor of such climb or descent. When rate of climb and altitude considerations enter the picture, they become the primary pitch instruments, and airspeed is once more the primary instrument for power requirements.

Airspeed would be primary for pitch, for example, when climbing at a fixed or desired airspeed with a constant climb power setting. This speed would then be controlled by the elevators and resultant pitch attitude. It is also primary for V_2 climb speeds after engine failure, engine-out climb speeds, normal descents at both high and low speeds, and emergency descents.

Before leaving the discussion of pitch control, trim should be mentioned. Elevator and stabilizer trims are important. The correct use of elevator trim is an integral part of pitch control, and good technique is very difficult without it. In high-performance jet aircraft with power-driven stabilizer trim, first make your correction with elevator control and then trim the stabilizer so that there is no control force or pressure left or required for the elevator.

Bank Control

I won't go into needle, ball, and airspeed. I'm assuming that all pilots know how to use the turn and bank indicator, or turn and slip as it is called now. It essentially shows the direction and rate of turn and slip or skid. But it is very useful for trimming the airplane, particularly after an engine loss or with an uneven fuel load. And the aircraft should be in proper trim, both rudder and aileron, for good bank control technique.

Turbojets are not as trim sensitive (not as critical in trim requirements) in yaw and roll as propeller driven aircraft. The rudder and aileron trims are centered before takeoff, symmetrical power is used throughout the flight, and very little trim about these axes is ever required except for an engine failure or an extreme fuel imbalance in flight. But turboprop and piston-engine aircraft, propeller driven, are extremely sensitive to rudder trim as they pass through a wide range of speed, power, and torque changes during a flight.

The Lockheed L-188, the Electra, is a good example of this rudder trim requirement. It is normally trimmed for 5° right rudder prior to takeoff. This setting is for an average speed at takeoff and maximum continuous power, ranging from V_2 to engine-out climb speed. As in most prop aircraft, the vertical fin is also offset for torque at cruise. As the speed and power are changed, the trim requirements change, and the aircraft must be retrimmed for yaw for maximum performance in all cases.

I've seen many landings in the Electra where the pilot closed the throttles to flight-idle in the flare just prior to touchdown, and the airplane invariably drifted to the right. Almost as invariably, the pilot didn't recognize the cause and corrected by lowering the left wing, thus landing with the left wheel slightly ahead of the right. This yaw or drift is caused by the vertical fin offset and the absence of torque at flight-idle. Pilots that are aware of this characteristic use a little left rudder as they reduce power, so that they land straight and on both wheels. The moral is: Be familiar with your aircraft and its trim requirements.

To trim for bank, as in pitch, you use all the instruments. An easy, quick trim may be accomplished by holding the wings level (or in a constant bank if necessary) and trimming the ball to the center with the rudder. Then trim off any pressure you may be using on the aileron with aileron trim to hold the wings level.

You have accomplished quick trim; fine adjustments of both rudder and aileron may then be necessary to trim for level flight (wings level and no turn indicated).

There are only three instruments to show the bank attitude of an aircraft—horizon, compass, and turn needle. The horizon shows an actual angle of bank; the compass and turn needle indirectly indicate a bank by turn rate. However, the turn needle is not necessary in full-panel attitude scan. Its primary use—rate of turn at 3° per second for a standard rate turn—may be performed by the horizon. You can take the first two digits of your *true* airspeed, add 9 to it, and use the sum as an angle of bank of 3° per second. As an example, with a true airspeed of 160 take the first two digits, 16, and add 9. This gives you a sum of 25, and a 25° bank at that speed will give you exactly 3° per second. This little trick may be used at any true airspeed with a high degree of accuracy.

Accurate interpretation of these instruments is essential for good bank control. As in all flight, errors in bank attitude should be noted quickly so that corrections will be small.

In bank control, as in pitch, there is always a primary instrument. *It is the one that should be stationary during the maneuver performed.* In straight flight, the compass should remain constant; therefore, it is the primary instrument. Simply stated, if you're not turning, you're not banking.

The approach horizon when used for heading mode (or computed heading in instrument approach) also becomes a primary bank instrument in relation to heading, selected heading, or computed heading. The horizon must then be scanned for pitch and bank information simultaneously and continuously. Correlate the information from all the bank instruments and correct any errors in bank and pitch that may be indicated.

A word of advice: The approach horizon receives its information from other instruments. It is wise to check these other instruments, such as the actual ILS course during approach, against the approach horizon. The approach horizon essentially tells you where you are going, and the "raw data" instruments tell you where you are. The approach horizon is a fantastic instrument, capable of making approaches very simple, but do not be lulled into a false sense of security and become dependent on it. Cross-check actual "raw data" information against the approach horizon at all times!

The horizon is, of course, primary for turns at a given angle of bank. It should remain constant, for example, while making a 30° banked turn.

When bank and pitch control are combined, at least two instruments become primary. For example, in straight and level flight the compass and altimeter are used as primary instruments to hold a constant heading and altitude. Combining bank and pitch control requires a more rapid instrument scan, and interpretations must be very accurate. An ability to scan, cross-check, interpret rapidly and accurately, and use the information the instruments present must be developed to allow more time for attention to other matters such as power changes, radio navigation, instrument approaches, etc.

Your heading control must be within 10°, plus or minus, and anything outside that tolerance is unsatisfactory. Heading control within 3° is excellent; 3–7°, average; 7–10°, below average. Incidentally, heading control is graded for smoothness as well as accuracy.

Before leaving bank control, there is a rule of thumb for correcting heading: When the actual heading deviates from the desired heading, the plane is to be returned to the desired heading by the use of a bank not to exceed the number of degrees to be turned. As an example, to correct your heading by 10° or to change course 10°, use no more than a 10° bank.

Speed Control and Power Changes

Power changes, which definitely affect the attitude of an airplane, require the proper use of elevator trim to maintain a constant altitude or pitch attitude and the use of rudder trim to offset torque in propeller-driven aircraft for bank and heading control.

In making power changes in attitude flight, good scan is essential; the primary instrument for pitch, bank, or speed is always the one that should remain constant. The instrument scan for attitude must be rapid and correct, since changing power tends to change the plane's attitude, particularly in pitch. An application of power tends to raise the nose, a decrease in power tends to drop the nose. This is almost immediate in multiengine propeller-driven airplanes. The propellers send either more or less air over the wing. High power for acceleration blows air over the wing behind the propellers, and low power lessens that flow of air. This change in airflow is rapid and immediately affects the lift and causes the nose to rise or drop, as the case may be. But power application does not affect airflow over the wing of a jet in that manner. The wing is just a board, either pushed or pulled through the air by the jet engines, and an artificial airflow faster than the speed of the wing itself is not blown over the wing as with propellers. The pitch attitude will change but only as a result of the actual speed of the wing. More power will move the wing through the air faster, causing it to create more lift and therefore affecting pitch attitude by raising the nose. Reducing power will have an opposite effect, slowing the wing and causing it to seek an angle of attack for the speed at which it is trimmed. But the jet reaction will be a little slower because acceleration or deceleration must occur to affect the lift and have a resultant pitch change.

The wing is the secret of flight. Without the necessary lift, we'd never get off the ground. But we also need power to move that wing fast enough to create the lift to sustain flight, and the two must be combined. Since this is true, the airspeed indicator is the primary instrument for power control. Speed, while affected by pitch attitude, is primarily a function of the amount of power being used. But never forget that speed and attitude are tied together inseparably; a change in one immediately causes a change in the other. You must learn to coordinate one with the other until it becomes instinctive. An airplane climbs or descends as a result of two things: (1) its angle of attack and (2) its speed through the air.

To decrease airspeed in level flight, or any reasonable pitch attitude for that matter, reduce the power to a setting below that required to maintain whatever the desired speed may be. Your power reference (depending upon what you are flying) might be manifold pressure, RPM, horsepower gauge, BMEP, engine pressure ratio, fuel flow, etc., but the power must be reduced to below the power required to give you the speed desired. The throttles should be moved smoothly and accurately in all power changes. It is not necessary to monitor your power readout except to be sure that the power is below what you estimate will be required to maintain the speed to which you are slowing but not so low as to cause piston slap in piston engines or to affect pressurization in turbine-powered aircraft.

As you begin to decelerate, proper control pressures or trim must be used to maintain the desired bank and pitch attitude, since the control pressures also change as the power is changed and the speed reduced. The pitch attitude must be changed as necessary to maintain a constant altitude, and the primary system of scan is required (the altimeter primary for level flight, the rate of climb primary in a constant climb or descent, etc.).

In any constant-speed maneuver (increase or decrease) the airspeed indicator is the primary instrument for power control; the engine power gauge is used solely as an aid in making the desired power changes to maintain a given speed. The airspeed indicator becomes primary for speed as it approaches the desired speed, and the power should be set to that which will maintain the desired airspeed a little less than 10 knots before reaching it. (This varies with different aircraft. Generally, props and turboprops should be led about 5 knots; jets from 5 to 10 knots depending on the particular aircraft.) Using the proper lead as the airspeed nears the desired speed, you have to look at the power gauge to reset the power. After setting it approximately, check the correctness of your power by airspeed indications and make small adjustments as necessary, using the airspeed indicator solely; don't even look at the power gauge. You're being graded, not on how well you can set 1.20 EPR, or 1,000 HP, or 30″ MAP, but on how well you can fly a given airspeed. The grading tolerance is ±10 knots using the same parameters as for heading.

To increase speed, use the same technique in reverse order; increase power to a setting greater than that estimated for desired speed; trim the nose down from the pitch change induced in acceleration; lead the power reduction the proper number of knots for your aircraft by observing the airspeed indicator; set the power to the approximate desired power setting; and then make any necessary small power changes from airspeed indications.

While maintaining level flight at given speeds (such as slow flight, holding, etc.), it is important that exact altitude and airspeed be maintained. It is done most easily if altitude is maintained with pitch control and airspeed maintained with power control. The need for a correction in power or pitch change is then indicated by a cross-check of the altimeter and airspeed. When the altitude is correct and the airspeed is off, a power change is necessary to attain the desired airspeed. When the altitude is low and the airspeed is high, or when the airspeed is low and the altitude is high, only a change in pitch attitude may be necessary to regain desired altitude and airspeed. When both altitude and airspeed are high or low, changes in both pitch and power are needed.

Summary

1. All flight may be defined as controlling an airplane's attitude and speed.

2. Attitude relates to the position of the wing, its angle of attack, and bank.

3. Attitude + speed = desired performance.

4. Attitude and speed are variables, each affecting the other.

5. The instrument that is to remain constant is the primary instrument, the starting point of scan for attitude.

6. Pitch attitude is seen in four instruments: horizon, altimeter, rate of climb, and airspeed indicator.

7. Bank attitude is seen in the compass for heading and the horizon for angle of bank.

8. Airspeed is usually the primary instrument for power control.

8

Takeoffs

■ The takeoff maneuver is one to which most instructors give very little time, especially when training experienced pilots for transition into new aircraft. All pilots have been performing takeoffs of one kind or another from the very beginning of their flying careers, and it is assumed that they know all there is to know about taking an aircraft off the ground. Other than the peculiarities of the specific aircraft, very little is said about the basics of the procedure. Perhaps this is as it should be. But the takeoff, while being essentially easy to perform, can be the most potentially dangerous part of any flight.

From a training standpoint, there are several types of takeoffs to consider: normal takeoff both from a standing start and rolling, rejected takeoff, takeoff with an engine failure and V_2 climb-out, and instrument takeoff with the hood up below 100′. The actual execution of these maneuvers is not hard, and most pilots in controlled flight training programs do them with little difficulty. But very few pilots know, or remember from their early training, all the factors involved in the takeoff.

One fact above all else must be considered. An aircraft is not a ground vehicle and is out of its natural element while on the ground. It handles very well in flight and at taxi speeds on the ground, but it is an extremely fractious and temperamental machine as it is accelerated from taxi speed to takeoff speed. It is during this period that it has the potential of getting away from the unwary.

This is the transitional period, from being groundborne to being airborne, that requires an abort if a malfunction occurs. Rejected takeoffs will be discussed in Chapter 9; in this chapter, we'll cover the normal takeoff, the takeoff with engine failure, and the hooded takeoff.

Before discussing the actual flight performance of the takeoff, we should review the factors involved (many of which pilots would have to obtain from performance charts for their particular aircraft), some of the terms related to takeoff, and the profiles of the various climb segments.

Minimum Takeoff Field Length

The minimum field length required for takeoff is certainly one of the pilot's first considerations and varies with gross weight, temperature, altitude, wind, and runway slope. These are the criteria used in certification tests of the aircraft, to which I strongly recommend adding *runway conditions.* Wet, slushy, icy, and slippery runways certainly affect both acceleration and stopping distance in an abort. Using the aircraft performance charts, the pilot should determine the effects of runway slush conditions and make the necessary correction to gross weight (about the only factor over which the pilot has control) for safety. However, no one but the military considers runway conditions as they might affect stopping distances in accelerate/stop performance. Airlines base runway requirements on the basic performance data obtained on a dry concrete or asphalt runway, adding only 15% for "wet runway" conditions—except for slush, which I'll discuss later.

An inoperative antiskid system will also add to your runway length requirements. Here again, you can't change the field length, altitude, temperature, or wind, so you must reduce weight according to maximum allowable for the runway. This will usually be found in the first few pages of your aircraft *Performance and Planning Manual* or in the *Runway Gross Weights Manual* (prepared for every runway and airport served by your airline by the engineering department for both takeoff and landing) and is a percentage reduction of weight for maximum allowable.

The takeoff field length required depends also on the V_1, V_R, and V_2 speeds, which in turn relate to weight, altitude, temperature, runway slope, and wind.

Not considering runway condition, inoperative antiskid, etc., the minimum runway length requirements for takeoff will be the longest of the following distances (shown in profile in Fig. 8.1):

1. Takeoff distance with an engine failure—The distance required to accelerate to V_1 with all engines

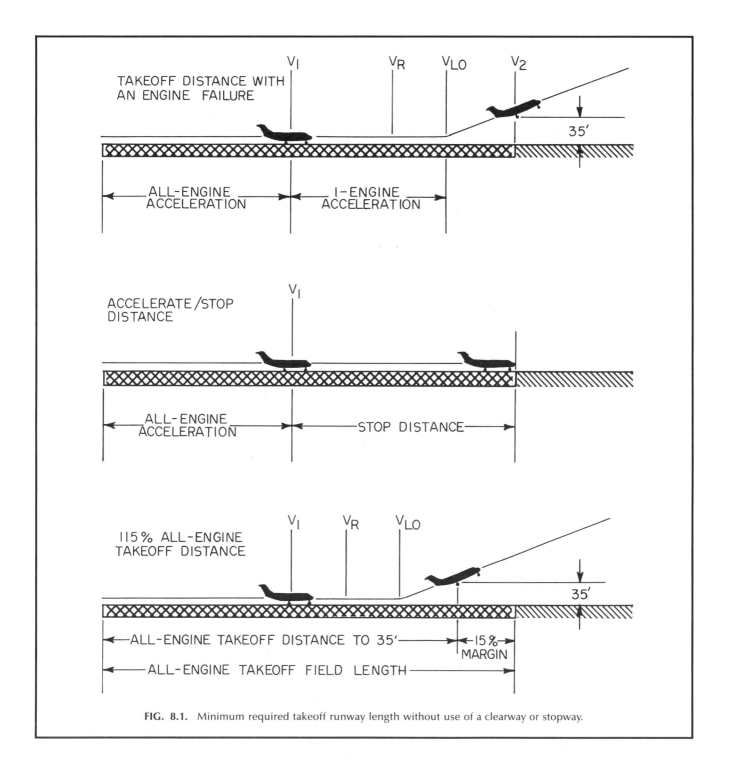

FIG. 8.1. Minimum required takeoff runway length without use of a clearway or stopway.

operating normally, experience the loss of an engine, and continue to accelerate on the remaining engines to V_R, at which time rotation is commenced to reach a height of 35' above the runway at V_2.

2. Accelerate/stop distance—The distance required to accelerate to the decision speed (V_1) with all engines operating normally, experience the loss of an engine at V_1, retard the operating engines, and bring the airplane to a full stop. (This will also be discussed in Chapter 9).

3. All-engine takeoff runway length—115% of the distance from brake release to a point where a 35' height above the runway is reached, with all engines operating normally at takeoff thrust.

Clearways and Stopways

In addition to the runway lengths just described, the runway lengths required may include the use of a clearway or stopway. Operations at the high gross weights commensurate with the increased performance capabilities of jet aircraft are enhanced by the incorporation of clearways and stopways. That's just a fancy way of making a high-weight takeoff legal without the actual runway pavement to meet the criteria we just discussed. The stopway is used to meet accelerate/stop distance requirements and the clearway to attain the 35' at V_2 *beyond* the runway end. Both are illustrated in profile in Figure 8.2 for two-engine aircraft; the same criteria apply to any multiengine T-category aircraft.

Note that the takeoff distance with engine failure shows the aircraft stop distance right at runway end and the V_1 takeoff attaining V_2 at 35' right at runway end. This is called the "balanced field concept," where accelerate/stop distance equals takeoff-with-engine-failure distance. Another more descriptive name for it is "critical field length."

A stopway is an extension *beyond the runway end* that may be used to decelerate an airplane in case of an aborted takeoff. It cannot be less in width than the runway and must be constructed so that its surface is adequate to support the airplane without inducing structural damage, but it is not intended for normal use.

A clearway is an area *beyond the runway* that is cleared of obstructions so that it provides an additional obstacle-free space for climb-out. It is also under control of the airport. Where a clearway exists, it is considered to start at the end of the runway regardless of whether a stopway is incorporated or not. It must be at least 500' wide with an upslope of not more than 0.0125%. There can be no protrusions within this area except runway threshold lights, and then only if they are not more than 26" in height and are to the side of the runway line.

Personally, I prefer longer runways!

Conditions Affecting Takeoff Performance

There are several basic conditions that influence takeoff performance. These conditions are taken into account in performance considerations, assuming that the takeoff is made in a normal manner and using the normal pilot technique for the aircraft. These conditions are: aircraft gross weight, temperature, pressure altitude, engine thrust, flap setting, stabilizer setting, wind direction and velocity, runway slope, and runway surface.

AIRCRAFT GROSS WEIGHT, TEMPERATURE, AND PRESSURE ALTITUDE

The takeoff field length is dependent on the speed to which the aircraft has to be accelerated and the acceleration available. Higher weight will reduce the available acceleration with the same engine thrust.

The true takeoff speeds will increase with decreasing air density, which is a function of temperature and pressure altitude. Therefore, lower density conditions will require more acceleration time to reach the higher true takeoff speed and will result in longer field lengths.

ENGINE THRUST

Engine thrust depends on ram air temperature, pressure altitude, aircraft speed, and engine bleed requirements. Higher temperatures and pressure altitudes normally result in lower thrust. Takeoff thrust also falls off with increasing speed. Reduced thrust causes a lower acceleration, which increases the field length. The thrust is also reduced when bleed air is required. Normally, all takeoff and climb-out performance is based on air conditioning systems being operative. However, corrections are necessary for use of anti-icing systems; the charts must be consulted for the correction where applicable.

FLAP SETTING

Takeoff performance naturally is based on the aircraft being in the takeoff configuration. In some aircraft this is one flap setting; in others, such as the DC-9-30 series, flap settings of both 15° and 5° can be used for takeoff. In aircraft with two flap settings, the higher setting is recommended for takeoff unless the maximum recommended takeoff weight is limiting in climb gradient at the higher setting. The lower flap setting requires longer takeoff distance than the higher setting but provides a better climb-out gradient capability.

STABILIZER SETTING

The stabilizer setting used for takeoff is determined by the aircraft center of gravity. This usually results in a stabilizer trim set for very close to engine-out en route climb speed, or about 1.35 over stall,

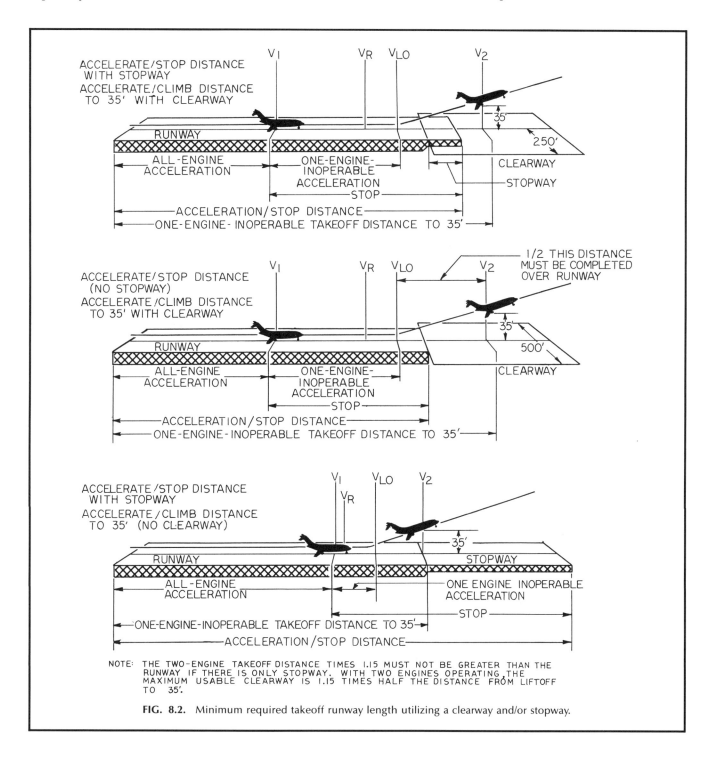

FIG. 8.2. Minimum required takeoff runway length utilizing a clearway and/or stopway.

which would also be a very good trim setting for a landing in the event of a malfunction rendering the stabilizer trim inoperative during takeoff.

WIND CORRECTION

The application of winds to takeoff and landing will be stated in considerations for performance chart construction. Here we'll define some terminology and recommend some methods of accounting for winds in determining runway requirements if you must compute takeoff and landing data and runway requirements yourself.

   1. Definitions of wind terminology:

     a. *Steady wind value*—Reported steady wind.

     b. *Gust increment*—Reported wind in excess of steady wind value.

     c. *Light and variable*—Winds of 5 knots or less; not to be applied to performance computations.

     d. *Variable at _____ knots*—Winds reported in excess of 5 knots; determine prevailing direction and apply most critical computation.

     e. *Component*—Effective wind parallel to or across runway.

     f. *Headwind*—Effective wind parallel to runway, determined from steady wind value.

     g. *Tailwind*—Effective wind parallel to runway, determined from steady wind value plus gust increment.

     h. *Crosswind*—Effective wind across runway, determined from steady wind value plus gust increment.

   2. Accounting for headwinds and tailwinds— Headwinds shorten takeoff ground run and improve obstacle clearance capability since they result in lower true ground speeds, which require less acceleration time on takeoff and less distance covered to reach a given height. Tailwinds have the opposite effect. Normally you should not take headwinds into account in runway length computations, using the zero wind reference on the chart, unless it is absolutely necessary to complete the flight. Tailwinds are always to be taken into account.

   3. Runway and crosswind component—The wind angle and wind component charts (Figs. 8.3 and 8.4) provide one method of determining runway and crosswind components of the reported wind condition.

RUNWAY SLOPE

The takeoff distance and time will be longer than usual if the takeoff is made on a slight uphill slope and will be shorter on a downhill slope. Corrections for runway slope are provided on all charts where applicable.

RUNWAY SURFACE

All performance involving stopping distance must be based on assumed coefficients of friction between the tires and the runway surface. The basic performance data assume a dry concrete or asphalt runway. Variations from the basic performance depend on the runway surface and the type and amount of runway surface covering.

Considerations in Performance Chart Construction

Runway length required for takeoff, takeoff performance, and takeoff field length are charted for pilot use in the performance section of the aircraft *Flight Crew Operating Manual*. These charts are constructed by the manufacturer from actual flight test data.

All takeoff performance is computed considering engine failure at the most critical time, V_1. If failure or malfunction occurs either before or after V_1, a further margin of safety is provided if proper procedures and techniques are used for stopping or continuing the takeoff.

To obtain the basic flight test data for the published performance, flight procedures are used that could be adopted by airline pilots or commercial operators to enable them to obtain similar performance. Some of the factors and procedures used in the flight tests and construction of the performance charts are as follows:

   1. Time delays are considered for the average reaction time that can be expected in airline operations.

   2. Reverse thrust is not used in establishing the ground stopping distances.

   3. Full altitude and temperature effects on performance are considered and taken into account where applicable.

   4. Performance charts are then constructed using only 50% of reported headwinds and 150% of reported tailwinds measured at a height of 50'.

To be more specific, the takeoff ground run distances are based on normal takeoff procedure using the stabilizer setting specified for the center of gravity, slats extended, and flaps set as specified. Takeoff and critical field length distances, climb performance, and minimum control speeds are based on maximum takeoff thrust settings. All stopping distances are based on takeoff flap settings, spoilers extended manually, no reverse thrust, *and maximum antiskid braking on dry concrete or asphalt*. Four seconds are allowed for transition from full takeoff power or thrust to maximum braking. This allows time to recognize the situation, make a decision to stop, and achieve the braking configuration.

WIND DIRECTION (DEGREES MAGNETIC)

RWY. NO.	10	20	30	40	50	60	70	80	90	100	110	120	130	140	150	160	170	180	190	200	210	220	230	240	250	260	270	280	290	300	310	320	330	340	350	360	RWY. NO.
1	0	10	20	30	40	50	60	70	80	90	100	110	120	130	140	150	160	170	180	170	160	150	140	130	120	110	100	90	80	70	60	50	40	30	20	10	1
2	10	0	10	20	30	40	50	60	70	80	90	100	110	120	130	140	150	160	170	180	170	160	150	140	130	120	110	100	90	80	70	60	50	40	30	20	2
3	20	10	0	10	20	30	40	50	60	70	80	90	100	110	120	130	140	150	160	170	180	170	160	150	140	130	120	110	100	90	80	70	60	50	40	30	3
4	30	20	10	0	10	20	30	40	50	60	70	80	90	100	110	120	130	140	150	160	170	180	170	160	150	140	130	120	110	100	90	80	70	60	50	40	4
5	40	30	20	10	0	10	20	30	40	50	60	70	80	90	100	110	120	130	140	150	160	170	180	170	160	150	140	130	120	110	100	90	80	70	60	50	5
6	50	40	30	20	10	0	10	20	30	40	50	60	70	80	90	100	110	120	130	140	150	160	170	180	170	160	150	140	130	120	110	100	90	80	70	60	6
7	60	50	40	30	20	10	0	10	20	30	40	50	60	70	80	90	100	110	120	130	140	150	160	170	180	170	160	150	140	130	120	110	100	90	80	70	7
8	70	60	50	40	30	20	10	0	10	20	30	40	50	60	70	80	90	100	110	120	130	140	150	160	170	180	170	160	150	140	130	120	110	100	90	80	8
9	80	70	60	50	40	30	20	10	0	10	20	30	40	50	60	70	80	90	100	110	120	130	140	150	160	170	180	170	160	150	140	130	120	110	100	90	9
10	90	80	70	60	50	40	30	20	10	0	10	20	30	40	50	60	70	80	90	100	110	120	130	140	150	160	170	180	170	160	150	140	130	120	110	100	10
11	100	90	80	70	60	50	40	30	20	10	0	10	20	30	40	50	60	70	80	90	100	110	120	130	140	150	160	170	180	170	160	150	140	130	120	110	11
12	110	100	90	80	70	60	50	40	30	20	10	0	10	20	30	40	50	60	70	80	90	100	110	120	130	140	150	160	170	180	170	160	150	140	130	120	12
13	120	110	100	90	80	70	60	50	40	30	20	10	0	10	20	30	40	50	60	70	80	90	100	110	120	130	140	150	160	170	180	170	160	150	140	130	13
14	130	120	110	100	90	80	70	60	50	40	30	20	10	0	10	20	30	40	50	60	70	80	90	100	110	120	130	140	150	160	170	180	170	160	150	140	14
15	140	130	120	110	100	90	80	70	60	50	40	30	20	10	0	10	20	30	40	50	60	70	80	90	100	110	120	130	140	150	160	170	180	170	160	150	15
16	150	140	130	120	110	100	90	80	70	60	50	40	30	20	10	0	10	20	30	40	50	60	70	80	90	100	110	120	130	140	150	160	170	180	170	160	16
17	160	150	140	130	120	110	100	90	80	70	60	50	40	30	20	10	0	10	20	30	40	50	60	70	80	90	100	110	120	130	140	150	160	170	180	170	17
18	170	160	150	140	130	120	110	100	90	80	70	60	50	40	30	20	10	0	10	20	30	40	50	60	70	80	90	100	110	120	130	140	150	160	170	180	18
19	180	170	160	150	140	130	120	110	100	90	80	70	60	50	40	30	20	10	0	10	20	30	40	50	60	70	80	90	100	110	120	130	140	150	160	170	19
20	170	180	170	160	150	140	130	120	110	100	90	80	70	60	50	40	30	20	10	0	10	20	30	40	50	60	70	80	90	100	110	120	130	140	150	160	20
21	160	170	180	170	160	150	140	130	120	110	100	90	80	70	60	50	40	30	20	10	0	10	20	30	40	50	60	70	80	90	100	110	120	130	140	150	21
22	150	160	170	180	170	160	150	140	130	120	110	100	90	80	70	60	50	40	30	20	10	0	10	20	30	40	50	60	70	80	90	100	110	120	130	140	22
23	140	150	160	170	180	170	160	150	140	130	120	110	100	90	80	70	60	50	40	30	20	10	0	10	20	30	40	50	60	70	80	90	100	110	120	130	23
24	130	140	150	160	170	180	170	160	150	140	130	120	110	100	90	80	70	60	50	40	30	20	10	0	10	20	30	40	50	60	70	80	90	100	110	120	24
25	120	130	140	150	160	170	180	170	160	150	140	130	120	110	100	90	80	70	60	50	40	30	20	10	0	10	20	30	40	50	60	70	80	90	100	110	25
26	110	120	130	140	150	160	170	180	170	160	150	140	130	120	110	100	90	80	70	60	50	40	30	20	10	0	10	20	30	40	50	60	70	80	90	100	26
27	100	110	120	130	140	150	160	170	180	170	160	150	140	130	120	110	100	90	80	70	60	50	40	30	20	10	0	10	20	30	40	50	60	70	80	90	27
28	90	100	110	120	130	140	150	160	170	180	170	160	150	140	130	120	110	100	90	80	70	60	50	40	30	20	10	0	10	20	30	40	50	60	70	80	28
29	80	90	100	110	120	130	140	150	160	170	180	170	160	150	140	130	120	110	100	90	80	70	60	50	40	30	20	10	0	10	20	30	40	50	60	70	29
30	70	80	90	100	110	120	130	140	150	160	170	180	170	160	150	140	130	120	110	100	90	80	70	60	50	40	30	20	10	0	10	20	30	40	50	60	30
31	60	70	80	90	100	110	120	130	140	150	160	170	180	170	160	150	140	130	120	110	100	90	80	70	60	50	40	30	20	10	0	10	20	30	40	50	31
32	50	60	70	80	90	100	110	120	130	140	150	160	170	180	170	160	150	140	130	120	110	100	90	80	70	60	50	40	30	20	10	0	10	20	30	40	32
33	40	50	60	70	80	90	100	110	120	130	140	150	160	170	180	170	160	150	140	130	120	110	100	90	80	70	60	50	40	30	20	10	0	10	20	30	33
34	30	40	50	60	70	80	90	100	110	120	130	140	150	160	170	180	170	160	150	140	130	120	110	100	90	80	70	60	50	40	30	20	10	0	10	20	34
35	20	30	40	50	60	70	80	90	100	110	120	130	140	150	160	170	180	170	160	150	140	130	120	110	100	90	80	70	60	50	40	30	20	10	0	10	35
36	10	20	30	40	50	60	70	80	90	100	110	120	130	140	150	160	170	180	170	160	150	140	130	120	110	100	90	80	70	60	50	40	30	20	10	0	36

THIS CHART PROVIDES RESULTANT <u>WIND ANGLE</u>. USE <u>WIND COMPONENT</u> <u>CHART</u> TO DETERMINE RESULTANT WIND COMPONENT VALUES.

WIND ANGLES IN CLEAR AREAS RESULT IN HEADWIND COMPONENTS.
WIND ANGLES IN SHADED AREAS RESULT IN TAILWIND COMPONENTS.

INSTRUCTIONS FOR USE OF CHARTS.

GIVEN: WIND DIRECTION 130° TRUE; MAGNETIC BEARING CORRECTION +10°; WIND SPEED 15 KNOTS; RUNWAY NO. 17.

1. ENTER <u>WIND ANGLE CHART</u> WITH CORRECTED WIND DIRECTION 140° MAGNETIC AND RUNWAY NO. 17 — READ WIND ANGLE 30°
2. ENTER <u>WIND COMPONENT CHART</u> WITH WIND ANGLE 30° AND WIND SPEED 15 KNOTS — READ RESULTING HEADWIND COMPONENT 13 KNOTS AND CORRESPONDING CROSSWIND COMPONENT 8 KNOTS.

FIG. 8.3. Wind angle chart.

FIG. 8.4. Wind component chart.

Speeds Associated with Takeoff

CRITICAL ENGINE FAILURE SPEED (V_{CEF})

This is a speed that the average airline or commercial pilot has never heard of, or believes is V_1. It isn't V_1, but it is related to it. Critical engine failure speed is the highest speed to which the aircraft can be accelerated, lose an engine, and then continue the takeoff or stop in the computed minimum field length. This would be when the distance to accelerate from V_1 to lift-off equals the distance required to stop.

DECISION SPEED (V_1)

Decision speed is the speed the pilot uses as a reference in deciding whether to continue or abort the takeoff.

The speeds given in the FAA-approved *Airplane Flight Manual* have been computed and selected so that (1) if engine failure is recognized at or above the V_1 speed, the takeoff may be continued on the remaining engines; or (2) a stop may be initiated at or prior to V_1 and completed within the distance specified in the FAA requirement for takeoff field length.

V_1 is the speed at which the pilot becomes committed to continue the takeoff. If a system emergency occurs before V_1, the takeoff is aborted. V_1 will occur after V_{CEF} or will equal V_R, depending on the runway available. To clarify the relationship of V_{CEF} to V_1, there is 1 second allowed for recognition time during which the aircraft will continue to accelerate to V_1 after engine failure at V_{CEF}.

ROTATION SPEED (V_R)

Rotation speed is that speed at which the pilot begins to rotate the airplane to the lift-off attitude. The rate of rotation can vary, but it should normally take about 2.5 seconds to rotate to lift-off attitude.

Rotation at the maximum practical rate will result in attaining the V_2 speed at or below 35′ with one engine inoperative or result in exceeding the V_2 speed at 35′ with all engines operating.

The criteria used for establishing rotation speed are as follows:

1. A speed that cannot be less than 5% above the minimum control speed in the air.

2. A speed that will result in at least the minimum required lift-off speed.

3. A speed that permits the attainment of V_2 prior to reaching 35′.

4. A speed that will not result in increasing the takeoff distance if rotation is commenced 5 knots lower than the established V_R during one-engine-inoperative acceleration or 10 knots lower than the established V_R during all-engine acceleration.

LIFT-OFF SPEED (V_{LO})

Lift-off speed is the speed at which the plane becomes airborne. If the airplane is rotated at maximum rate, the minimum lift-off speed must be at least 5% above the one-engine-inoperative minimum unstick speed (V_{MU} one-engine-inoperative) and 10% above the all-engine minimum unstick speed (V_{MU} all engines).

TAKEOFF SAFETY SPEED (V_2)

This is the speed attained at 35′ above the runway with engine failure at critical engine failure speed (V_{CEF}) or a speed at least 20% above the stall speed, whichever is greater, and is the minimum recommended climb-out speed. It cannot, however, be less than 10% above the minimum control speed in the air (V_{MCA}). The correct V_2 is a result of proper rotation and lift-off procedures and allows the airplane to maintain a specified gradient in the climb-out flight path.

MINIMUM CONTROL SPEED—GROUND (V_{MCG})

The minimum control speed with the critical engine inoperative must be considered for both ground and air. This again is a variable. Some aircraft with tail-mounted center-line thrust engines have such a low V_{MC} that it is hardly worth consideration for flight and is useful only in computations of some of the other speeds just discussed. This is particularly true of aircraft with nosewheel steering through rudder control for minimum control speeds on the ground. But there are others, such as the L-188 Electra or any other propeller-driven aircraft, that may be very critical for V_{MCG}.

Ground minimum control speed is the minimum airspeed at which the aircraft can lose an engine during takeoff roll with the remaining engines at takeoff thrust and can maintain directional control by use of full rudder deflection with no nosewheel steering. V_{MCG} is not affected by runway slope or headwind component.

Until this speed is reached, it would be proper and wise to lightly monitor nosewheel steering and maintain directional control with flight controls throughout the takeoff just as if nosewheel steering didn't exist; but it would be readily available if needed.

Improper use of nosewheel steering on an aborted takeoff with engine failure at too high speeds or during skids after landing has caused more damage to aircraft than you might imagine. Nosewheel steering is designed for taxi operations—making large and sharp turns at low speeds, turning off the runway, and parking at the ramp.

When you're taking off on wet, icy, or slippery runways, the nosewheel begins to hydroplane between 70 and 90 knots (depending on tire pressure and depth of water or slush) and has very little steering effect.

V_{MCG} is always lower than V_1.

MINIMUM CONTROL SPEED—AIR (V_{MCA})

Air minimum control speed is the minimum speed at which an engine can be lost after lift-off and directional control maintained. It is not critical in aircraft with tail-mounted engines. Aircraft with near-center-line thrust (DC-9) have such low V_{MCA} speeds that they are not limiting to flight, since they are below stall speed. The loss of an engine still produces adverse yaw and requires considerable control, but it may be considered as having negligible effect on takeoff directional control after lift-off. The yaw may be offset with much less than full deflection of the rudder.

Engine failure in wing-mounted engines, either props or jets, will cause an adverse yaw, and there will be definite minimum speeds at which directional control can be maintained. This minimum control speed in the air, of course, must be less than V_2; but it can sometimes become a critical factor, and this is where "flying the wing" can be helpful.

As an example, the V_{MCA} for the Lockheed L-188 with the critical or number one engine failed and the other three at takeoff power is 113 knots at sea level on a standard day. This V_{MCA} jumps to a minimum of 145 knots with two engines on the same side failed and the wings level. Merely raising the wing with the dead engines and banking into the good engines only 5° reduces two-engine V_{MCA} to 128. The lift of the wing turning into the good engines and against the yaw reduces the rudder force necessary for directional control and hence the minimum control speed. Having less induced drag from the rudder, the airplane will also accelerate better.

MINIMUM UNSTICK SPEED (V_{MU})

The minimum unstick speed is the minimum speed with which the airplane can be made to lift off the ground without demonstrating hazardous characteristics while continuing the takeoff. This speed is established with all engines operating and with one engine failed.

This speed is very rarely found in the pilot's *Airplane Flight Manual*. For all practical purposes, it must be higher than stall speed and minimum control speed and may in some instances be about V_1 speed. With some of the longer fuselage aircraft, it may also be limited by tail clearance; the aircraft could become airborne and the tail would strike the pavement in the high angle of attack and rotation necessary to take off. The only real purpose of such a speed is to establish minimum lift-off speeds, V_{LO}.

Takeoff Flight Path Performance

The FAA climb segment gradients required for two-engine turbine-powered aircraft are: (1) for the first segment, a positive rate of climb; (2) for the second segment, a climb gradient of not less than 2.4%; (3) for the transition segment, a climb gradient of not less than 1.2%.

The takeoff flight path is considered to begin when the airplane has reached a height of 35' above the surface and continues to a point 1,500' above the surface, or to the point where the single-engine en route climb speed of 1.38 V_S (138% of stall speed) is reached, whichever point is higher.

Most jet engines are approved for continuous operation at takeoff thrust for 5 minutes. The takeoff flight path data with one engine inoperative are based on the use of takeoff thrust for the full time allowed. If the takeoff thrust is reduced by the use of airplane deicing or engine anti-icing, a reduction from airplane performance must be taken into account. (This is usually a gross weight reduction in order to meet climb requirements.)

Figure 8.5 shows the performance requirements and profiles for various takeoff flight path configurations. The airplane must meet the required climb gradient performance with one engine inoperative at all approved operating gross weights, temperatures, and altitudes for which it is certified.

GRADIENT REQUIREMENTS

The gradient method of calculating climb performance during the various climb flight path segments is basically a percentage of the horizontal distance traveled with zero wind. For example, if a 2.4% climb gradient is required, for every 1,000' the airplane travels horizontally it must climb 24'.

TAKEOFF OBSTACLE CLEARANCE

Capability to meet the three climb segments is required assuming an engine failure at V_1. In that condition, FAA regulations require that the airplane *weight* allow a net takeoff flight path that clears all obstacles, either by a height of 35' vertically or by at least 200' horizontally within the airport boundaries and by at least 300' horizontally after passing the boundaries. All obstacles must be cleared by a specified amount until the airplane reaches final segment climb speed or 1,500' above the airport elevation, whichever is greater.

The aircraft cannot be banked before reaching 50', and thereafter the maximum bank angle is 15°. The gradient loss in a steady turn at a 15° bank angle

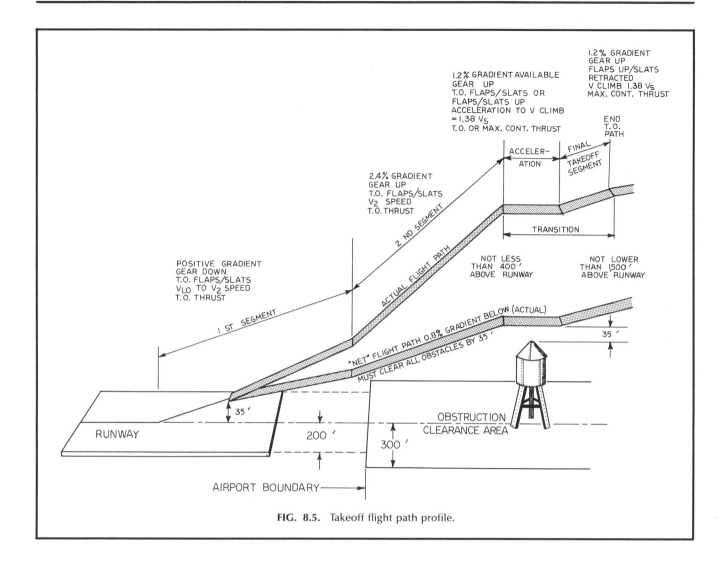

FIG. 8.5. Takeoff flight path profile.

is 0.57%, with a linear variation to 0° bank angle. As a rule of thumb, the height *not* obtained as a result of a 15° bank angle is 15′ for each 5° heading change, with a linear variation to 0° bank angle.

The obstacle clearance requirement is an expanding type, which requires that the *net* flight path must clear all obstacles by 35′. The net path is determined by reducing the 0 wind gradients by 0.8% and then correcting for wind. With 0 wind, this gives an expanding clearance equal to 0.8% of the distance from the end of the required takeoff field length. Figure 8.5 shows this as "net flight path," which is 0.8% below actual flight path to any given point.

Since local obstacle clearance is dependent upon the height of the obstacles and their distance from the end of the runway, each airline (or airplane operator) must make its own flight path profile analysis for the airports from which it operates. This results in various obstacle clearance altitudes for acceleration after V_2 climb. If such takeoff flight path profiles are not available, it would then be necessary to extend second segment (V_2) climb to 1,500′.

To obtain the clearance required for engine-out operation, the second segment of the takeoff flight path must be flown at V_2 speed.

With both engines operating, the climb-out gradient at a pitch attitude of 15–16° is approximately four times steeper than the one-engine-out gradient flown at V_2. This climb capability assures obstacle clearance with all engines operating, provided the thrust available for climb is not used excessively for acceleration, thereby inadvertently failing to achieve the flight path that provides obstacle clearance. The maximum pitch attitude should not exceed 16°.

TAKEOFF CLIMB

1. First segment—This is the climb segment just after lift-off and continues until the landing gear is retracted. The speed will vary from V_{LO} to V_2 at 35′. The allowable gradient is a positive climb, and I don't know of any aircraft that are limited in this segment. However, all climb performance requirements are based on gear retraction initiated within 3 seconds after lift-off, with the aircraft accelerated to a speed of V^2 minimum at 35′.

2. Second segment—This segment starts at the time the landing gear is fully retracted and continues until the airplane reaches an altitude of at least 400′ above the runway. It is flown at V_2 speed (with an inoperative engine) and must have a gradient of at least 2.4%.

3. Transition segment—The takeoff flight path profile will vary during the transition segment depending upon obstacle location and height. If obstacle clearance requirements warrant, the climb to 1,500′ could be made in the second segment configuration (gear up, flaps at takeoff setting, and takeoff power). On the other hand, without any obstacles the transition segment may start at 400′, as shown in Figure 8.5. Thus the second segment altitude depends on obstacle clearance requirements; it is *never* lower than 400′ in jet aircraft but may be as low as 200′ in propeller-driven aircraft.

At the beginning of the transition or acceleration segment, as the flaps and slats are retracted and the airplane accelerates, the indicated airspeed should never be less than 120% of the stall speed. This is the basis for the establishment of a speed schedule for flap/slat retraction. The *available* gradient must be at least 1.2%, with the flaps at the takeoff setting or retracted and with the thrust at takeoff or maximum continuous. When the airplane reaches the final takeoff flight path point at 1,500′, it must have a climb gradient of 1.2% while using maximum continuous power, with gear up, flaps and slats retracted, and a speed of at least 38% above the stall speed.

In a normal takeoff, if the takeoff weight has been limited by obstacles in the flight path, climbing out at the maximum pitch angle for your aircraft (usually no more than 15–16°) after an all-engine takeoff will assure obstacle clearance as long as takeoff flaps/slats and power are maintained. However, distant obstacles must also be considered in accelerating to en route climb speeds, configuration, and thrust settings.

Normal Takeoffs

The entire length of the runway should be available for use, especially if the precalculated takeoff performance shows the airplane to be limited by runway length or obstacles. An FAA solution to the traffic delays being experienced these days is a directive allowing the tower controller to instruct you to make an intersection takeoff. When you do so, they are also required to give you the length of the runway available from the intersection. However, this is a violation of the operating specification of every airline operations manual, which requires the pilot in command to use the full length of the runway for every takeoff. I've never felt that a compromise of safety, which an intersection takeoff is, is an adequate method of expediting traffic.

After taxiing into position at the end of the runway, the airplane should always be aligned in the center of the runway, allowing equal distance on either side for recovery from a swerve caused by a blown tire or engine failure. Runway available for takeoff is actual runway length less the aircraft lineup distance, and a nominal lineup distance is considered to be 100′.

Figure 8.6 shows the steps in a normal takeoff; they are discussed in detail below.

After runway alignment in takeoff position in a jet, hold the brakes and advance the throttles to a power setting beyond the bleed valve range (this may be either an EPR setting or a percentage of N1 RPM) and allow the engine to stabilize. Check the engine instruments for proper operation. Most failures occur in the range of bleed valve closing and may be noted at this time. This procedure assures you of symmetrical thrust during the takeoff roll. You may hold the brakes until takeoff thrust is set, but this sometimes causes the brakes to drag and clatter from heat, and the aircraft will lurch forward abruptly as brakes are released. A smoother takeoff roll without diminishing performance would be a "rolling" takeoff, accomplished by fully releasing the brakes after engine stabilization from first power application, noting engine instruments after bleed valve closing, advancing the throttle smoothly to takeoff thrust, and setting takeoff thrust before reaching 60 knots.

A slight forward pressure should be held on the control column to keep the nosewheel rolling firmly on the runway. If rudder pedal nosewheel steer is available, the pilot need not use the nose gear steering wheel to maintain runway alignment and directional control. However, in an airplane with a critical V_{MCG}, the pilot should monitor the nosewheel steering to about 80 knots (or V_{MC} for the particular aircraft), with the copilot applying slight forward pressure and crosswind aileron correction, and then bring the left hand up to the wheel after V_{MC} is attained.

The pilot flying the aircraft and making the takeoff should be on the throttles at least until reaching V_1. With the application of takeoff power, he or she gives the command, "Takeoff power," but doesn't turn the throttles loose. The copilot (flight engineer in some

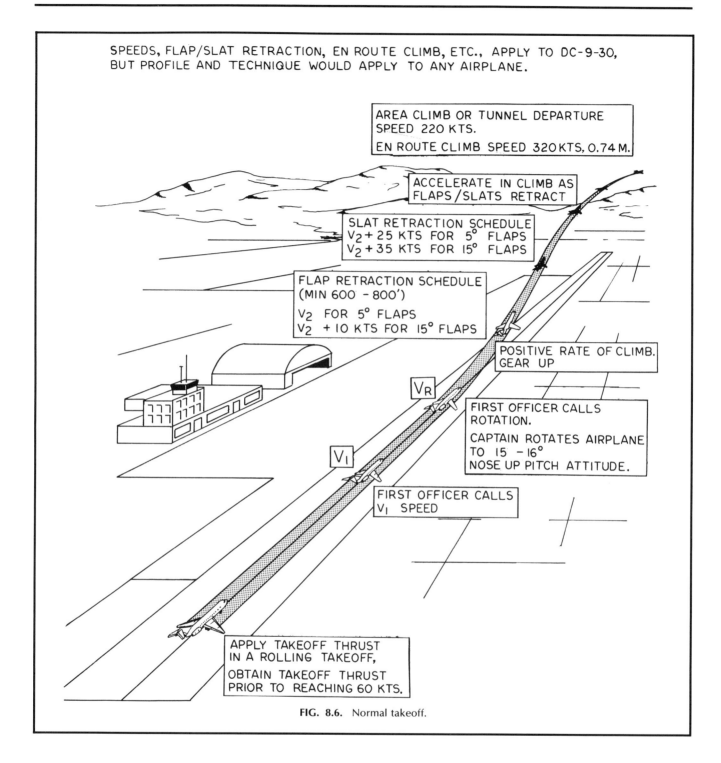

SPEEDS, FLAP/SLAT RETRACTION, EN ROUTE CLIMB, ETC., APPLY TO DC-9-30, BUT PROFILE AND TECHNIQUE WOULD APPLY TO ANY AIRPLANE.

AREA CLIMB OR TUNNEL DEPARTURE SPEED 220 KTS.
EN ROUTE CLIMB SPEED 320 KTS, 0.74 M.

ACCELERATE IN CLIMB AS FLAPS/SLATS RETRACT

SLAT RETRACTION SCHEDULE
$V_2 + 25$ KTS FOR $5°$ FLAPS
$V_2 + 35$ KTS FOR $15°$ FLAPS

FLAP RETRACTION SCHEDULE
(MIN 600 - 800')
V_2 FOR $5°$ FLAPS
$V_2 + 10$ KTS FOR $15°$ FLAPS

POSITIVE RATE OF CLIMB. GEAR UP

V_R

FIRST OFFICER CALLS ROTATION.
CAPTAIN ROTATES AIRPLANE TO $15 - 16°$ NOSE UP PITCH ATTITUDE.

V_1

FIRST OFFICER CALLS V_1 SPEED

APPLY TAKEOFF THRUST IN A ROLLING TAKEOFF,
OBTAIN TAKEOFF THRUST PRIOR TO REACHING 60 KTS.

FIG. 8.6. Normal takeoff.

aircraft) should then set the power by trimming it up evenly, monitor the engine instruments, observe the general airplane condition and performance, and immediately call any malfunction to the attention of the pilot. The copilot should also call out V_1 and V_R speeds.

After passing V_1 speed on the takeoff roll it is not mandatory for the pilot to keep a hand on the throt-

tles; the point for abort has passed, and both hands may be placed on the control wheel if desired. At V_R speed the pilot should fly the nosewheel off the ground, with a pressure of 15-20 lb. of pull force on the wheel, to start the rotation to lift-off attitude.

I'm sure you've seen jet aircraft seemingly pitch up sharply at this point. That isn't necessary at all!

You must establish a positive angle of attack to fly off the ground, but a smooth rotation (requiring 2–3 seconds) should result in a lift-off at an angle of 7–9°. The rotation should be continued until reaching a pitch attitude of 15–16° (requiring another 2–3 seconds). The use of a smooth and proper technique assures a clean lift-off; at 35′ above the surface, the speed will invariably be above V_2 when the 15° pitch attitude is established. Actually, the airspeed will stabilize substantially above V_2 depending on gross weight, temperature, and altitude.

The smooth continuation of the rotation to at least 15° (or the correct rotation attitude of your particular aircraft) is necessary to get desired and planned performance. If the rotation is started too late or is too slow or if the nose is not rotated high enough, the ground roll will be increased. A pitch attitude in excess of 15–16° is not considerate of passenger comfort. Also, an engine failure right at this point with too high an angle of attack in the rotation will immediately put you on the backside of the power curve. You'll have to drop the nose to make the airplane fly, and this will result in a sacrifice of altitude.

Landing gear retraction should be accomplished upon establishing a positive rate of climb. The climb attitude (maximum of 16°) should be held and the airplane allowed to accelerate to flap/slat retraction speed. However, the flaps/slats should never be retracted until after the minimum altitude of the second segment climb (400′) has been passed. Ground effect and gear drag reduction results in rapid acceleration to the desired speed in this phase of the takeoff flight path.

Prior to takeoff, the stabilizer trim is set according to the center of gravity position in percentage of mean aerodynamic chord. This setting is for 0 elevator force at 135% of stall. Because of this, a diminishing back pressure will be experienced from initial rotation to about engine-out climb speed.

Crosswind Takeoffs

Crosswind takeoffs in sweptwing aircraft are not much different from those in other types. *The primary objective during the takeoff roll is to keep the wings level.* The upwind wing will have a tendency to rise soon after brake release, and the aileron must be held into the crosswind to maintain wings level. This will not materially affect the takeoff performance. Nosewheel steering will maintain directional control, and the rudder becomes aerodynamically effective at about 50 knots. As airplane speed increases, the aileron required will diminish but will never reduce to zero during the ground roll.

Takeoff on Icy or Slippery Runways

One of the considerations in takeoff performance is the ability to accelerate to a decision speed, recognize a malfunction or engine loss at that point, and bring the aircraft to a stop. All performance involving stopping (and stopping distance) must be based on assumed coefficients of friction between the tires and runway surface, and all performance data are based on a dry concrete or asphalt runway. Obviously, wet or icy surfaces will produce a lower coefficient of friction between the tires and the runway surface. Slippery runway surfaces will increase stopping distances and consequently should increase required field lengths.

If icing conditions are present or anticipated during the use of takeoff power, engine anti-ice should be turned on *before* the application of takeoff thrust. The takeoff technique may also vary a bit. Nosewheel steering will not be too effective in maintaining directional control after about 70 knots on wet, icy, or slippery pavement. Once the plane is moving, the nosewheel doesn't do much except turn sideways and skid. Also, with tail-mounted engines in appreciable water or slush, you might not want too much forward pressure on the yoke. Holding the nosewheel down into such conditions in an attempt to improve its steering capability slows acceleration and has a nasty habit of throwing a spectacular "rooster tail" that may be ingested into the engine and possibly cause a flameout. I like to lock the nosewheel in the center, holding the nose steering wheel firmly to override nosewheel steering demands that may be a result of rudder action, and then have the full and unrestricted use of the rudder for directional control. In every airplane I've ever flown, the rudder is fully sufficient above 50 knots IAS for normal directional control up to a maximum crosswind condition. The copilot can hold the aileron into the wind for you until you're ready to rotate.

Takeoff in Slush, Standing Water, and Snow

Experience has proved that standing water, slush, and snow affect takeoff performance and must be taken into consideration. The ground roll is extended from slower acceleration, and there is also a possibility of damage to the aircraft from flying water and slush, particularly in the flap area. The condition of the runway should be determined as near departure time as possible. To make a decision of go/no-go, be sure the depth is measured by checking a number of places along the runway, particularly the section where the high-speed portion of the takeoff run will occur. No adjustment is required for standing water, slush, or wet snow up to 0.2″ or dry snow up to 4″. Make certain that the depth limits of your aircraft will not be exceeded for takeoff, be sure to adjust your takeoff runway lengths (or your gross weight for the runway) for the condition, and be very wary of deep slush or water in the high-speed portion of your takeoff roll.

The depth limitations for takeoff vary for different aircraft, but no aircraft has a limitation of more than 1″ of water, slush, or wet snow—usually ½″. The limitation usually reads, "Takeoff should not be attempted on any runway when the *average* depth of standing water, slush, or wet snow over an appreciable portion of the runway is in excess of ½ inch."

Taking off on a snow-covered runway is another consideration. Runways covered with snow have a variable braking coefficient somewhere between wet and slippery and icy. When there is a variable such as this, you are much better off to apply the most critical factor; I recommend that you consider braking on snow to be poor to nil. Dry snow is defined as snow that cannot be readily formed into a snowball; generally, takeoffs are permitted in dry snow depths ranging from 3 to 6″.

The total effect of water- or slush-covered runways may be summed up in two statements: (1) The depth of the surface covering of a runway can cause a significant reduction in takeoff performance due to the retarding effect of the tires displacing the covering, plus the additional drag effect of the material being sprayed and consequently striking the aircraft surfaces. (2) This retarding effect will no longer be a factor when the aircraft reaches hydroplaning speed, but then the braking coefficient is reduced to almost 0 at speeds above hydroplaning speed.

There are two ways of making allowances for the effect of slush-covered runways: (1) reduce your gross weight for the runway length, wind, and temperature or (2) use a runway requirement chart and then add a percentage factor for additional takeoff ground run. The weight reduction is most commonly used by the airlines and perhaps offers the most safety for an engine loss, but no method yet devised takes increased stopping distance for an abort into consideration, although the weight reduction does reduce the V_1 and lift-off speeds. But all accelerate/stop data are still based on dry runway performance.

To use the weight reduction method, the airline pilot refers to the *Gross Weights Manual*, finds the maximum weight for the wind and temperature for the runway, and then reduces the gross weight further according to runway conditions., For example, let's

assume that we're taking off in a DC-9-31 from a runway with an effective field length of 5,004′, 80°F temperature, 0 wind, and ½″ of standing water on the runway. Table 8.1 shows a wind/temperature adjusted takeoff gross weight of 91,500 lb. for the runway. The next step, for the ½″ of water, is to refer to Table 8.2

TABLE 8.1. Takeoff Gross Weights, DC-9-31 (effective field length 5,004′, 15° flaps)

Temp. (°F)	Wind Velocity					
	−10 Knots	0 Wind	+10 Knots	+20 Knots	+30 Knots	Climb Limit
			(lb.)			
−20	92,400	101,100	102,800	104,500	105,000	100,000
−10	91,500	100,100	102,000	103,900	105,000	100,000
0	90,700	99,200	101,300	103,400	105,000	100,000
10	89,700	98,100	100,500	102,900	105,000	100,000
20	88,800	97,100	99,700	102,300	104,900	100,000
30	87,800	96,100	98,700	101,400	104,000	100,000
32	87,600	95,900	98,500	101,200	103,800	100,000
34	87,500	95,600	98,300	101,000	103,700	100,000
36	87,300	95,400	98,100	100,800	103,500	100,000
38	87,100	95,200	97,900	100,600	103,300	100,000
40	86,900	95,000	97,700	100,400	103,100	100,000
42	86,700	94,800	97,500	100,200	102,900	100,000
44	86,600	94,700	97,400	100,100	102,800	100,000
46	86,400	94,500	97,200	99,900	102,600	100,000
48	86,300	94,300	97,000	99,700	102,400	100,000
50	86,100	94,200	96,800	99,500	102,200	100,000
52	85,900	94,000	96,700	99,300	102,000	100,000
54	85,800	93,800	96,500	99,200	101,800	100,000
56	85,600	93,700	96,300	99,000	101,700	100,000
58	85,500	93,500	96,100	98,800	101,500	100,000
60	85,300	93,200	95,900	98,500	101,200	99,900
62	85,100	93,100	95,700	98,400	101,000	99,900
64	84,900	92,900	95,500	98,200	100,800	99,900
66	84,800	92,700	95,300	98,000	100,600	99,900
68	84,600	92,500	95,200	97,800	100,400	99,900
70	84,400	92,400	95,000	97,600	100,200	99,900
72	84,300	92,200	94,800	97,400	100,100	99,900
74	84,100	92,000	94,600	97,200	99,900	99,900
76	84,000	91,800	94,400	97,100	99,700	99,900
78	83,800	91,600	94,300	96,900	99,500	99,900
80	83,600	91,500	94,100	96,700	99,300	99,900
82	83,500	91,300	93,900	96,500	99,100	99,900
84	83,400	91,200	93,800	96,400	99,000	99,900
86	82,900	90,600	93,200	95,800	98,400	99,000
88	82,400	90,100	92,700	95,200	97,800	98,200
90	81,900	89,600	92,100	94,700	97,200	97,400
92	81,400	89,000	91,600	94,100	96,600	96,500
94	80,900	88,500	91,000	93,500	96,100	95,700
96	80,400	88,000	90,500	93,000	95,500	94,900
98	79,900	87,400	89,900	92,400	94,900	94,000
100	79,400	86,900	89,400	91,800	94,300	93,200
110	77,000	84,200	86,600	89,000	91,400	89,100

Note: The maximum allowable takeoff weight for any runway is the least of the following:

1. The weight derived from runway tables for the appropriate combination of existing wind and temperature.

2. The weight permitted by the climb limit of the airplane at the existing limits.

3. The maximum weight permitted by airplane structural limits.

4. The weight limited by specific performance and/or operating conditions described in the performance information section of the *Gross Weights Manual*.

TABLE 8.2. Takeoff Weights with Standing Water, Slush, Wet or Dry Snow (DC-9-31)

Takeoff Gross Weight*	Slush, Water, Wet Snow			Dry Snow 4–6″
	¼″	⅜″	½″	
	(lb.)			
Under 75,000 lb.	1,500	4,500	9,500	9,500
75,000–80,000 lb.	2,000	5,000	10,000	10,000
80,000–85,000 lb.	2,300	6,000	12,000	12,000
85,000–90,000 lb.	2,500	7,000	13,000	13,000
90,000–95,000 lb.	3,000	8,000	14,000	14,000
Over 95,000 lb.	4,000	9,000	15,000	15,000

*Adjusted for wind and temperature. Include effects of anti-icing if used.

where we find that we must reduce our weight another 14,000 lb. This gives us a total allowable takeoff weight of 77,500 lb. adjusted for wind; temperature; and standing water, slush, or snow—which rather limits either payload or fuel for the takeoff.

As a rule of thumb: Compute your maximum weight for the runway with the wind and temperature prevailing and then reduce your gross weight 15% for ½″ of water, slush, or snow.

Usually, however, takeoff field length requirements are based on the distance required to either stop or take off with a recognized engine failure at V₁. This is a critical field length, and the effect of slush on the two-engine takeoff for most twin-engine jets operating today may be compensated for by adding the additional ground run required to the field length. An average of additional ground run requirements for two-engine jets is shown in Table 8.3.

TABLE 8.3. Increase in Takeoff Ground Run

Depth of Slush	Tailwind, 15 Knots	0 Wind	Headwind, 25 Knots	Headwind, 50 Knots
			(%)	
¼″	30	30	35	40
½″	75	80	100	120

To use Table 8.3, you would need a chart enabling you to determine the takeoff ground run for a dry runway, interpolate a percentage increase from the above figures for wind component, and then adjust your runway requirements accordingly. I favor the method of reducing your weight according to runway length available, adjusting your maximum takeoff weight for that runway according to wind and temperature, and then correcting for runway condition by a further weight reduction as in Table 8.2.

But you must be aware that the phenomenon of hydroplaning makes it difficult if not impossible to determine critical field length for slush-covered run-

ways. (Anytime your computations for runway requirement for takeoff amount to maximum weight for the runway or minimum runway required for weight, wind, temperature, etc., you're operating at critical field length.) The lift-off speed is associated with aircraft flight characteristics and must be generalized in terms of indicated airspeeds, which are dependent on gross weight, altitude, temperature, and wind. Aircraft acceleration and the hydroplaning condition are dependent on true ground speed and tire pressure and therefore cannot be correlated with lift-off speed at all temperatures, altitudes, and weights. For these reasons, the method of increasing takeoff ground run distance considers only all-engine operation. The method of weight reduction from maximum wind/temperature adjusted weight affords the maximum safety. This can be seen by comparing the climb limit weight for the example we used earlier (99,900 lb.) against the 77,500 lb. we found to be the total maximum allowable weight for ½″ of standing water. The following information will give some guidance as to the effects of slush on takeoff performance in the event of an engine failure. In fact, it will provide you with the information necessary to either produce your own weight reduction chart, such as shown in Table 8.2, or allow you to figure the correct weight reduction for every takeoff in slush.

At speeds below hydroplaning speed, the slush drag is large enough to have a significant effect on acceleration after engine failure. Acceleration on the runway after engine failure is obtained by dividing engine thrust minus total drag by the aircraft weight on the ground and subtracting the runway slope. Since the initial climb-out gradient is the engine thrust minus total drag in the air divided by the aircraft weight, the initial climb-out gradient chart can be used to obtain an approximate calculation of ground acceleration. It is not recommended that a takeoff be attempted where loss of an engine would result in a ground acceleration of less than 1′ per second per second.

The maximum takeoff weight in slush that would provide a minimum ground acceleration of 1′/sec./sec. for most two-engine jet aircraft in the event of an engine failure can be estimated by using the Climb Gradient Chart in the performance section of your aircraft's *Pilot Operating Manual*. It's just like using the chart to obtain a limiting weight for engine-out performance, except that we are going to use a higher minimum gradient to offset the effect of slush on ground acceleration as follows: (1) Use a minimum gradient of 3% for ¼″ of slush and 7.5% for slush in excess of ¼″ but no more than ½″; (2) add the runway slope to the minimum gradient for corrected gradient; and (3) enter the chart with the corrected gradient, altitude, and temperature and obtain the limiting weight.

Engine Failure at V₁ and One-Engine Inoperative Takeoff

For all practical purposes, consider a takeoff mandatory for an engine failure recognized *after* V₁ unless the actual runway greatly exceeds the runway length required—by at least 50%. This is especially true on anything other than a dry runway and after reaching hydroplaning speed. In actual practice, recognizing an engine failure or malfunction at V₁ will result in an abort being initiated at a speed between V₁ and V_R, and this is perfectly safe on a dry runway of the longest of the several distances that must be considered for takeoff field length. But *never* attempt an abort after initiating rotation. In other words, if the engine failure occurs *after* the decision speed is reached, the takeoff must be continued (Fig. 8.7).

The airplane will yaw *toward* the failed engine. Use whatever rudder is necessary to maintain directional control, usually one-third to one-half deflection when the rudder is also tied in with nosewheel steering, and keep the wings level with the ailerons. This rudder requirement will diminish as speed increases after lift-off. In performing the takeoff with an engine failure, be firm enough with the controls to let the airplane know you're the boss and in full control, yet be cautious of overcontrolling.

The rotation to takeoff attitude at V_R should be accomplished in exactly the same manner as in the normal all-engine takeoff. However, if you have nose steering through the rudder pedals, you may need a slight amount more rudder, since you lose the steering effect from the nosewheel in maintaining directional control.

At lift-off, due to asymmetric thrust, the rudder and aileron must be used smoothly and with discretion to avoid unnecessary rolling and yawing tendencies. The yaw damper, magnetically held in the ON position, will probably snap off with the electrical interruption as the failed engine generator drops off the line and will not be able to assist in lateral stability. Overcontrol, therefore, may cause Dutch roll, which is highly undesirable.

As the nose comes up to the normal climb angle and you no longer have a good visual reference for directional control, the compass will become the primary instrument for bank. You should be able to hold your heading within 5° by holding the ball in the center with rudder smoothly applied and maintain your heading and directional control by using slight bank angles. Remember though, the airplane must not be banked below 50', and then no more than 15° during V₂ climb. However, it is possible to make the wing work for you and lessen some of the rudder required to overcome yaw by banking slightly into the good engines. The rudder and aileron control forces required to maintain direction in most airplanes are relatively light and need not be trimmed out immediately.

As soon as the engine loss is noted, call out the failure and then take no further action except to fly the airplane until the takeoff is accomplished.

Gear retraction should be initiated (by command of the pilot) when a definite rate of climb is noted. Second-segment climb requirements, when operating at limiting climb weights, require gear retraction initiated within 3 seconds after lift-off. Second-segment climb begins at V₂ (35') with takeoff flaps/slats and gear up; but acceleration is such that obtaining V₂ and 35', even with one engine out, is not difficult with proper rotation and lift-off speed.

The normal or average pitch attitude for most aircraft to maintain V₂ climb is 15–16°; however, airspeed is the primary consideration.

After you're established in second-segment V₂ climb, command—step by step and waiting until each step has been accomplished and verified before commanding the next—the immediate-action items for a failed engine. Engine failure at V₁ is no real emergency unless you panic and make it one. Just take your time, fly the airplane, and command what you want done one step at a time.

The exception, of course, is in prop aircraft where there is additional drag from a windmilling prop. The prop should be feathered as soon as possible after failure, and the engine shutdown and feather should be commanded as soon as the engine failure is noted. The pilot should *command* and continue to fly the airplane.

When the immediate-action items have been accomplished, take no further action until after the second-segment climb is accomplished and acceleration to engine-out en route climb has been established. Fly right up to the obstacle clearance altitude at V₂, level off, and *maintain* that altitude as you accelerate and retract the flaps/slats on a normal speed schedule. The airplane will probably have a tendency to settle as the flaps come up. Don't let it! Make whatever pitch change is necessary to compensate for the flaps and hold the altitude smoothly.

Maintain obstacle clearance altitude, or acceleration altitude, until engine-out en route speed is reached and normal engine-out climb is begun. Takeoff thrust should be used until reaching this final-segment climb speed or your takeoff thrust time limits, whichever comes first. Most jet engines have a 5-minute takeoff thrust limitation, and most piston engines a 2-minute limitation. However, it is important to thoroughly understand the 2-minute limitation of the piston engine. It is related to cylinder head temperature and merely means that the manufacturer doesn't guarantee the engine for more than 2 minutes of operation at its maximum limiting cylinder head temperature. You may use the power for more than 2 min-

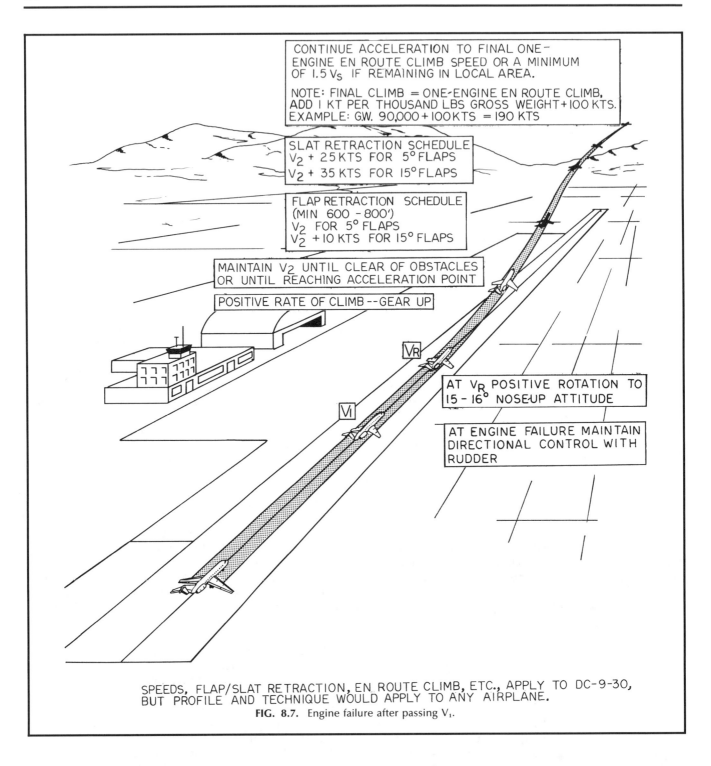

CONTINUE ACCELERATION TO FINAL ONE-
ENGINE EN ROUTE CLIMB SPEED OR A MINIMUM
OF 1.5 V_S IF REMAINING IN LOCAL AREA.

NOTE: FINAL CLIMB = ONE-ENGINE EN ROUTE CLIMB,
ADD 1 KT PER THOUSAND LBS GROSS WEIGHT+100 KTS.
EXAMPLE: G.W. 90,000+100 KTS = 190 KTS

SLAT RETRACTION SCHEDULE
V_2 + 25 KTS FOR 5° FLAPS
V_2 + 35 KTS FOR 15° FLAPS

FLAP RETRACTION SCHEDULE
(MIN 600 - 800')
V_2 FOR 5° FLAPS
V_2 +10 KTS FOR 15° FLAPS

MAINTAIN V_2 UNTIL CLEAR OF OBSTACLES
OR UNTIL REACHING ACCELERATION POINT

POSITIVE RATE OF CLIMB--GEAR UP

AT V_R POSITIVE ROTATION TO
15-16° NOSE-UP ATTITUDE

AT ENGINE FAILURE MAINTAIN
DIRECTIONAL CONTROL WITH
RUDDER

SPEEDS, FLAP/SLAT RETRACTION, EN ROUTE CLIMB, ETC., APPLY TO DC-9-30,
BUT PROFILE AND TECHNIQUE WOULD APPLY TO ANY AIRPLANE.
FIG. 8.7. Engine failure after passing V_1.

utes without experiencing detonation in a well-maintained engine as long as the cylinder head temperature can be kept down.

After establishing engine-out en route climb, set your power and command, "Max continuous power, engine-out checklist!" And after that, "Climb check!"

The next consideration is where to land as soon as practical. You might, weather permitting, return to the field you just took off from, dumping fuel if necessary (fuel-dumping time required should be computed prior to takeoff), which might necessitate a clearance to a dumping area; or you might need a clearance to a takeoff alternate if the field of departure is below landing limits.

Air traffic control should be notified that you have an engine shutdown and what your intentions are. But this is the last item in the takeoff and climb-out profile and is to allow the tower (or departure or approach control—whichever would be appropriate) to assist you by giving you priority to expedite your requests. If IFR, you would merely tell departure control, "I've lost an engine and am returning to land." If VFR, just turn downwind and tell the tower you're downwind with an engine out and returning to land. But if you need a clearance to a takeoff alternate, you'll have to specify your routing, altitude, and time en route when requesting your clearance. You are, in effect, requesting a completely new clearance and, at the same time, asking them to expedite your routing and traffic separation because you have an engine out.

Instrument Takeoffs

In actual practice there are no 100% instrument takeoffs in airline operation. There are certain weather minimums (generally no ceiling and ¼ mile visibility) that require center-line lights or markings to provide sufficient visual reference for the takeoff roll. However, after rotation and lift-off when taking off with low ceiling and visibility, you are very shortly on instruments. The best method of accomplishing the takeoff under such conditions is to maintain alignment with visual reference and then go on the instruments at rotation.

To simulate this condition in training, the hood is raised just after lift-off and below 100′. You're already airborne, and this makes it too easy. Let's talk about the old-fashioned way, the full hooded takeoff and the scan technique. In flight training, though it's not required, we'll do at least one; it makes the takeoff with the hood up at 100′ much easier.

Be sure you're lined up straight and in the center of the runway. Advance the throttles to 70–80% power, let the power stabilize, check the closing of the bleed valves, and then release the brakes and smoothly apply takeoff thrust. The technique is exactly the same as for normal takeoff, but now the compass rather than visual reference is primary for directional control. Check your heading closely before beginning your takeoff roll, while aligned in the center of the runway, and maintain an exact heading during acceleration.

The most common error is drifting to one side and then correcting back to runway heading to parallel the runway center. A sense of timing is required. If you drift right 3°, correct back 3° to the *left of the runway* heading for about the same length of time that you drifted right. If you merely correct back to runway heading, you'll be to the right of the center line, and another error to the right may put you very close to the runway's edge. By doubling your original error in correcting and then returning to runway heading, you'll bracket the center of the runway very closely.

Speed is primary for beginning rotation; then the horizon becomes primary for pitch attitude for lift-off and also for bank, and the compass becomes secondary for bank and directional control.

After lift-off, the gear should *not* be retracted until a positive rate of climb is indicated. Two instruments, the rate of climb and the altimeter, should show an indication of positive climb before retracting the gear.

After gear retraction, it becomes a routine takeoff climb. Hold the proper climb attitude (about 15° nose-up pitch) and climb out straight ahead. Use a normal flap/slat retraction schedule, not below the minimum altitude of second-segment climb and on the proper speed schedule for your aircraft. This may also be modified by noise abatement procedures. But I've always felt that a pilot's first responsibility is to the passengers and their safety, and noise abatement is secondary. Therefore, I never use a noise abatement procedure if in precipitation, turbulence, or on actual instruments. Anyway, don't forget to use your climb check after establishing normal climb.

To this point we have discussed takeoffs on minimum runway lengths and normal takeoffs, which flight training will use and stress. However, there are certain conditions involving weather phenomena that could modify normal takeoff procedures and considerations. See Chapter 30.

Wake Turbulence

Flight tests have proved that vortices may be generated with tangential (turned aside from a straight course or digressing) velocities of 150′ per second (about 90 knots—88.81 to be exact). They are generated by all aircraft, but the greatest vortex (and most dangerous) strength occurs when the generating aircraft is heavy, slow, and in clean configuration. Its angle of divergence is such that the turbulence moves to the right of the generating aircraft at a speed of 4.93399 knots or .08223 knots per minute. That's about 500′ per minute (499.99986 to be exact). In addition to the angle of divergence when the aircraft is airborne, the vortices move downward at 500′ per minute for about 900′. They range from 25′ to 50′ in diameter. I've never seen any information on how long it takes for them to dissipate, but, from incidents I have reviewed, I would estimate from 3 to 5 minutes.

Wake turbulence is related to an aircraft in a high angle of attack and producing a high lift coefficient in any configuration. Also aircraft with shorter wings (a low aspect ratio) will produce more wake turbulence than those of the same weight with a longer wing (a high aspect ratio).

I have related most of the questions in the next section to takeoff because wake turbulence encounters are most frequent and dangerous in this aspect of flight. However, they have been encountered in approach to landing and other phases of flight as well.

WHAT IS YOUR WAKE TURBULENCE IQ?

Listed below are 10 multiple choice questions that reflect findings by the FAA and NASA concerning wake turbulence. Check your answers against those at the end of the test. Credit yourself with 15 points per correct answer; if you have a perfect score of 150, you have the necessary high wake turbulence IQ. If your score is less than 135, missing only one, you'd better learn *all* the answers.

1. When departing behind a large cargo aircraft, or any aircraft that may be designated as "heavy," which of the following types of wind would result in the most persistent runway turbulence:

a. Calm winds.
b. Direct headwinds.
c. 5-knot crosswind component.
d. 10-knot crosswind component.

2. During a calm wind condition, a "heavy" jet aircraft departs on runway 36L. When should a pilot expect the turbulence to reach runway 36R if the distance between the two runways is 1,000′?

a. ½ minute.
b. 1 minute.
c. 1½ minutes.
d. 2 minutes.

3. When does a departing aircraft start producing wingtip vortices?

a. At the start of the takeoff roll.
b. At an approximate speed of 60 to 90 knots.
c. At lift-off.
d. When the nose is first rotated for takeoff.

4. What conditions of airspeed, weight, and configuration would generate the greatest amount of wake turbulence?

	Airspeed	Weight	Configuration
a.	Slow	Heavy	Flaps down
b.	Slow	Heavy	Clean
c.	Fast	Heavy	Flaps down
d.	Fast	Heavy	Clean

5. At what rate and to what altitude will the vortices generated by an aircraft descend?

a. 500′ per minute for 900′.
b. 500′ per minute for 500′.
c. 1,000′ per minute for 2,000′.
d. 1,000′ per minute to ground level.

6. The major danger associated with the high velocities of large jet aircraft would be present during which type of operation?

a. Landing.
b. Takeoff.
c. All flight operations.
d. Ground operations.

7. When taking off behind a departing jet aircraft, a good technique would be to:

a. Lift off prior to the point of rotation of the jet and stay above or away from its flight path.
b. Delay lift-off as long as possible to create excessive airspeed for penetration of the vortices.
c. Climb to 500′, level off, and turn so as to cross the vortex path at a 90° angle.
d. Adjust your flight path so as to penetrate the vortex core 500′ below the generating aircraft.

8. Generated vortex cores range in diameter from 25 to 50′. How are the two vortices of an aircraft affected by time?

a. The cores rapidly expand until they overlap and dissipate.
b. They stay very close together with little expansion until dissipation.
c. They gradually reduce in size until dissipation.
d. Depending on the atmospheric conditions, they sometimes increase or decrease in size.

9. Which of the following tangential velocities would approximate those of a very heavy jet such as the C-5 or Boeing 747?

a. 500′ per minute.
b. 5,000′ per minute.
c. 9,000′ per minute.
d. 15,000′ per minute.

10. Which of the following encounters with wake turbulence would probably result in the greatest loss of control of the penetrating aircraft?

a. Crossing the wake at a 90° angle.
b. Climbing through the wake at 90° angle.
c. Climbing through the wake on the same heading as the generating aircraft.
d. Flights 1,000′ below the generating aircraft.

Answers

1. The answer is (c). I stated that the angle of divergence is to the right at a speed of 500′ per minute. With a calm wind (a) or a headwind (b) it would be moving to the right and away from the runway and/or the generating aircraft's flight path. With a 10-knot crosswind (d) it would still be moving to the right at 500′ per minute or to the left at 1,500′ per minute. A 5-knot crosswind from the *right* would make it persist on the runway for a longer period of time.

2. The answer is (d). The aircraft taking off on runway 36R may expect the vortices to move the 1,000′ from runway 36L in 2 minutes.

3. The answer is (c). The aircraft is at a high angle of attack and producing a high lift coefficient.

4. The answer is (b). The generating aircraft is slow, heavy, and in a clean configuration. Without flaps it has a higher angle of attack than it would with flaps.

5. The answer is (a). They descend 500' per minute for about 900'.

6. The answer is (d), ground operations. Small aircraft may be affected even on the taxi strip alongside the runway. Takeoffs and landings are also, in essence, ground operations. This is not to say, however, that other flight operations cannot be affected; they can and have. There have been accidents caused by wake turbulence with aircraft near the ground after takeoff or on final approach, and upsets have occurred in other flight operations.

7. The answer is (a). Even though a time delay is provided for takeoff behind a "heavy" jet, smaller aircraft should consider any air carrier or aircraft much larger than the one being flown as capable of causing wake turbulence. Take the time delay and lift off before the preceding aircraft's point of rotation and, considering wind and its effect on the vortices, establish your flight path away from its flight path.

8. The answer is (b). They stay very close together until dissipation.

9. The answer is (c). Isn't 150' per minute 9,000' per second?

10. The answer is (c), climbing through the wake on the same heading as the generating aircraft.

Summary

I learned about wake turbulence the hard way; fighting it and prop wash (prop wash is also a form of wake turbulence) in formation flight, following an L-1011 too closely in approach with a DC-9 during a training flight and seeing two accidents occur in wake turbulence. As a result I thoroughly researched the subject, flew in such a manner as to avoid an encounter, and taught the above to all my students, who have also been successful in avoiding the phenomenon. Wake turbulence, like any other flight hazard, is something to be avoided. To avoid it takes knowledge. Good pilots, professional pilots, devote their entire careers, every minute in flight, to learning all they possibly can to make flight safer. You can learn the hard way through experience, from others, and from observation. I hope you learn from the above and from all in this book. Remember, *never stop learning!*

9

Rejected Takeoffs

The FFA regulations pertaining to training and check rides refer to "rejected" takeoffs and "rejected" landings as synonyms for "aborted" takeoffs and landings, meaning the arrest of the development of the takeoff or landing during an attempt and before it is actually accomplished.

Flying heavily loaded bombers during World War II, we found that takeoff at maximum load was the most dangerous part of any mission. The terms V_1, V_R, and V_2 were unknown at the time; we had to use "Kentucky windage" on every takeoff. You'd look down the runway—assessing your expected acceleration considering your load and weight, field elevation, wind, temperature, etc.—and pick a reference point. If any mechanical or engine failure occurred prior to reaching that point, you'd abort and try to stop before going off the end of the runway. Past that point, you knew you couldn't stop and had to fly! This was known as Kentucky windage, and it wasn't a very exact way to determine actual performance criteria for aborting a takeoff. It was an educated guess at best, but from it was born the more scientific accelerate/stop performance of V_1 used today. One fact became apparent, however, and still applies: It may be safer to go off the end of the runway and plow through rough terrain at 50 knots or so, attempting to stop, than to struggle into the air at too low an airspeed and then come to a stop a little farther out and from a higher altitude and speed. It pointed out the need for adequate runway length and, if actual runway length was not available, for an overrun area at the runway's end.

Kentucky windage, intuition, and luck have long since been replaced with the velocities V_1, V_R, and V_2. These are speeds, developed and proved by flight test, that are used as a guide by the pilot in making an abort/fly decision and for performance in the takeoff maneuver. The decision speed of V_1, the accelerate/stop speed at which a stop may be executed and beyond which flight is mandatory on minimum length runways, is the governing speed in determining whether to continue the takeoff or to abort.

When you are operating a jet that is loaded to the maximum weight allowable for the runway in use, whether it be a large and heavily loaded four-engine jet on an intercontinental flight or a smaller jet operating from a smaller airport, it begins to border on a marginal operation. In fact, there are many who are of the opinion that it is extremely hazardous, especially when runway surface conditions reduce brake effectiveness.

The most hazardous parts of any flight are the takeoff run and deceleration after touchdown on landing. The aircraft, designed to fly, is hard to control during these periods of acceleration and deceleration on the ground, especially when there is a sudden and unexpected yaw as may be caused by a blown tire or engine failure. Compound these malfunctions with tailwind, crosswind, wet and slippery or icy runways and there are a few seconds when control is rather touchy. Also, there seems to be a point of transition on minimum length runways where neither abort nor flight seems desirable from a standpoint of safety in the event of a malfunction. This is right at V_1. For a split second, you're damned if you do and damned if you don't. You may either abort or continue the takeoff, at the captain's option, but the correct decision must be made and action initiated instantly.

Much can be done to make takeoffs safer. The most obvious solutions are longer and wider runways, with ample overrun at each end, and a better method of computing accelerate/stop distance, with more realistic V_1 speeds and runway lengths as related to runway surface conditions. The Air Force mechanically examines braking coefficient capability for varying runway conditions and takes these findings into consideration for both takeoff and landing performance. But this is not the case thus far in civil aviation or airline operation; until something similar is accomplished there, better training in the abort of the takeoff is the only alternative.

The FAA has established criteria for takeoff performance by which a pilot must abide. The numbers vary for different airplanes, but they are established by actual flight test for all aircraft. Basically, they establish the speeds associated with takeoff—V_1, V_R, V_2, and *minimum runway length.*

Minimum takeoff field length requirements vary with gross weight, temperature, altitude, wind, and runway slope. The minimum length must be the greatest of the following:

1. 115% of the all-engine takeoff distance to a height of 35'. This is called the *all-engine takeoff field length.*

2. The distance to accelerate to the decision speed (V_1) with all engines operating, recognize the loss of an engine at V_1 speed, retard the operating engine or engines, and bring the airplane to a full stop. This is called the *accelerate/stop distance.* (Note! No mention is made of any other malfunction that could affect aircraft performance and require an abort—blown tires, striking an object, control malfunction, instrument failure, fire of any nature, or any mechanical failure other than an engine—but they must also be considered by the pilot.)

3. The distance to accelerate to V_1 speed with all engines operating, lose an engine, continue to accelerate on the remaining engine or engines to V_R speed, at which time rotation is commenced so as to reach a height of 35' *above the runway* at V_2 speed. This is called the *takeoff distance with an engine failure.*

The last two are the most critical, and they must balance. To make V_1 a decision speed, affording an alternative, there must be the ability to either stop (as in case 2 above) or continue (as in case 3). Therefore, there is also a concept called the *balanced field length.* This is when a V_1 speed is selected so that the accelerate/stop distance is equal to the takeoff distance with an engine failure.

The takeoff decision speed, V_1, which for all practical purposes may be considered the critical-engine-failure speed (though this speed actually is 1 second prior to V_1 to allow for recognition of failure), is the speed the pilot uses as a reference in deciding whether to continue or abort. In actual practice, at max gross for a particular runway creating a condition of critical field length, an abort is mandatory for speeds at or below V_1; to continue the takeoff is mandatory at speeds above V_1. The split second the aircraft is exactly at V_1 speed is the split second in which the pilot,

if he or she experiences and recognizes a malfunction, must correctly decide to abort or to continue the takeoff.

One airline, being aware of the complex dangers of this moment of decision, put out a bulletin to its pilots stating, "The decision to abort a takeoff involves a thought process fraught with potentials for trouble. It necessitates a sudden reversal in plans from the inherent intent of the takeoff maneuver and a shift in gears in the use of the controls." There is not the slightest doubt that this factor causes inordinate delays in execution of the abort procedure, and occasional omissions in the procedural steps. At least one airline admits that there may be a moment of hesitation while recognizing a failure, shifting gears mentally, and beginning the abort procedure when some danger does exist in an abort. There is a time delay factor considered for accelerate/stop, but this time delay can be exceeded in experiencing an unexpected engine failure recognized at V_1.

Considering the minimum runway requirements, it would appear that the aborted takeoff is not as critical or dangerous as some believe. But before forming an opinion and to more thoroughly understand the aborted takeoff and the hazards that may be associated with it, it is necessary to further explore V_1. We know that it may be considered as the critical-engine-failure speed, the recognition-of-failure speed, the go/no-go speed, or the decision speed of takeoff. But could this speed be even better defined? And how is it arrived at in the first place in aircraft certification tests?

The takeoff decision speed, V_1, given in the FAA approved Airplane Flight Manual has been selected so that: (1) if engine failure is recognized at or before the V_1 speed, the takeoff may be continued on the remaining operating engine (engines) to a 35' height; or (2) a stop may be made *on the runway* at V_1, or at speeds less than V_1, without the aid of reverse thrust; and without, in either case, exceeding the FAA takeoff field length. These minimum takeoff field lengths are based on stopping if engine failure is recognized before reaching V_1 and on continuing the takeoff if engine failure is recognized after V_1.

The stopping performance is based on retarding the power levers to idle *at the time of engine failure recognition, maximum braking on a dry runway* with antiskid operative (unless otherwise noted) and with ground spoilers extension initiated after the power levers are moved to idle. *The effect of reverse thrust is not included.* [Eastern Air Lines Douglas DC-9 Flight Manual]

The V_1 performance figures for certification relating to stopping distance are derived from the flight test. Thoroughly trained test pilots prove maximum capability of the aircraft on a dry runway and without bad weather or other problems, and from that performance the V₁ for accelerate/stop will be established. They fly a plane carefully prepared for this test; its engines, brakes, and tires are perfect. Every step of the maneuver, including the 3-second delay for application of complete abort procedure, has been carefully rehearsed. They begin the takeoff with hands on the power levers, knowing they are going to abort; they are prepared to immediately retard the takeoff thrust to idle; toes are on the brake pedals prepared for brake application; they know the exact speed at which the simulated failure will occur; so they abort the takeoff perfectly.

The actual conditions under which you may have to abort are far different. Your plane has not been specially prepared; your tires may not be brand new; your brakes may not have been adjusted and checked for this one takeoff; your takeoff data may not have been so carefully computed; the runway may be wet and slippery, causing you to hydroplane; the runway may be hot asphalt, causing slow acceleration; a tire may blow, causing slow acceleration, vibration, and structural damage from flying pieces of rubber, including engine failure from ingestion of rubber. All your malfunctions and failures are unplanned surprises requiring the mental shift of gears to the abort. These routine conditions make an abort to the degree of proficiency demonstrated by the test pilot extremely difficult to duplicate.

However, the rejected takeoff maneuver must be practiced in training and demonstrated in a proficiency check. Both the captain and copilot are trained to abort. Normally, things that may be required and safely performed in an aircraft are performed just as they may occur in line flying while carrying passengers. Since V₁ is a decision speed, you would expect to be required to demonstrate proficiency for both alternatives. You *are* required to accelerate to V₁, experience a simulated engine failure, and continue the takeoff with an engine at idle. It should also be required to perform the other side of the maneuver, the abort from V₁. But this just isn't done! In fact, it's specifically prohibited by regulation!

Part 121 of Federal Air Regulations, Appendix F, has set forth the proficiency check requirements for air carriers. In relation to rejected takeoffs the following is stated, requiring one rejected takeoff in the flight check: "The proper procedures for the rejected takeoff: (1) may be performed in an airplane after a normal takeoff has been started at a *reasonable* speed that has been determined by giving due consideration to runway length, surface conditions, brake heat energy, and other pertinent factors (but in no event at a speed greater than 50% of V₁ speed); (2) may be performed in an approved simulator with an approved visual system or may be demonstrated in an airplane other than during an actual takeoff; or (3) may be waived."

Apparently the airlines and/or the FAA don't want to have a training airplane (with only a crew and perhaps an FAA inspector on board) involved in an abort, even under controlled and expected conditions, though you may have one on your next takeoff with a full load of passengers. Instruction is given in the simulator, the procedure is practiced in the simulator, and the maneuver is waived on the flight check. Maximum braking puts some wear on the aircraft and raises the cost of training, but I think the real reason for not practicing the aborted takeoff is that it's dangerous on minimum field length.

That's all we have to work with at the present time. It's up to the flight instructors to figure out how to teach the procedures outlined in the flight manual and to give you reliable information about a maneuver they are prohibited from demonstrating and that you can't practice under real conditions. The following are my own suggestions, not necessarily those of the FAA, manufacturer, airline, or ALPA. They are offered as a supplement to the rules (which you must follow) set forth by the above parties and are not intended to replace company procedures.

First, the mechanical process: Close the throttles, apply brakes simultaneously, actuate ground spoilers, and then use reverse thrust as desired. This procedure will conform to the procedure outlined in your flight manual. As an alternative procedure, you might close the throttles and pull reverse thrust while your hands are still on the throttle, apply maximum braking at the same time, then pull the ground spoilers and come back to monitor reverse thrust. But if you use the reverse before the spoilers, you may have to explain why you used a device not considered in stopping performance before using the spoilers, which are. In all cases the yoke should be hard forward to destroy lift and provide drag from a negative angle of attack as well as to assist in steering. Directional control must be maintained by any means available—nosewheel steering, brakes, flight controls, or even differential reverse thrust if necessary—but it would be well to remember our discussion about skids.

There is little opportunity to practice aborted takeoffs, and fortunately they occur infrequently. But it is necessary for the operating crew to mentally run through the abort procedure (and all other possibilities related to the takeoff) before every takeoff. It should become a habit for the pilot to command, "Call my speeds—V₁, V_R, and V₂—and any malfunctions" before beginning the takeoff run.

The captain should consider the following every time the position to begin takeoff is taken:

1. V₁ = critical-engine-failure speed on a minimum length runway. What is the minimum field length for weight, temperature, etc., or is the plane at maximum weight for the runway?

2. If not at critical field length (operating at

weight less than maximum for the runway), how much *surplus* runway length is available in the form of paved runway, approved overrun, clearway, or stopway?

3. Consider again the headwind component. Is it the same as that used in determining the maximum weight and critical field length? Is it more, increasing safety of takeoff performance, or is it less?

4. What are the adverse factors to acceleration (slush on the runway, hot asphalt pavement)?

5. What are the adverse factors to stopping (crosswind or downwind components; wet, slippery, or icy runway surfaces; slush on the runway; hydroplaning speed; worn tires; brakes)?

All these are factors that will affect takeoff performance, specifically accelerate/stop distances; we'll talk about how to use them as "Kentucky windage" at the end of the chapter.

Deciding whether to abort or continue the takeoff remains a matter of judgment, the *captain's judgment,* and it is the captain's responsibility to determine whether to abort or continue, regardless of whether the captain or the copilot is making the takeoff. The captain, therefore, should be mentally alert to the aforementioned pertinent conditions; all other crew members should be alert to engine and aircraft performance during takeoff that might necessitate aborting and should call malfunctions to the captain's attention *immediately!*

It follows then that during all takeoffs the crew will coordinate their functions in the following manner:

1. All crew members must be alert and in position to act.

2. Any observed abnormal condition or malfunction must be reported to the captain immediately.

3. The captain must make the decision, if necessary, to abort or continue the takeoff.

4. The captain must announce that decision loud and clear and, when flying, take the proper action plus calling for any assistance needed. (It may vary with the different airplanes due to cockpit layout and procedural differences. As an example, the captain might reduce the power levers to idle, control yaw tendencies by using rudder and nosewheel steering as necessary, actuate ground spoilers, and apply maximum braking and reverse thrust; the copilot might be required to maintain forward pressure on the wheel, shut down an engine, actuate ground spoilers [in some airplanes], discharge fire agent if necessary, etc.)

5. The captain must be even more alert when the copilot is flying in order to make and announce decisions and to take over the controls immediately.

6. The copilot will remain on the controls when flying, executing the abort or takeoff according to the captain's decision and command, and keep the aircraft under control as well as possible until the captain has taken over the controls and announced that fact loud and clear. The copilot shall then relinquish control to the captain and carry out such duties as requested or as might be pertinent to the situation.

7. Having made the decision to abort the takeoff, the captain shall immediately command the same, immediately taking over the controls to maintain directional control and making this fact known in a loud voice. The captain shall then proceed to use the proper deceleration and stopping procedures, calling for assistance as necessary from other crew members.

We've covered virtually everything normally considered about aborted or rejected takeoffs, but everything we've said so far about minimum runway lengths, V_1 speeds, and stopping the aircraft has been for dry runway conditions. To make the discussion complete, let's clarify some terms and consider the effects of runway conditions on stopping.

Runway Surface Condition

All performance involving stopping distance must be based on assumed coefficients of friction between the tires and runway surface. The minimum field length for takeoff (the basic performance data including the accelerate/stop distance) assumes a *dry* concrete or asphalt runway. Variations in this basic performance depend on the runway surface and the type and amount of runway surface covering.

Wet or icy runway surfaces will produce a lower coefficient of friction (depending on the amount of water or ice on the runway), and slippery runway surfaces will increase stopping distances, which *should* increase required field lengths.

The pressure between the fluid on the runway and the tires increases as the speed increases until the tires are entirely supported on top of the fluid. The speed at which this occurs is called hydroplaning speed and may be computed by multiplying the square root of your tire pressure by 9.

REJECTED TAKEOFFS

The braking coefficient is reduced to almost 0 at speeds above hydroplaning speed. This would make the determination of the minimum field length (computing the actual distance required to accelerate to V_1 and then stop within the remaining runway length) very difficult. Any retarding effect of slush is virtually eliminated when the aircraft reaches hydroplaning speed.

Because of the effects of slush upon acceleration, the ability to accelerate to V_R after losing an engine is accomplished by reducing gross weight (see Table 8.2). But no weight reduction or any other correction factor is taken for reduced brake effectiveness. Therefore, since V_1 is, by definition, the highest speed to which the aircraft can be accelerated, lose an engine, and then take off or stop in the computed field length, V_1 should never exceed hydroplaning speed on any runway conditions where it may be a factor affecting braking coefficient.

Critical Field Lengths

The major consideration in determining the minimum runway required for takeoff is selecting the distance required to accelerate to V_1 as a critical-engine-failure recognition speed, experience an engine failure, and then either continue to accelerate and lift off and attain V_2 at an altitude of 35′ over the runway or to stop on the runway. The minimum distance in which this may be accomplished is called critical field length.

Critical field lengths are also computed assuming dry runway conditions, since they too are part of the basic performance data. When using a gross weights manual, the maximum weight corrected for wind and temperature for a runway would indicate that you are taking off on a runway's critical field length dry. Just as there are weight reductions for takeoff in slush, there should also be weight reductions for any takeoff where runway surface condition will reduce braking coefficients, but there aren't. No adjustment is required for standing water, slush, wet snow depth up to 0.2″, or dry snow depths up to 4″. *Absolutely no correction at all is made for runway conditions as related to stopping performance!*

This being the condition in air carrier operation, the pilot should actually compute a minimum field length requirement for every takeoff on any runway condition other than dry. Then, to conform to the letter of the regulations requiring an ability to get to a certain speed and then either be able to abort or take off, the pilot will have to *reduce* V_1 speeds and, at the same time, *increase* minimum field length.

The combination of a decrease in V_1 and an increase in critical field length provides (1) a more realistic decision speed in relation to stopping distance as affected by runway surface conditions and (2) adequate runway for a stop within the runway limits from the reduced V_1 as well as adequate runway to continue the takeoff, accelerating from the lower V_1, and to accomplish the takeoff safely.

The correction factors in Table 9.1 provide the only information that assures a safe stop on anything other than a dry runway. You can always take off, assuming normal acceleration and proper weight reductions for slush, but you'll never be able to stop on a wet and slippery or icy runway or at hydroplaning speed in water when operating on a minimum field length based on dry stopping data.

Runway slope is one of the factors to be considered, since takeoff distance and time will be longer than usual on an uphill slope and shorter on a downhill slope.

These correction factors are calculated figures and are not approved by any agency. I'm sure that more realistic V_1 speeds should be required for braking conditions on some runway surfaces and that adding runway to the basic requirement or, as an alternative, reducing the allowable weight for the runway condition should also be required.

As a final word of advice, the correct abort procedure is contained in the emergency operating section of your aircraft manual; this procedure, along with all emergency procedures, should be reviewed frequently.

TABLE 9.1. Correction Factors for Takeoff Runway Conditions

Runway Surface	Braking Action	Decrease in V_1	Increase in Critical Field Length	Decrease in V_1	Increase in Critical Field Length	Decrease in V_1	Increase in Critical Field Length
		(−0.2 downhill slope)		(0 slope)		(+0.02 uphill slope)	
Wet concrete or macadam	Good to medium	10%	16%	8%	17%	7%	25%
Wet and slippery	Fair	20%	30%	15%	30%	9%	35%
Icy	Poor to nil	33⅓%	35%	25%	40%	15%	50%

Note: The critical field length is the *dry* runway distance required to accelerate to V_1, lose an engine, and either stop or continue to accelerate with a failed engine to V_R and to attain V_2 at an altitude of 35′ within the confines of the runway. V_1 may be limited by V_{MCG} in some cases on a downhill slope.

10

Climb, Cruise, and Descent

■ Climb, cruise, and descent are en route consider-ations and are usually given little time in a training program, since the professional pilot is expected to have the ability to fly an airplane cross-country, which of course requires climb, cruise, and descent. The pro-fessional pilot transferring to jet equipment is nor-mally just briefed on climb and descent speeds, and actual in-flight training is confined to the check ride required to obtain the rating for the aircraft. This is one of the weaknesses of training in a new aircraft. Pilots are left to their own devices in learning to oper-ate the airplane normally in an en route situation. However, we're not restricted by time and operating cost here, so a discussion of these three phases of flight may prove helpful, even in flight training.

Climb and Climbing Turns

In a jet operation there are at least six different climb regimes: (1) one-engine-inoperative en route climb, (2) climb in a "tunnel departure," (3) 250 knots indicated airspeed climb to 10,000', (4) best rate of climb or best time to altitude, (5) long-range climb or best fuel to altitude, and (6) normal high-speed en route climb or best distance to altitude.

ONE-ENGINE-INOPERATIVE EN ROUTE CLIMB

The one-engine-inoperative climb is the only ab-normal situation that we'll have to consider. For most airplanes, the best rate of climb with an engine in-operative is the same speed used as the final segment of climb (see Fig. 8.5). This speed represents 138% of stall speed for gross weight, with the landing gear and flaps retracted and the airplane in the clean configura-tion. It provides the best combination of climb gra-dient and highest limiting gross weight at all tempera-tures and altitudes. However, after obstruction clearance is no longer a factor, the speed should be increased to at least the minimum recommended clean maneuver speed of 150% of stall for a better lift/drag ratio and more safety in maneuvering.

TUNNEL DEPARTURES

There are many instances where it is necessary that your flight be held to a low altitude in departure to avoid inbound IFR traffic arriving over you. This is the "tunnel departure" where you depart in a "tun-nel" under the inbounds. This situation may be per-plexing due to the FAA regulation restricting operat-ing speeds to 250 knots indicated speed below 10,000' and the possibility of turbulence requiring operation at turbulence penetration speed. In most jet carrier aircraft, 250 knots indicated at weights below maximum landing weight and at low altitudes away from Mach considerations is a safe turbulence speed. However, at weights above maximum landing, it may be necessary to accelerate to turbulence penetration speed and then inform the controller of your speed and its necessity.

Where turbulence is not a factor, the rapid acceler-ation and high rates of climb are the factors that make climb and speed control difficult. Climbs in excess of 6,000' per minute are not uncommon. And rapid ac-celeration, not only far beyond 250 knots indicated but right on up to the maximum operating speed, is possible if you are really precise with power and atti-tude control.

To make the tunnel departure easier, weather con-ditions and turbulence permitting, I'd recommend es-

73

tablishing the desired aircraft attitude of about 10–15° nose up and control your speed and rate of climb in this constant attitude with power to a speed of 150% of stall. In some aircraft this may be as high an indicated speed as 220 knots. In any case it will be a slow-flight type of procedure at a speed below 250 knots indicated. This will give you better speed, rate of climb, and power control and will result in lower fuel consumption at low altitude.

250 KNOTS INDICATED AIRSPEED CLIMB BELOW 10,000′

When the climb restriction of the tunnel departure is lifted, set climb thrust to attain 250 knots indicated airspeed. Then power becomes primary for pitch attitude until a maximum attitude of 15° nose-up is attained.

Climbs in excess of 15° nose-up are uncomfortable and disconcerting to your passengers. Therefore, if climb thrust and 15° nose-up will give you faster than 250 knots indicated airspeed, a power control of the speed is necessary.

BEST RATE OF CLIMB

If reaching altitude in a hurry is your prime consideration, climb at the speed above 10,000′, which is the best average lift/drag ratio speed for your aircraft. This speed will vary from aircraft to aircraft and will also vary according to gross weight for a particular aircraft. You'll have to dig it out of your particular aircraft manual, along with its corresponding Mach number at altitude.

I've found, for rule-of-thumb and practical application, that V_B speed (turbulence penetration speed) is most nearly the best average lift/drag ratio speed to use for indicated airspeed in every case.

LONG-RANGE CLIMB

As its name implies, long-range climb is at a speed that is the best combination of climb, fuel burn-off, and distance covered. It is somewhat faster than the best rate of climb speed and slower than the normal high-speed climb. This is another figure you'll have to obtain from your aircraft manual, but it is usually halfway between best climb and high-speed climb.

The DC-9-10, for example, has the best rate of climb at 260 knots indicated, the long-range climb at 290 knots indicated to Mach 0.72, and a normal high-speed climb of 320 knots indicated to Mach 0.74.

Since long-range climb is the best combination of all factors, I like to use it for long flights or anytime I know I will experience a good tailwind in climb and at altitude.

NORMAL HIGH-SPEED CLIMB

The recommended en route climb speed is a generally high indicated airspeed to be used at all weights and temperatures. This is the speed on which your "top-of-climb" performance is computed, either from tables in your aircraft performance manual or contained in your computer flight plan. It is the speed that provides the highest practical true airspeed in climb and therefore the best ground distance covered in climb.

General Comments

The airspeed indicator is the primary instrument for pitch attitude all during the climb with a constant power, and the desired indicated speed should be maintained until it crosses the recommended Mach number for Mach climb regime. At this point, the Mach number becomes primary for pitch attitude and should be maintained. As altitude increases, a constant Mach indication results in a continually decreasing indicated airspeed.

For the most economical and efficient operation, it is desirable to establish en route climb speed as quickly as possible after obstacle clearance. In establishing en route climb, maintain a positive climb indication of at least 500′ per minute while accelerating to the desired speed.

The aircraft should be trimmed about all three axes and the en route climb speed adhered to as closely as possible. The airspeed is always primary for pitch while climbing at a constant thrust when rate of climb is not a consideration; small pitch changes result in rapid speed changes.

If you have to level off at any point during climb, monitor your thrust closely because the thrust accelerates the aircraft and you may find yourself at maximum speed very quickly. To resume climb, set climb thrust and establish the climb speed with the elevator control.

During the climb, as long as the temperature lapse rate remains standard with altitude increase, the climb thrust setting will remain reasonably normal and correct with a constant throttle setting and position. However, the thrust schedule should be checked at each 5,000′ of climb to ensure the correct thrust setting and not overboost the engines.

Maneuvering should be held to a minimum, and climbing turns should not exceed 25–30° bank angle. Banks of more than 25° materially affect climb performance, reducing rate of climb through loss of vertical lift in relation to gravity; banks of 30° or more may cause passenger discomfort due to the load factors imposed at high speeds.

To climb the last 1,000' to assigned altitude at 500' per minute, the rate of climb becomes the primary instrument for pitch attitude and climb. Set it smoothly on the desired rate with elevator control. Your airspeed will probably increase now and becomes the primary instrument for power control.

During level-off at cruise altitude, lead the rate of climb by 10% (lead the 500' per minute by 50' below the desired altitude) and roll the nose right over onto your desired altitude smoothly with the elevator. The altimeter now becomes the primary instrument for pitch attitude to hold altitude. Leave the climb thrust on until cruise speed is attained, thereby attaining the fastest acceleration to the desired speed. As your Mach number and indicated airspeed increase, the ram air temperature also increases, so it is essential that cruise thrust is set as soon as cruise speed is attained to ensure accurate thrust settings.

Instrument Scan in Climb

The instrument scan and control technique in the climb break down to the following:

1. The airspeed indicator is the primary instrument for pitch attitude in climb at a constant power setting and a constant indicated speed.

2. The rate-of-climb instrument is the primary instrument for pitch in climb, controlled by the elevators, to climb at a fixed rate; the airspeed indicator is the primary instrument for power to maintain a constant speed if desired.

3. The level-off from climb is most easily accomplished if led by 10% of the rate of climb below the desired altitude. This is elevator action alone if you wish to level off and accelerate; it is a combination of elevator and stabilizer trim to level off and a power reduction to maintain a constant speed if acceleration is not desired.

4. Anytime you are maintaining a constant altitude in level flight, the altimeter is controlled by the elevator and is the primary instrument for pitch attitude. The airspeed is controlled by power and the airspeed indicator is the primary instrument for power control.

Cruise

Since jets have become so numerous in both airline and corporation fleets, the term "jet service" seems to be synonymous in most people's minds with long range and high speed. Both are characteristics of jet operation, especially in transcontinental and transoceanic flights; but the vast majority of jet aircraft, particularly the small business jets and the smaller

airline jets, are designed for the purpose of offering the passenger the obvious benefits of jet flight in speed and comfort for short and medium trips as well. Basically, they are designed with a versatility for economical operations ranging from 50 to 1,500 nautical miles.

As you might imagine, a cruise regime on the shorter flights is virtually nonexistent. The flight profile could, and does in some cases, consist of takeoff, climb, descent, and landing, so consideration of cruise is not a factor. From a practical standpoint, considering both fuel burn-off and air traffic control, the optimum altitude selection for short distances may be easily determined by multiplying the route distance times 100. For example: a flight of 60 nautical miles $\times 100 = 6,000'$ altitude; 150 nautical miles $\times 100 = 15,000'$ altitude.

For cruise on longer stages of flight, the distance times 100 works very well for most aircraft until stage lengths of more than 300 miles are reached. For distances greater than 300 miles, cruise above 30,000' or even higher may be optimum. A Lear jet, for example, may find a 35,000–40,000' altitude more efficient for a 300-mile trip, due to its light weight and climb capability; a DC-9 would be more economical at 30,000–31,000'. Familiarity with your own aircraft's performance may vary the rule of thumb as to where the distance times 100 breaks off in selection of optimum altitude for cruise.

There are actually three cruise procedures that may be used in jet operation: (1) high-speed cruise at maximum indicated airspeed or constant Mach number, (2) a recommended cruise of 10 knots below maximum airspeed and a lower constant Mach number to prevent inadvertent and nuisance sounding of the overspeed clacker warning, and (3) long-range cruise varying Mach number in relation to gross weight and altitude. Each procedure has advantages either in actual flight time or fuel used; it is up to the individual operator to determine the procedure best suited to the routes and operation.

Most operators, weather and turbulence permitting, normally fly at high-speed or recommended cruise, since this permits minimum direct flight time and average operating costs. When using high-speed or recommended cruise, it is important that you adhere to optimum altitudes for operation in order to achieve the ultimate in both speed and efficiency. If cruise is made at an altitude less than optimum, the penalty will be increased fuel consumption. More accurately stated, a compromise must be made at lower than optimum altitude by reducing speed for the sake of economy or sacrificing fuel in favor of faster flight time.

Always remember that cruise altitudes must also consider weather, turbulence, buffet boundaries, etc., and that altitudes higher than optimum may sometimes be desirable for exceptionally long flights or in consideration of wind.

Three terms related to cruise—fuel flow, miles per pound, and range constant—will be discussed here in general terms, since there is a wide variance in actual numbers with different aircraft. I strongly suggest that you become thoroughly familiar with the cruise performance charts and tables of your particular aircraft. Defining these terms will be helpful in enabling you to use the charts and your aircraft more efficiently.

1. Fuel flow—Pounds per hour of fuel consumption. The fuel flow of turbojet engines is a function of temperature, engine pressure ratio, altitude, and speed or Mach number. With so many variables it is not easy to generalize. However, I've found that specific fuel flow is closely related to indicated airspeed and that the fuel flow for any indicated speed will be very nearly constant at any altitude. For example, 250 knots indicated airspeed at 30,000′ requires very nearly the same fuel flow as 250 knots indicated airspeed at any altitude. The difference in true airspeed for any given indicated airspeed at altitude and the same indicated airspeed at lower levels is great; therefore, the range and ground speed is better at altitude.

2. Miles per pound—Nautical miles per pound of fuel. This is a function of weight, altitude, and speed and is essentially independent of temperature. The speed of sound increases with an increase in temperature, but the efficiency of turbojet engines also decreases with an increase in temperature. In general, the specific fuel consumption for given values of thrust, altitude, and speed or Mach number increases with temperature at approximately the same rate as the speed of sound.

3. Range constant—Miles per pound of fuel times the weight of the fuel. The greater the range constant, the more efficient the cruise procedure. This is particularly true in the long-range operation and should point out the absolute necessity of adhering strictly to optimum altitudes to obtain the ultimate in speed and efficiency.

Long-Range Cruise

As the name implies, long-range cruise is used when fuel economy is the prime consideration. The engineers that design and build airplanes, knowing the lift required for various weights; the lift/drag ratio at various speeds; and the thrust required for the most efficient combinations of fuel flow, miles per pound of fuel, and the resultant range constant, are able to prepare long-range cruise charts to give the most efficient operation.

From a pilot standpoint, it boils down to an indicated airspeed or Mach number. The long-range speed is defined as the greater of the two speeds at which 1% of range is sacrificed, and it varies with gross weight and altitude. It is a speed to produce flight at 99% of the absolute maximum nautical miles per pound of fuel.

To use long-range cruise, the pilot should determine gross weight and refer to the long-range cruise chart for the indicated speed and engine pressure ratio power settings required. The chart must be used with regularity, since the weight decreases with fuel burn-off, and the proper speed schedule must be maintained with power reductions.

It's not as easy as it sounds. The airplane must be flown smoothly, and any error in setting the thrust to the correct engine pressure ratio or selecting the wrong speed or Mach number will result in a large speed or power discrepancy. You'll be going either too fast or too slow and in either case not getting the range you desire. It might also result in an unstable flight condition if you overcontrol pitch attitude while making power changes. It is important that you be smooth, deliberate, and accurate in setting thrust and maintaining the desired speed.

En Route Holding

In jet aircraft, holding is normally done in the clean configuration, with the slats and flaps retracted, at an indicated airspeed of 1.53 V_s. This may be the minimum recommended speed for maneuvering clean (an average speed for normal operating weights) or may, at the pilot's option, be computed for an actual 1.53 V_s speed in relation to actual gross weight. You may either compute the stall speed clean for your aircraft weight and then add 53% to it for a holding speed, or it may be done by rule of thumb if you know the index number for such computation for your aircraft. For example, in the DC-9-31, you should take 120 knots and add 1 knot for every 1,000 lb. of gross weight.

A holding speed of 153% of stall, 1.53 V_s, is a compromise considering stability, safe margin above stall for maneuvering, and fuel economy. But the pilot must still exercise good judgment in the determination of holding speed and the configuration to be used. If conservation of fuel is critical, holding should be done in the clean configuration (less drag, less thrust) and at the highest practical altitude. If pattern size or lower holding speed is the primary concern, as at lower altitudes prior to approach, then the flaps and slats should be extended to a takeoff setting and the airspeed should be reduced to the recommended maneuver speed for the configuration (that is, 1.53 V_s).

Descents

The technique of getting a jet aircraft down from cruising altitude to traffic pattern altitude is important. Beginning descent too soon results in more flight at low altitude, which carries the penalty of additional fuel burn-off. Beginning descent too late makes it difficult to control speed and rate of descent simultaneously in the traffic pattern or on approach. A proper descent procedure requires planning to arrive at the desired maneuvering speed and altitude *before* the final instrument approach fix or downwind leg in the case of a VFR approach.

There are actually two ways to descend—high-speed descent or long-range descent—depending on weather, turbulence, etc. The actual indicated speeds vary, of course, for different aircraft, but the high-speed descent is accomplished at the high-speed cruise Mach until V_{MO} minus 10 (indicated airspeed for high-speed cruise) is reached. The DC-9, for example, should be descended at Mach 0.80 until 350 knots indicated airspeed is reached. However, I find it easier to use a speed of 20–30 knots below V_{MO} minus 10 and descend the DC-9 at 330 knots indicated when below Mach 0.80. The long-range descent uses less fuel but requires more time than the high-speed descent and is accomplished at the maximum-range indicated airspeed for your aircraft. The speed for this performance varies widely with different aircraft; even in the same type of aircraft it varies according to gross weight and altitude. You will just have to learn the appropriate speed to use in your aircraft. However, the average speed used for long-range descent is 250 knots indicated even though the best glide ratio could be obtained at a speed equal to 1.38 V_s or the best engine-out climb speed for your weight.

Either type of descent requires a decision for beginning of descent at the proper time, and time always involves distance as related to ground speed. One sure way of selecting the proper time to begin descent is to estimate the time to the final fix (usually the outer marker) and establish a rate of descent that will assure you of reaching the desired altitude at a reasonable distance prior to reaching the fix. In any case, you must consider the requirement of maintaining 250 knots indicated airspeed below 10,000' and plan the descent accordingly. Also, the use of the speed brake should not be anticipated in the planning except when excessive altitude or speed must be lost.

I have found, through actual flight experience, that the easiest way to select your beginning of descent (if ATC requirements will allow you to do so) is to take 4 times your altitude, add 8 miles, and vary your speed and rate of descent according to your actual ground speed and distance remaining during the descent. This consistently works with sufficient accuracy and pilot control of the variables involved, without sacrificing performance either in time or fuel requirements, that it may be used as a rule of thumb procedure in virtually every case.

Normal Descent

Changes in ATC procedures have resulted in high rates of descent near the airport. To accomplish the "get down and slow down" request, I recommend that speed brakes be used where practicable.

Normal descents and even profile descents are predicated upon a rate of descent of 300' per mile. The rate of descent required to attain this distance may be determined by dividing ground speed by 2 and adding a 0 or by multiplying ground speed by 5. For example, 450 knots ground speed divided by 2 = 225(0) or 2,250' per minute; 450 × 5 = 2,250 required descent rate.

To prove the above, 450 knots = 7.5 miles per minute; 2,250 divided by 7.5 = 300' per mile.

As your ground speed changes (reduces) simply use the above rule of thumb to adjust your rate of descent. For example, 400 knots ground speed divided by 2 = 200(0), or 200 × 5 = 2,000. It's simple, therefore, by being aware of your ground speed at any given moment, to apply the rule of thumb and adjust your rate of descent accordingly.

In the newer aircraft, with their sophisticated navigation systems and ground speed readouts, it is easy to establish the beginning of descent and descent speeds for fuel savings. For aircraft that do not have such systems, or if they cannot be used, the following may be considered in order to develop an efficient rate of descent and speed profile: Without ground speed readout, distance measuring equipment change per minute times 60 equals ground speed.

Whenever practical for fuel economy, descend at the minimum thrust required to maintain cabin pressurization. From cruise altitude, descend at cruise Mach until reaching 280 knots indicated airspeed (crossover altitude), then descend at 280 knots to 10,000'. At 10,000' level off and decelerate to 250 knots indicated airspeed; this will require approximately 2 miles. At any point, speed brakes may be used to increase the rate of descent if necessary.

Emergency Descent

This is really an emergency procedure caused by a loss of pressurization or rapid decompression of the aircraft. At the high altitudes of jet operation, people don't live too long without oxygen, and the airplane must descend to a breathable atmosphere as rapidly as possible. Prompt and positive reaction to an emergency of this type is essential; however, the few times that rapid decompression or total loss of pressurization has occurred in airline jet operation have proved that "hurried" actions are not necessary as a general rule and in some instances may create confusion.

The need for a rapid descent is predicated on the actual need for oxygen. It is apparent that a rapid decompression (caused by a rupture of the pressure envelope of the aircraft or by the unlikely total loss of pressurization through malfunction within the air conditioning and pressurization system) at a high altitude would require an immediate descent. And rapid decompression is readily recognized. You will notice the rapid pressure change before getting any mechanical warning of failure.

However, a simple mechanical failure of a pressure regulator or a system failure causing a slow climb in cabin pressure does not necessarily indicate the need for an emergency descent. If the cabin altitude can be maintained at 14,000' or below (where you still have a life-sustaining breathable atmosphere) there is no need to expose yourself to the hazard of descending through the altitudes of unknown traffic.

From a training standpoint or an actual *rapid* decompression, immediate action is required. Every airplane has a recommended procedure for emergency descent, but I can recommend a sequence of action that is a good basic procedure in virtually every case. The idea is to get down rapidly and stay alive by going on oxygen. Know the procedure for your own aircraft and consider the following comments and procedure as explanatory for some of the phases in emergency descent.

1. Crew on oxygen and emergency interphone immediately. Because of the minimum time that may be available, this will keep the crew functioning and with the ability to communicate.

2. Close your outflow valve. This will maintain as much cabin pressure as possible in the event of loss of pressurization from system malfunction and may be accomplished by locking your manual pressurization control in the closed position. In some aircraft it would also be desirable to go to ram air, bringing some air into the cabin from the unpressurized source. You would also close your pneumatic crossfeeds, if open for using wing anti-icing, because of the possibility of bleed air loss through a ruptured duct robbing air from pressurization.

3. Emergency descent, if necessary. The magnitude of the pressurization loss is the determining factor. Check your cabin altitude and rate of climb, and make your descent according to requirements. Use your judgment!

4. If emergency descent is necessary, accomplish the following simultaneously: (a) autopilot off (the airplane should now be flown by the pilot), (b) throttles idle, (c) speed brake extended, (d) begin descent.

5. Evaluate descent requirements. The rate of descent necessary depends on cabin altitude and rate of climb. Consider the terrain. (You don't want to dive into a mountain!) You should also be aware that the emergency descent profile—full power off, speed brake extended, at maximum speed—is based on loss of pressurization with no loss of structural integrity of the aircraft and assumes smooth air conditions. Known failures that affect structural integrity and/or turbulence might dictate other speed profiles and entry techniques into the emergency descent.

If your speed brakes are the spoiler type and tied into your ailerons where they will materially affect your roll rate when extended, either nose straight over or use aileron gently to establish a bank of no more than 30°. If your airplane has a normal tuck characteristic that begins at a speed less than V_{MO} (maximum operating speed) or very near your normal cruise Mach, do not exceed a 10° nose-down attitude.

If your speed brake provides drag only and doesn't affect roll characteristics, roll into a vertical bank to lose vertical lift and reduce the angle of bank about 10 knots prior to desired speed. It is possible to control your speed then with bank angle and using very little pitch change.

Adequate descent rate will be obtained by descending at your normal high-speed Mach number for cruise and maximum normal airspeed below the Mach regime.

You might also inform ATC of your descent and direction of turn off the airway. They will provide all the assistance possible and give you an altimeter setting.

6. Check your passenger oxygen requirements. If the cabin altitude has reached the altitude where the masks should be automatically deployed (usually 11,500'), check to see that they have done so. Oxygen should be available automatically at 14,000', but if the masks have deployed, you might as well go to the manual position and provide oxygen.

It may seem at this point that you have done a lot, but you haven't really. In fact, you've barely begun descent. It's now time to turn on the seat belt and no smoking signs, reset your master warning, and clean up any remaining checklist items.

7. Below 14,000', or 2,000' above your desired altitude, gradually reduce your rate of descent. Upon reaching the desired altitude, retract the speed brake and resume normal flight. It may now be necessary to reevaluate your fuel requirements for your unpressurized flight at a lower altitude and plan your flight accordingly.

11

Steep Turns

■ Steep turns are those requiring more than a 30° bank and are normally practiced with a bank of 45°. This is a steep angle of bank while carrying passengers but is included in the check ride to assess the pilot's ability to maneuver the aircraft at high bank angles safely and with good heading and altitude control. Proficiency in performance is judged on the pilot's smoothness, coordination, and ability to perform within the prescribed tolerance parameters (100′, 10 knots, 10°) for altitude and heading. Therefore, the pilot is checked on (1) correct entry into the turn, (2) maintaining a constant bank angle and speed in the turn, (3) altitude control, and (4) recovery to level flight.

Banked turns of no more than 45° are safe at any normal operating airspeed, from maximum operating down to minimum maneuvering speed for any configuration, but you should always remember the G load imposed in a 45° bank. Approximately 1½ Gs are induced, raising the equivalent gross weight 1½ times; the stall speed is also increased about 17%. When operating at normal maneuvering speed (1.5 V_s or 1½ times the stall speed for a given weight), such as in the traffic pattern, there would be an adequate and safe margin over the stall; steep turns at those altitudes and speeds are rarely required and are considered poor judgment on the part of the pilot. Operating in the Mach structure of high altitude, steep turns at high gross weights at high altitude are more likely to induce a high- or low-speed buffet. A gust load of 1½ Gs, imposed either by bank angle or turbulence, may very well bring about the phenomenon of the Mach buffet when operating at or near the buffet boundaries for the aircraft. However, we're more concerned with the steep turn as a training maneuver here, and I refer you to Chapters 12 and 25 for other considerations.

The speed at which steep turns are practiced varies widely with different aircraft but is usually V_{NA}, the maximum maneuvering speed at which full deflection of the controls may be used. One exception is the Sabreliner, with max V_{NA} speed of 200 knots at sea level ranging to 220 knots at 40,000′; but the Sabreliner flight training syllabus calls for steep turns at 250 knots. In the DC-9, steep turns are practiced at 220 knots (V_{NA}); in the Electra, 200 knots. You may be assured that the speed used for practice will be the speed at which the maneuver is most difficult. I'd suggest that you learn the speed your particular flight training curriculum calls for or use V_{NA} if practicing in your own airplane.

In the execution of every maneuver, there are some common errors that the experienced instructor or check pilot looks for. It is most helpful to trainees if they are aware of these common errors, analyze the maneuver completely while in training, and then endeavor to eliminate these errors in their performance.

The most common errors made in steep turns are: improper rate of roll into the bank, either too fast or too slow; poor altitude control on entry; varying the angle of bank in the turn; poor altitude control in the turn and pumping the nose up or down rather than holding altitude smoothly; poor speed control in the turn, usually losing airspeed and gaining altitude; poor recovery by starting the rollout either too soon or too late and either overshooting or undershooting the desired heading; poor altitude control on recovery, usually gaining altitude; poor speed control on recovery, usually gaining airspeed.

No one can fly right on heading, altitude, and speed all the time. Flying has been defined as the constant correction of mistakes. This is certainly a true statement. But the more quickly you recognize a mistake, the more quickly and smoothly you can correct it. The best pilots can anticipate errors, have a good instrument scan, and know what effects are going to be caused aerodynamically as they use the controls to maneuver. Steep turns are a good example. Using the primary system of instrument scan and utilizing the theory of flying the wing, making it work for you by positioning it to the desired attitude and controlling its speed for the desired performance, make steep turns an easier maneuver.

Since you are being graded in the check ride on the performance of individual maneuvers, doing them one at a time, it is desirable to have the aircraft trimmed up and stable on the desired speed, altitude, and heading prior to entry into the maneuver. From that point, let's completely break down a steep turn.

Let's assume we want to do a steep turn using a 45° bank, a full 360° to the left. The technique is the same in any airplane (only the roll rate and speed vary), but to have some numbers to work with, we'll use a DC-9. So we're trimmed at 220 knots, at 10,000′, and on a 360° heading to begin the maneuver.

To enter the turn, a slightly faster rate of roll than that normally used for turns is used. It is desirable in high-performance aircraft to use a roll rate that will result in the aircraft turning one-third of the angle of bank (15°) by the time the angle of bank is established. In aircraft with heavier control pressures and slower roll rates (Convair 440, Lockheed Constellation, DC-6, and DC-7), a roll rate turning the aircraft 22–25° or about half the angle of bank should be used. The entry roll, even though at a slightly faster rate than normal, should still be executed with smooth, steady control pressures. This technique makes altitude and speed control in the entry easier and gives you a heading reference to begin your recovery.

This roll rate results in an aerodynamic effect rarely considered by many pilots and causes some difficulty in altitude control on entry. This may be eliminated by understanding what is going on out on the wings and knowing what makes the plane roll and bank. Not taking into consideration the aerodynamic forces being induced and their effect on the wing is the primary cause of the pilot's getting behind in altitude and applying control pressures incorrectly in pitch attitude. An improperly executed entry may result in the pilot's porpoising the nose up and down to bracket the desired altitude for as much as 180° of turn. We'll talk about this at greater length in a moment.

During the steep turn, the horizon is the primary instrument for bank angle, heading is the primary reference for recovery, and the altimeter is primary for pitch and altitude control.

You are also aware of the necessity of creating an increased angle of attack during the turn, by applying back pressure on the elevators, to maintain altitude. You must create more lift to overcome the loss of vertical lift from the angle of bank extending the length of the column of air supporting the weight of the aircraft and the resultant centrifugal forces and increased equivalent gross weight as this increased angle of attack is applied; with this increased lift will be increased drag that will reduce your speed.

These factors are basic, learned prior to first solo, and the instrument scan is also basic, using the primary instrument scan technique. But understanding control effect will make the maneuver much smoother and error-free.

The first tendency of the aircraft as you roll into the turn is to climb! This is an effect induced by the ailerons and is rarely taken into consideration. Most pilots begin the turn, using the horizon to check bank angle and applying a slight back pressure as the bank increases, and then glance back at the altimeter about halfway through the roll. The altimeter will read high (50–100′), so they relax back pressure on the elevator and go back to the horizon to stop the bank on 45°. When they have established the bank, they go back to the altimeter and are surprised to find they are as much as 100′ low! From that point on, through the first 180° of turn, they have altitude control problems. This is caused by control effect, but pilots usually aggravate it by applying back pressure too soon if they don't know about control effect and don't know at what point the loss of vertical lift will cancel this effect, and back pressure for an increased angle of attack will be required.

To turn to the left, left aileron is used, causing the aircraft to bank and turn left. The left aileron comes up into an area of high-speed but lower-pressure air, causing the airflow around that area of the wing to change its speed and pressure. The airflow over the top of the wing is slowed down and its pressure slightly increased; the airflow on the bottom is speeded up and its pressure slightly decreased; this results in a slight loss of lift on the left wing, causing it to go down.

The aileron on the right wing goes down and acts very much like a flap. Its effect on the right wing is just the opposite of the effect on the left wing and of a greater magnitude. In fact, in some aircraft the aileron effect is a ratio of approximately 6:1: the wing rising will rise 6′, while the wing descending will go down 1′. The right aileron, deflected downward into a relatively low-speed but high-pressure airflow, changes the camber of the wing in that area. The airflow across the top of the wing speeds up, causing an even lower pressure; the airflow on the bottom of the wing is slightly reduced in speed, causing greater pressure; this results in a tendency to climb. So the wing with the aileron down climbs and at a faster rate than the wing with the aileron up descends. This causes the aircraft to have a slight tendency to climb as ailerons are applied to roll into a turn.

The right wing is also on the outside of the turn, moving through the air just a little faster during the turn, and is on the high side of the turn. Its greater lift as it climbs in the beginning of the turn will cause the plane to climb slightly until the loss of vertical lift begins, at which point the aircraft will start to descend unless the angle of attack is increased to create greater lift.

There is very little loss of vertical lift until the bank angle exceeds 15°; it is really apparent beyond 22.5°. This is why bank angles must not exceed 15° in V_2 climb-out. It is also the reason that climbing turns are done at 25° bank; the loss of vertical lift is not great enough to materially affect climb performance, and yet the angle is steep enough to cause a reasonable rate of turn at climb speeds.

As you enter a steep turn, be aware of this characteristic. Either use a slight amount of forward pressure initially or ignore the initial climb completely and begin applying a slight back pressure at as near 22.5° bank as possible. As you pass about 20° bank, check the altimeter and use steady back pressure to stop the altimeter right on the desired altitude.

Stop the bank, using the horizon, on 45° and keep it constant. Varying the angle of bank varies the vertical lift, causes the airplane to climb or descend accordingly, and makes the back pressure requirements change rapidly. Altitude control is very difficult under such circumstances.

Control the altitude with the elevator, using the altimeter as the primary reference (a sort of nose-on-horizon indicator), and use steady pressures rather than large control movement. Use just enough pressure to positively control the altimeter needle. Try to catch its movement as quickly as possible—no more than 20–40′ off desired altitude—and smoothly stop its climb or descent tendency. As soon as the needle stops moving, use no more or less pressure and it should remain constant.

If corrections are necessary to climb or descend, use small pressure changes. The rate of climb (except with instant vertical speed indicators that are affected by G forces) may be taken into scan to establish the rate of climb required to correct altitude.

However, with practice you will find the nose-on position on the horizon for a given speed. Below this point the aircraft will descend; above this point it will climb. In the DC-9 at 220 knots, this position is about 5° above the horizon. This horizon position may also be used for altitude corrections. Using a constant bank and speed, the altimeter for primary pitch control, smooth control pressures, and combining the horizon

with rate of climb and altimeter result in steep turns with an altitude variance of no more than 50′ as average performance.

It is also necessary to control speed with a tolerance of no more than plus or minus 10 knots. This too can be done within 50% of maximum parameter, or 5 knots. You know that an increased angle of attack also increases drag. At about 25° angle of bank, when the loss of vertical lift must be offset with back pressure, add a slight amount of back pressure and simultaneously apply a slight increase in power. In a jet use about 0.05 EPRs; in a prop aircraft, about 2″ of manifold pressure. But do this with throttle movement and do not take your eyes from the flight instruments to check a power setting. You are being graded, not on how well you can set power, but on how well you can control attitude and speed. Looking at the power instruments and attempting to set an exact power invariably results in an interruption of instrument scan and flight errors. You just have to learn that to move the throttles an X amount increases or decreases power and speed an X amount.

The airspeed is primary for power. Consider that the airspeed needle is tied to and controlled by power and adjust it accordingly after arbitrarily increasing power in the entry. You'll be surprised at how easily and well it can be done.

During the turn the altimeter is primary for pitch attitude; the rate of climb and horizon are secondary to position the nose for altitude corrections; the horizon is primary for constant bank angle; and the airspeed is primary for power control.

The compass should come into scan on entry after the bank is established. As soon as the bank is at 45°, glance at the compass and check the number of degrees turned. You should turn the number of degrees equal to one-third the angle of bank, or 15°, but there may be some slight error in your roll rate. You may turn slightly more or less, but the number of degrees you have turned will be the same number you will lead your desired heading to begin recovery at the same rate of roll used in the entry.

The compass comes into scan again after 270° of turn. When it is the same number of degrees you turned on entry before the desired heading, roll out, using the same rate of roll as the entry. Then the compass becomes primary for bank when the heading is reached and wings are level. Using this technique, you should be able to roll out with an error of no more than 5° consistently.

Remember that as you roll out, the vertical lift increases. Hold the altimeter right on the desired altitude smoothly with forward elevator, relaxing the back pressure smoothly as the bank decreases. The increase of vertical lift as the bank decreases must be counteracted with a corresponding decrease in angle of attack to maintain altitude. This decreases drag, and the power must again be reduced to the amount required for the desired speed in level flight. Stay right on the flight instruments, using the primary scan system, and reduce the power smoothly as the bank decreases and the back pressure is reduced.

Think the maneuver through before execution; be trimmed on altitude, heading, and speed before beginning the maneuver. With just a little practice, you can learn the proper rate of roll, the aerodynamic tendencies of control effect, and power requirements for speed control. It takes smoothness, steady pressures at the proper time, and power adjustments in small increments from throttle position to roll into and out of a steep turn exactly on heading, altitude, and speed.

12

Stalls

There are several types of stalls—accelerated, full, and an approach to a stall, which isn't a complete stall at all. Beginning with primary flight training, every pilot has practiced stalls and should be completely familiar with the stall and the proper recovery from it. Pilots should practice the maneuver in every airplane they fly until they recognize the specific flight characteristics of the stall and approach to stall of the particular airplane and are able to execute the proper recovery procedure quickly and smoothly.

In air carrier flight training, full stalls aren't practiced. Pilots, if properly trained to recognize the approach to a stall, should never get their planes into a stalled condition while carrying passengers and would take the correct recovery procedures before a full stall develops.

The execution of practice stalls and recovery is one maneuver where we really get to utilize the theory of flying the wing. The stall is an aerodynamic function of the wing, induced by too high an angle of attack (Fig. 12.1).

The whole matter of lift is concerned with the smooth flow of air over and under the wing. As the angle of attack of the wing is increased, the air particles are forced to make sharper and sharper changes in direction in order to follow the "up-over-and-down" contour of the upper surface of the wing. At moderate angles of attack the particles can negotiate the initial turn of the leading edge but cannot follow the curve of the wing contour completely, so separation occurs near the trailing edge. As this occurs, the wing is in an approach to a stall.

As the angle of attack is increased, the point at which this separation occurs moves forward. If the high angle of attack is maintained or increased, the separation point moves rapidly forward toward the leading edge. This amount of separation is too much

for the wing to tolerate. Lift, the greatest portion of which is produced near the leading edge, is destroyed; the sustaining flight characteristics of the wing disappear with the formation of turbulence; and the wing is in a full stall.

Since the stall is caused by an excessive angle of attack—the angle of the wing's relative forward motion to the surrounding air—it may be induced at any speed and in any attitude by suddenly causing too high an angle of attack with rapid movement of the elevators. This is called an accelerated stall and is one of the first steps in executing a snap roll in aerobatics.

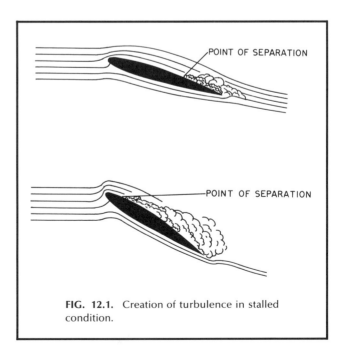

FIG. 12.1. Creation of turbulence in stalled condition.

Accelerated Stalls

Accelerated stall characteristics are essentially like those for normal, unaccelerated stalls. However, buffet warning begins considerably above normal stall speed and is more sudden in its appearance and recognition, but it doesn't increase significantly after it is induced. Recovery from an accelerated stall is accomplished by merely relaxing the back pressure you're holding on the elevator.

Before my first flight in any airplane, I look at the performance charts and explore the graph of stall speeds very thoroughly. I select a weight (usually either near 0 fuel weight or max landing weight) and find the stall speeds for this weight in all configurations. This weight, with its various stall speeds, then becomes a reference index for comparison of stall speeds at other weights and configurations. From then on, for any given weight and flap setting I can mentally compute the stall speed within 2–3 knots.

Table 12.1 gives the stall speed numbers for the DC-9-14 obtained by research of the stall performance graph in the aircraft flight manual. The reference weight and stall speed immediately become apparent, as does the effect of flaps: 70,000 lb., and its stall speed clean. The only speed to remember is the 0 flap speed at the reference weight of 70,000 lb.—121 knots. This is the actual, wings-level stall speed and *not* the speed at which the stall warning sounds. Then, by averaging the stall speeds at various weights and flap configurations, a rule of thumb for stall speed computation in this airplane is also apparent. It isn't exact, of course, but it gives a pilot a good knowledge of the effect of weight and flap settings on stall speeds.

I find that I may add or subtract 0.8 knots for every 1,000 lb. difference from 70,000 lb.; 85,000 lb., for example, is 15,000 lb. more than 70,000, and 15 × 0.8 = 12 knots. This 12 knots added to the memorized index speed of 121 knots shows a stall speed clean of 133 knots.

I also find that 10° flaps, as an average, reduce clean stall speed by 10 knots; 20° another 8 knots, or 18 knots; 30° another 5 knots, or 23 knots; and full flaps another 6 knots, or 29 knots.

Suppose you had no way of computing a V_2 speed or a V_{ref} speed, but you knew your weight. Let's take max takeoff weight, 90,700 lb.: 21 × 0.8 = 16.8, rounded to 17, added to 121 for a clean stall speed of 138; 138 − 18 = 120 knots for a stall speed with 20° flaps and a V_2 of 1.2 over stall = 144 knots; 138 − 29 = 109 stall speed with full flaps and 1.3 approach speed = 141.7, or 142 knots.

This basic knowledge of stall speeds for my aircraft and the effect of flaps makes me aware of my margin over stall at all times in normal operation simply by looking at my airspeed indicator. It also makes stall practice and stall recovery easier during training and on check rides.

Full Stalls

Full stalls are not commonly practiced in airline types of aircraft. If you were ever in the tail end of the fuselage when a transport was stalled, you'd know why. The horizontal stabilizer and tail surfaces, usually the weakest points of a large aircraft structure, are being struck by the turbulent air caused by the separation of the airstream around the wing. It seems the larger the aircraft, the more pronounced this disturbance.

However, if your aircraft is stressed for such maneuvers, full stalls should be practiced to familiarize you with the characteristics of the plane in a deep stall. Common sense dictates the necessity for practice at a safe altitude and clear of other traffic—somewhere above 10,000′ and below 14,000′ where you will not be affected by Mach characteristics.

Precompute your stall speed for weight and flap configuration and then set your power to a value below cruise that will allow you to decelerate but high enough so as not to cause damage to your engine or allow a jet to unspool. For piston aircraft, I would suggest a power setting just sufficient to make sure that the engine will be driving the prop at all times rather than the prop driving the engine. This will be a low power setting of approximately 100 BMEP and will prevent piston slap. In turboprop aircraft, a setting of

TABLE 12.1. Actual Stall Speeds of DC-9-14

Weight (000 lb.)	Flap Settings (gear up or down does not affect stall speeds)				
	0	10	20	30	50 full/down
90.7 max takeoff	138	128	120	115	109
85	133	123	115	110	103
81.7 max landing	131	121	113	108	102
80	129	119	112	107	101
75	125	116	109	103	97
71.8 zero fuel	123	113	106	101	96
70 (reference weight)	121	112	105	100	94
65	116	108	101	96	91
60	112	103	97	93	87
55	107	99	93	88	83
52 empty weight	104	96	90	86	81

100 horsepower will prevent the negative torque system from giving you an uncomfortable yaw, which you would experience at flight-idle, and will provide a continual thrust of 100 horsepower to the prop. In jet aircraft adjust the power levers to a setting just above the starting bleed strap or bleed valve setting. Usually 55–60% N1 RPM is used for stalls in the clean and takeoff flap configuration, and 70% power for stalls in the full dirty landing configuration. In some of the light business twins the ignition comes on automatically with the stall warning. But this isn't usually the case with the larger engines, and I would suggest placing the ignition switch to the continuous position. The power setting will prevent compressor stall at the stall angle of attack and will also keep the engine unspooled; the continuous ignition will protect you from flameout during the power application at stall recovery or from an inadvertent compressor stall.

Maintaining altitude is not important and would be extremely time-consuming in the entry, so pull the nose up to approximately 25° (15–20° nose-up attitude is sufficient in the landing configuration) and hold it. Or you might hold your altitude until the speed reaches a predetermined value, usually the minimum trim speed, and then use about 1 knot per second speed bleed-off attained with smooth back pressure on the elevator until the stall occurs. This latter method will result in the stall occurring at the expected speed and is the method I would use if checking the stall warning devices of a new aircraft.

For a full stall with either entry method, you would keep reducing your airspeed at a steady rate with smooth elevator or hold your attitude smoothly with back pressure through the stall warning and the first mild buffet of an approaching stall. The buffet will progress slightly as the stall progresses (as you might expect from looking at the diagram in Fig. 12.1), and then an abrupt break will occur as the separation of airflow moves rapidly forward toward the leading edge of the wing and the wing stalls completely. However, there will still be effective rudder control and reduced aileron effectiveness, and the aircraft will still have lateral control.

This isn't always true of T-tail aircraft. The Boeing 727, BAC-1-11, and Lockheed Jetstar, if deep stalled at such a nose-high attitude, may drop tail first for several thousand feet before pitching over—if it pitches over at all. The DC-9-14 may also do this if stalled at a high altitude, though it normally pitches over straight ahead; the DC-9-30 series may snap rapidly one way or another or possibly hang in the tail-down attitude. The aerodynamics causing these aircraft to remain with the nose high and the elevator ineffective is very simply a stalled horizontal stabilizer and elevator. The ailerons are effective down to a speed about 40% less than the full stall speed; the aircraft may be rolled over vertically from the full stall

with the aileron, and the nose will very definitely fall through the horizon. At the first sign of an approach to a stall in a T-tail in a nose-high attitude, roll off on a wing, establishing a steep bank while relaxing any pressure you may have on the elevator, and let the loss of vertical lift cause the nose to drop smoothly back to the horizon before rolling back level at a safe speed.

Most aircraft other than the T-tail can be controlled with the wings level well below the stall speed to a minimum speed, which in most conditions is limited by available longitudinal control.

Stall recovery should be initiated as the nose pitches through the horizon (the level flight attitude) and accomplished with coordinated flight control usage and smooth application of power by lowering the nose of the aircraft smoothly (more by a relaxing of back pressure than the traditional violent forward elevator) and increasing engine thrust symmetrically to takeoff power as necessary to build up airspeed. (Note: Use max continuous power for training, although you command takeoff power, to be sure you do not overboost the engine or exceed its power limitations.) Recovery is completed by stopping your descent when a safe airspeed is attained and with a minimum loss of altitude.

Approach To Stall

Air carrier flight training usually practices a more sophisticated type of stall. It seems far more desirable to train pilots to recognize the approach to a stall and to smoothly initiate a proper recovery at that point than to do full stalls. At the same time it allows them to practice maximum performance at reduced speeds by maintaining a constant altitude on entry, a constant angle of bank, or a constant heading. Only one stall or approach to a stall is required in a type rating ride or proficiency check (one stall is the minimum requirement under FAR 121), and it is usually in the takeoff flap configuration. But to be thoroughly familiar with the stall characteristics of your aircraft, you should practice approaches to a stall in the clean configuration with takeoff or approach flaps and in the landing configuration with gear and full flaps extended.

The speeds for entry and stall vary from aircraft to aircraft, and the manner in which the maneuver is accomplished may vary slightly in different training programs, but the procedure is essentially the same. For practicing stalls (since an inadvertent stall would most likely occur while operating at maneuvering speed rather than in cruise), the aircraft should be prepared by the use of the appropriate "in range" or "descent" checklist in order to have all of the systems in full operation for the use of the flaps and landing gear, just as if you were approaching an airport control zone prior to landing. As a second consideration, the initial entry speed should always be at the minimum recommended maneuvering speed for your configuration. And third, the most effective recovery is made by applying power for speed increase, setting flaps in the takeoff configuration for best lift, and retracting the gear after establishing a positive climb.

The execution of the approach to a stall and recovery then becomes a perfectly natural procedure: Start at the proper entry speed for your configuration; reduce the power as required for deceleration; hold your altitude until the first physical signs of a stall or stall warning appear; execute the proper recovery procedure for a safe speed; climb as near that speed as possible back to your original altitude; and level off and accelerate to the minimum recommended maneuvering speed for your configuration.

Approaches to a stall are done in level flight, in all configurations, and in banks of no less than 20° and no more than 30°. A turn with a 25° bank is commonly used for approach to a stall in the clean and takeoff flap configuration, but the full dirty landing configuration stall is usually done straight ahead in level flight since you are rarely expected to be maneuvering with 20° banks on the final approach to a landing.

Not everyone is flying jets, at least not yet, so we'll describe a stall in a piston aircraft, a turboprop, and a T-tail jet.

DC-7 Clean Stall

The DC-7 is an example of piston aircraft. The stall speeds will be different for a DC-4, DC-6, Connie, Martin, or Convair, but the procedure will be typical for all piston airline aircraft. Let's do an approach to a stall, clean, in level flight. You are all trimmed out at 8,000′ on a 270° west heading at a speed of 175 knots. The hood is up and you are on instruments.

Since you are in a multiengine prop-driven plane, you must be aware that there are two different stall speeds—one with power off and another with power on (lower because of the increased airflow across the wing from the propeller slipstream). This is to serve as a reminder that immediate power application in stall recovery is highly desirable.

The throttle (control of power) is a vital flight control that should be handled as much as possible by the pilot flying the aircraft. I recommend that you set your own power, initiating a power reduction for deceleration, and then command the flight engineer to set 100 BMEP.

We'll assume that at your weight the clean stall should occur at about 118 knots indicated and that V_2

speed with flaps 20° is 122 knots. It is apparent then that extending the flaps to takeoff as quickly as possible upon recognition of the approach to a stall will be highly beneficial in recovery.

After reducing the power, hold your altitude and allow the aircraft to slow down. The altimeter is the primary instrument for pitch; the compass is primary for bank and heading. Watch the airspeed for deceleration only. To maintain altitude, trim down to about 140 knots (about 120% of clean stall speed) and then trim no further. You are now trimmed for a safe recovery speed even if the flaps didn't extend or there was a control malfunction and are very near the final speed of your recovery in a normal execution of the maneuver.

As the speed slows below 140, take rate of climb into your scan, along with the altimeter for pitch control, and keep the needle just barely above the 0 index for a very slight indication of climb. This serves two purposes: (1) It assures a positive altitude control rather than settling before the stall buffet, and (2) it gives you the first indication of the approach to the stall. A split second before the actual buffet of the approach to a stall is felt, the rate of climb needle will suddenly drop and show a descent. At that instant apply power, commanding, "Takeoff power! Flaps 20!" Ease off the back pressure you're holding (remember you're trimmed for 140) and allow the nose to drop to the horizon only. *Do not push the nose beyond the horizon!* At this point the artificial horizon is your primary instrument for pitch. As soon as the nose is on the horizon, check the airspeed indicator. If it is V_2 or more (it should be since there are only 4 knots difference in stall speed and V_2), go to the altimeter and make any corrections necessary back to your original altitude. When on altitude, let the airspeed build up and make a power reduction to stop it on the minimum recommended maneuver speed for 20° flaps, 145 knots.

All through the maneuver the compass is primary for bank and for maintaining your west heading. You should be able to stay within 3–5° of your desired heading.

I emphasized not lowering the nose beyond the horizon in the recovery because this is all the attitude change necessary to break the stall angle of attack regardless of speed. Airplanes stall at different airspeeds and at different weights and flap settings because this is the absolute minimum speed at which the wing will furnish enough lift to support the plane. But a stall is not a function of speed, it is a function of angle of attack. The most basic knowledge of aerodynamics, as illustrated in Figure 12.1, proves that the stall is a function of angle of attack. Since a stall is a separation of airflow caused by angle of attack, it follows that a stall or buffet from a separation of airflow may be induced at any speed by an abrupt change in angle of attack. Conversely, in a stall recovery the buffet and separation of airflow may be caused to disappear entirely by reducing the demand for lift by changing the angle of attack at speeds even less than the minimum speed required to support the plane. If, by a change in angle of attack, the wing is not required to support the weight of the plane or is required to support only a portion of the weight, then the buffet will disappear because there is no longer a separation of airflow over the wing.

The angle of attack is the angle of the wing in relation to the aircraft's motion through the air. Just as we can induce greater lift by increasing angle of attack, so can we lose lift by diminishing angle of attack. It is by making the wing produce lift less than the weight of the aircraft that we descend, and this is the characteristic used in flying the wing to recover from a stall.

By allowing the nose to drop to the horizon, we've really raised the tail instead of lowering the nose; we've also placed the wing in a position parallel to the earth and gravity. We've lessened our demand for lift; the wing is once again producing lift in proportion to its speed with a smooth airflow and is not being required to support the weight of the aircraft. By using just enough elevator control, applied smoothly, to hold the nose right on the horizon, we have already broken the stall and are in a safe attitude to accelerate to a minimum safe speed that will support our weight. We will be descending, but we will not experience a stall buffet again unless we raise the nose above the horizon before a safe speed is attained.

Just as a stall may be induced at any speed, it may also be induced in any attitude. For example, if you were to dump the nose over violently and well below the horizon and then attempt to correct back to the horizon abruptly, you'd probably experience a second stall more violent than the first one. This would be a result of placing a high lift demand upon a wing already below level stall speed by abruptly changing its angle of attack, even though the nose may still be well below the horizon.

You should lower the nose to the horizon smoothly by releasing back pressure, hold it there gently and smoothly until your minimum safe recovery speed is attained, and then raise it smoothly above the horizon to stop your descent and to climb back to your original altitude. By using the technique of applying power for acceleration, thus simultaneously breaking the separation of airflow by dropping the nose to the horizon, and applying takeoff flaps to reduce the stall speed by increasing the lift capacity of the wing, you will effect the most rapid stall recovery possible while using the absolute minimum altitude loss every time.

Electra Stall

The technique in a turboprop is essentially the same, but let's fly an L-188 Lockheed Electra through a stall in a 25° bank and with takeoff/approach flaps. Electra flap settings are in percentage and in stages of 30, 60, and 100%. The 30% setting is a takeoff setting very rarely used except for operation at airports with high field elevations; the 60% position is the most effective takeoff position at airports nearer sea level and is also normally used as the approach setting; the 100% position is the landing position. They are full Fowler flaps; the first 60% extension is out and back from the trailing edge, materially increasing the wing area; the remaining 40% extension is downward, increasing lift by changing the wing curve but also adding a high drag factor.

The approach speed for an Electra is 1.4 over stall with flaps 60% and 1.25 with flaps 100%. The approach speeds for various weights are placarded just above the pilot's head, so be aware of your gross weight and take a look at your approach speed for that weight. Let's say that the 1.4 chart showed an approach speed of 138 knots. You know right away that your stall speed is about 99 knots. Add about 7 knots to this for the 25° bank, and you could expect a stall at 106 knots indicated.

The maneuver speed of an Electra with the flaps at the 60% takeoff/approach position is 150 knots; in this example, it is almost exactly 1.5 over stall. Have the aircraft stabilized on the correct speed and altitude and then begin your turn to establish bank angle *before* reducing your power. *After* the bank angle is established, reduce your power to 100 horsepower. Now all you have to do is maintain a constant bank and altitude as the aircraft decelerates until the very first indication of a stall is recognized. I stressed reducing your power after establishing your turn. It is much easier this way to control bank and altitude. If you reduce power as you roll into the turn, you have the diminishing lift from deceleration and loss of prop airflow over the wings as well as the increased angle of attack required to maintain altitude to contend with in control of pitch attitude. This makes altitude control more difficult, so always establish the turn first, taking care of the pitch requirements for angle of bank; then reduce the power and take care of the additional pitch requirements from the power change.

Make a mental note at this point of V_2 speed (approximately 120 knots) and trim the aircraft down to about the 1.4 approach speed of 138. All of the instrument scan techniques used earlier in the DC-7 apply here the same way. At about 5 knots over your safe recovery speed (V_2 or about 120 knots), take the vertical speed indicator into your scan—both for a more positive altitude control and for a definite indication to the first buffet of an approach to a stall to follow *immediately* after the rate of climb needle breaks sharply to a downward indication.

The recovery technique is also the same except for correcting the bank angle to regain lift by recovering with the wings level. At the first indication of the approach to a stall, smoothly apply power, commanding, "Takeoff power!" At the same time, allow the nose to drop smoothly to the horizon and simultaneously level the wings to stop at whatever heading you may roll out on.

The primary instrument scan throughout the entire maneuver would be: (1) horizon for angle of bank and altimeter for pitch in entry; (2) horizon for angle of bank, altimeter for pitch, airspeed for trim, and rate of climb secondary for pitch and altitude control and warning of impending stall at speed below recovery speed; (3) horizon primary in the recovery for nose to the horizon and wings-level, airspeed primary for safe recovery speed and climb back to altitude after recovery, compass primary for bank, and altimeter primary for pitch while accelerating back to 150 knots on altitude and in level flight.

Stall Series in DC-9-14

A jet is entirely different from a prop-driven airplane. The wing is just a big board pushed through the air by the jet engines, and it is more honest in a stall. First of all, there is no torque to be concerned with; second, the stall speed is constant for a given weight and configuration, since there is no additional airflow over the wing from prop blast. For these reasons the training maneuver of approach to a stall is performed more easily in a jet than in a prop-driven aircraft.

The approach to a stall and recovery are strictly training maneuvers in airline operation. It is highly unlikely that anyone with sufficient experience to be an airline captain would inadvertently stall the aircraft. But performance of the training maneuvers gives you an opportunity to get some slow flight experience in the aircraft as well as a degree of proficiency required to pass the rating ride. However, since inadvertent stalls are a possibility and would occur when operating at slow speeds such as maneuvering for a landing, we'll begin the stall sequence by preparing the aircraft for the stall and recovery.

I'm going to conduct this training in the same manner I would conduct the rating ride as a check pilot. I'll set up the maneuvers, and you fly them. I'll give you all the help, constructive criticism and demonstration necessary. My purpose is to instill in you from the very beginning the understanding that it is necessary for you to take command, to fly the aircraft

as if I were your copilot, and to use your own judgment.

As an example, I may say, (or the check pilot on your ride may say), "Show me a stall, in a turn to the left clean," or simply, "Let's get set up for stalls." That's your cue to take command.

To prepare for the stall series:

1. Command, "Pumps to high," or "In-range check." This will assure your having hydraulic pressure to operate the gear and flaps.

2. Place the ignition to continuous. This will ensure against inadvertent flameout.

3. Check your stall warning. The DC-9 has two systems, Number 1 on the left side of the aircraft and Number 2 on the right. They must be operable for dispatch, and both should be checked before practicing stalls.

4. Request the stick-shaker speed. I (or the check pilot) will give it to you, and you are to set the computed stick-shaker speed on your "bug" and then recover from the stall when the stick shaker goes off or your speed gets down to the bug speed, whichever comes first.

Let me reemphasize that last part. Prior to executing the approach to a stall, we'll compute the stall buffet speed and the stick-shaker speed for the aircraft weight and configuration we'll use in the maneuver. There are two reasons for this: (1) The aircraft shouldn't be allowed to encounter the full stall buffet, so it will only be slowed to the initial buffet or stick-shaker speed, and (2) I want to impress upon you the accuracy with which the stall and stick-shaker speeds are computed and the reliability of the stall warning system.

Begin each approach to a stall from a recommended maneuver speed for your configuration as a minimum entry speed and be all trimmed and stable before beginning the power reduction for the approach to stall. The recommended maneuver speed clean is 190 knots; with 20° flaps, 160 knots. Be on altitude and establish the bank for the stalls in a turn before reducing your power.

Reducing power is something else you can command. Concentrate on your altitude and attitude (if you're on the correct speed, it makes it easier) and then pull the power levers back slightly and command your copilot to set the desired power. Use 55–66% N1 for the approach to a stall clean, or with 20° flaps use 70% N1 for the approach to a stall with the gear down and full flaps.

After your power is set and the aircraft begins to decelerate, trim down to what I call a secondary entry speed and then trim the stabilizer no further during the entry to the approach to the stall. This secondary entry speed is a very generous safe speed for recovery in the entry configuration and is considerably above the actual minimum safe speed. Trimming to this speed will also assist in the performance of the maneuver by requiring you to hold your altitude and control pitch with the elevator as you slow beyond these speeds and will give you some feel of the approach to the stall in reducing elevator effectiveness. There should be no further trimming on the entry below 170 knots clean, 150 knots with flaps 20°, and 135 knots with flaps full.

As you begin the approach to a stall, mentally add 25 knots to your preset stick-shaker or bug speed. This is the speed at which you may raise the nose above the horizon during the recovery—the speed at which you would climb back to your assigned altitude. Note this speed as you slow through it in entry, taking your IVSI (instant vertical speed indicator) into your scan at this point to assist in maintaining your altitude; for recovery at stick-shaker speed plus 25 this would be the speed reference telling you to rotate the aircraft to stop descent.

Stalls will be practiced in clean, maneuvering, and landing configurations in both level flight and with 25° bank. However, we'll do the landing stall straight ahead only. You're always graded on altitude, heading, and speed control, but in this maneuver you're also graded on smoothness of entry, with a zero rate of climb or a very slight drift up, and recognition of the warning of the approach to a stall with prompt, smooth recovery without inducing a secondary stall warning.

In the performance of the stalls, I want you to compare the stall-warning stick-shaker speed to the computed speeds I'll give you. In level flight the stick shaker should go off almost exactly on the computed speed, but a bank component would have to be added to it for increased angle of attack in a turn.

For comparison of stick-shaker speeds to actual full stall speeds, I've prepared Table 12.2. The stall warning is at a slightly higher speed than the full stall and is very nearly the speed at which the initial separation of the airflow occurs in the approach to a stall. These speeds are for level flight; thus the stall warning may be caused to occur at faster speeds. You should, therefore, think of a stall warning as an angle of attack warning to take corrective action. The lower speeds are the full stall speeds, and the faster speeds are those at which the stick shaker should give its warning. The weight would never exceed the maximum

listed on a training flight. To use this chart, go to the weight nearest your actual weight and add or subtract 1 knot per 1,000 lb. difference between your weight and those listed. For example, the stick-shaker speed for 67,000 lb. with 20° flaps would be 108 knots (the stick-shaker speed for 65,000 lb. plus 2).

The approach to stall should be practiced by trimming for the proper entry speed (the secondary entry speed) for the configuration and decreasing the airspeed until the stick-shaker or precomputed stall speed is noted. Recovery will be effected by lowering the nose and adding thrust to maximum takeoff power. However, in training we'll use max climb power in the interest of conserving the engines except for the stall in the landing configuration.

Particular care should be used to be sure that the airspeed has increased and the stall warning has ceased before the airplane is rotated nose up for the recovery. Raising the nose too soon may cause an accelerated stall, and the stick shaker will go off again. The only thing that will cause the stall warning to go off at speeds faster than the correctly computed stick-shaker speeds is flight load acceleration; this may be caused by gusts, bank angle, or control action. Make sure your control action is smooth and applied at the right time.

Earlier in this chapter I showed you an analysis of stall performance and a manner of determining a reference speed and weight for rule of thumb computation of stall speeds (Table 12.1). I would also suggest that you take the stick-shaker and full stall speed chart for your aircraft, similar to the one for the DC-9-14, and analyze some stall problems during flight training. Note the margins between normal recommended maneuver speed, stall warning or stick-shaker speed, and full stall speed. Compare the speeds for different weights and note how they vary for both weight and flap configurations.

A turbine-powered aircraft will not accelerate from the stalled or nearly stalled condition without decreasing the angle of attack, so some loss of altitude may be expected. But the amount lost depends on the method of entry toward the stall, the power application, timing, and the delicacy and smoothness with which you make attitude changes on recovery.

The best methods for handling power, flaps, and gear for stall recovery in the various configurations follow.

CLEAN

At stick-shaker speed, apply takeoff thrust, simultaneously lowering the nose smoothly to the horizon or just below with 20° flaps; then accelerate to stick-shaker speed + 25 knots before raising the nose above the horizon. The recovery speed will be very

TABLE 12.2. Stick-Shaker and Full Stall Speeds, DC-9-14

| | Gross Weight (000 lb) | | | |
	60	65	70	75
0 flaps	122/112	127/116	132/121	137/125
20° flaps	102/97	106/101	111/105	115/108
50° flaps	93/87	97/91	101/94	102/97

close to the recommended maneuver speed with 20° flaps, and you should note the trim change as the flaps extend and see that little if any altitude is lost.

MANEUVERING, 20° FLAPS

At stick-shaker speed, apply takeoff thrust, simultaneously lowering the nose to the horizon or slightly below. After stick-shaker speed stops, during acceleration with the nose on the horizon, use a *gentle* back pressure to stop loss of altitude at about 5 knots before stick-shaker speed + 25; lead stick-shaker speed + 25 by 5 knots, and you will be able to climb back to initial altitude exactly on recovery speed. You should note the relatively slow acceleration rate and pitch tendency, the moderate loss of altitude if recovery is delayed or improperly done, and the reactivation of the stick shaker if the recovery is attempted too soon or an excessive load factor is applied in recovery.

LANDING, 50° FLAPS

At stick-shaker speed, apply takeoff thrust, simultaneously lowering the nose to the horizon with 20° flaps. Accelerate to stick-shaker speed + 25, then stop descent. When descent has stopped and the aircraft is climbing (the altimeter stops descent and the IVSI comes through 0 to the climb indication) with a positive rate, gear up and climb back to the initial altitude. Note the loss of altitude on recovery (considerably more than recovery from the previous configurations), the lateral instability as the speed nears and is at the stall warning, and the reactivation of the stick shaker if you raise the nose too soon or abruptly.

General Comments

Ailerons should be used to keep the wings level the rudder is not normally used as positive control, since the yaw damper function is available without it for coordination. Further, the rudder is so aerodynamically effective even in the stall that it is easy to overcontrol, thus inducing a yaw that aggravates the lateral instability with Dutch roll.

I recommend leading the recovery speed by 5 knots, with smooth and gently applied back pressure, to allow the aircraft to stabilize on stick-shaker speed + 25 knots to climb back to initial altitude if necessary.

In the recovery from the approach to a stall with 50° flaps and gear down, the altitude loss will be an average of 300'. This may be reduced by about 150' by leading the recovery speed 5 knots and smoothly pulling the nose up to about 10° above the horizon to climb back to altitude.

When doing stalls in 25° bank, expect the stall warning about 7–8 knots before the computed speed due to the increased angle of attack to maintain altitude in the bank. Some instructors add this to the stick-shaker speed and set it on the bug before entry into the stall. I prefer to leave the bug speed at the basic wings-level speed, since you should recover wings level; an additional 7–8 knots in raising the nose will result in unnecessary loss of altitude in the recovery.

As you can see, the old basic stall recovery is still appropriate: apply thrust while dropping the nose to the horizon, rolling wings level, and extending the takeoff flaps all at the same time; then raise the nose to climb after a safe speed is attained.

13

Unusual Attitudes

The FAA has deleted the necessity for a pilot to demonstrate ability to recognize and recover from unusual attitudes in transport aircraft. Perhaps they reason that today's modern aircraft, with its inherent stability and performance capabilities coupled with excellent instrumentation, will never inadvertently be placed into an unusual attitude; or that the pilot's experience has been such that recovery from unusual attitudes has been taught; or that the practice of such maneuvers may impose undue stress upon the aircraft. Whatever the reason, it is still required that the pilot be exposed to "jet upset" in high-speed aerodynamics, warned of wake turbulence behind large aircraft, told of the correct techniques for turbulence penetration, and given a refresher in the mountain wave if routes are over mountainous terrain.

Until 1967, recovery from unusual attitudes was a part of air carrier flight training programs, though it was rarely required in the check ride. Even then, it was unrealistic. The pitch attitude was restricted to plus or minus 15°, the bank not to exceed 30°, and the speed not to exceed the maximum speed for use of full deflection control. However, according to FARs, unusual attitudes begin when the airplane exceeds a 60° bank or 30° of pitch above or below the horizon.

There are several possible causes for an airplane to enter an unusual attitude: severe turbulence, wake turbulence, high-speed Mach characteristics, or a violent maneuver to avoid a collision. When the unusual attitude occurs, the pilot's recovery reaction must be instantaneous and correct. Every pilot should review the procedure from time to time.

There are five basic unusual attitudes that can occur in an airplane:
1. Spins.
2. Inverted positions.
3. Vertical dive attitudes.
4. Vertical climb attitudes.
5. An extreme stalled attitude.

These are the basic positions or attitudes, and more complicated attitudes can develop from these. From an extreme stalled attitude a spin or tail slide can develop; the inverted position can lead to a split S and damaging airspeed buildup. The unusual attitude should be recognized and prompt recovery action initiated before the attitude becomes too extreme.

Spins

Spins are rare for transports, but they can occur from delayed or improper recovery from a stall, from a vertical climb, or from cross-controlling in slow flight. If your plane starts to spin, first stop the rotation with opposite rudder and then break the stall. Very little stress is imposed on the aircraft during the spin, but the speed will build rapidly after recovery, so straight and level flight should be attained as quickly as possible.

The modern instrumentation of today's aircraft, with full roll capability, was developed after extensive evaluation of human factors when it was found that the ground/sky reference is a major influence on instinctive, correct recovery reaction. The three-dimensional flight director indicator—the artificial horizon—provides this full-time ground/sky reference and becomes the primary instrument in recovery from unusual attitudes. Regardless of aircraft attitude, the pilot quickly sees any corrective action that might be required. This instrument capability is the primary reason that recovery from unusual attitude is no longer practiced.

Figure 13.1 shows the flight director indicator for an aircraft with wings level and in a 15° nose-up attitude. The light portion indicates the sky and the black the ground. The long pitch reference marks on this attitude tape (marked in number of degrees nose up or down) and the delta wing and horizon bar furnish the mental picture required to cause instinctive reaction.

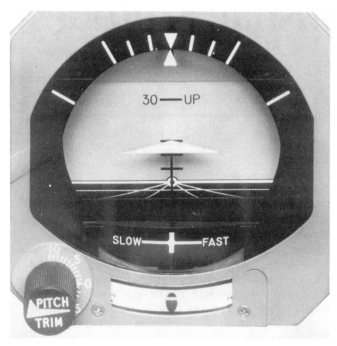

FIG. 13.1. FD-109 with pitch and bank markings. (Courtesy Collins Radio Company)

Inverted Position

The inverted position would be rare but can happen in wake turbulence or severe thunderstorm turbulence. *Rolling* is the best means of recovering from the inverted position, and promptness in rolling out of this unexpected condition will prevent great loss of altitude or the high-speed buildup associated with non-precision inverted flight.

Any kind of roll will do. It's not necessary to use opposite rudder from the aileron or to hold a heading; just wrap in the aileron and roll. If the nose drops during the roll, as it no doubt will, ease off the power to prevent excessive speed buildup and recover as gently as possible from the dive.

It is possible to use a split S as a recovery technique from the inverted position, but I don't recommend it. Hauling back on the elevator should be avoided. High speeds will build rapidly, far beyond the limiting speed of your aircraft, and excessive load fac-

tors are sure to result. Also, altitude will be lost very quickly.

Pushing forward on the elevator is also not recommended. If you happened to also be very nose high, this could result in an inverted stall followed by an inverted spin or spiral.

To recover from the inverted position, roll immediately and avoid the use of any elevator control until past the vertical position. When the aircraft passes through the vertical, the tape display of the flight director will rotate 180° in the roll axis and continue to display the correct ground/sky and attitude reference during the inverted portion of the maneuver. From that statement, you would naturally assume that the flat or bottom portion of the delta wing would be toward the sky portion of the tape, and you'd be right. The index pointer, shown at the top of the instrument in Figure 13.1, would also be at the bottom of the instrument case.

You may tend to pass this too quickly. The DC-9-31 in a full-stall condition will probably snap to at least a 30° inverted bank position but can be recovered quickly and safely by rolling rapidly.

Vertical Dive

The vertical dive would also be rare. Much more common would be the diving turn, known as the "graveyard spiral" during the days of needle, ball, and airspeed. This attitude is illustrated in Figure 13.2,

FIG. 13.2. FD-109, nose down and rolling left. (Courtesy Collins Radio Company)

which shows the aircraft in a 50° nose-down attitude and a 40° bank. This is the easiest position to recover from, and recovery should be accomplished before the speed becomes excessive and takes you into a high-speed Mach buffet or exceeds your speed limitations. The aircraft would be accelerating rapidly from the dive and turning from the bank while the altimeter would be unwinding the lost altitude. Pulling back on the elevator would only tighten the turn. (This is what used to happen to those who tried to control speed with pitch attitude on primary instruments.)

To recover, you should reduce the power to slow the aircraft, roll the wings level to recover the lift lost in the angle of bank, and pull the nose up gently to stop the descent. After this has been accomplished, you're back in normal flight and will recover your lost altitude in normal climb.

Vertical Climb

The tailslide and whipstall—the airplane actually falling and sliding backward until it whips over to a vertical nose down position, or even past the vertical to inverted—are the most dangerous possibilities of the vertical climb attitude.

One way to recover from the vertical climb is to rudder the nose off to one side as in a hammerhead turn while you still have speed. The best technique is the rolling wingover. A coordinated turn into a roll with a slight amount of back elevator pressure will result in a falling rolling wingover, allowing the nose to drop toward the horizon, at which time power is reduced and the aircraft rolled to level flight.

A combination of steep climb attitude and bank is more common and is illustrated in Figure 13.3.

The fly-the-wing system provides the easiest and best recovery from this attitude. With the extreme nose-up climb, you don't want additional lift to make it tend to climb even higher. Just roll the aircraft over into an even steeper bank or maintain the present angle of bank and let the nose drop of its own accord by applying no elevator pressure. Then at or just before reaching the horizon, roll the wings level and continue your flight—just like doing a wingover.

Stalled Attitude

The approach to a stall is practiced in air carrier flight training (see Chapter 12). But the impression a pilot may get from practicing stalls may cause trouble in recovery from a stalled attitude that may also be an unusual attitude. When the stalled attitude is not extreme, the fastest and surest way to recover is to promptly lower the nose to break the stalling angle of attack and regain flying speed and to increase power

FIG. 13.3.   FD-109, nose up and rolling right.
(Courtesy Collins Radio Company)

to accelerate to a safe flying speed faster. But in the attitude illustrated in Figure 13.3, this would be entirely incorrect.

Some aircraft (the BAC 1-11, Lockheed Jetstar, and other T-tail jets with aft-mounted engines) may also hang in a deep stall, with the elevator not effective at all until the whipstall occurs.

This condition can be corrected by quickly rolling the airplane with the aileron and rudders with the elevator neutral. The rudder may also be blanked out and ineffective, but this would make little difference in the recovery. The ailerons can roll it off, and with simultaneous reduction of power the nose will promptly fall toward the horizon. This can occur with a steep bank or even from the inverted position, but in any event the nose of the airplane will head down toward the pull of earth's gravity, and then a normal recovery can be initiated.

It would be absolutely unthinkable for a pilot to pull back on the elevator from an extreme stalled attitude, but few recognize that it is equally important to avoid the instinctive reaction to push forward on the elevator in the expectation of lowering the nose. The elevator control may no longer be aerodynamically effective in some aircraft; in others, pushing forward on the elevator may cause an inverted spin or a whipstall. Rolling with the aileron rather than pushing with the elevators will bring the nose to the horizon every time, safely and smoothly, even from speeds below the normal stalling speed.

The recovery from unusual attitudes then boils down to the following rules of thumb: (1) With the nose well above the horizon, use a steep bank to lose vertical lift, let the nose drop smoothly to the horizon, and roll the wings level. (2) With the nose well below the horizon, roll the wings level to recover vertical lift, reduce the power to control speed, and gently pull the nose up. (3) If inverted, roll to the nearest vertical index and proceed to recover from the vertical position. Just fly the wing, position it for the lift factor you want, and control speed with power.

General Comments

Note the index pointer on the instruments in Figures 13.1, 13.2, and 13.3. It is in the inner scale of the instrument and in the center of the light portion of the sky/ground tape. Normally this pointer is used to determine bank angles by comparing it to the bank angle marks on the top portion of the instrument, but it is also an excellent reference in unusual attitudes. Using the center of the glide slope indication on the left of the instrument and the fast/slow indication on the right as vertical bank indication, when the index pointer is toward the top of the instrument (between these vertical indications and the top of the case), you are right side up; if the pointer is below these vertical indications and in the bottom half of the instrument case, you're inverted.

It should be apparent now that with the instrument roll and pitch capability you can normally reorient yourself by establishing a level flight indication on the attitude horizon and resuming a normal cross-check of the other instruments. But in cases where the attitude of the aircraft is more extreme, the sequence and manner in which attitude and power are controlled become important. Applying back pressure in an inverted dive prior to reducing the bank to less than 90° will cause a great loss of altitude and could result in exceeding the speed limitations of the aircraft. This could be very bad at an altitude where Mach characteristics and compressibility effects would come into play and at lower altitudes might cause you to apply severe G forces in the recovery—if indeed you could recover at all.

When recovering from any steep climbing attitude, a banked attitude should be established and maintained to allow the nose to come smoothly to the horizon. This will avoid the stall in an extreme pitch attitude (the deep stall), the possible tailslide and whipstall, and the resultant negative Gs of pushing the nose hard over.

To recapitulate: If you are *inverted*, always roll *toward* the index pointer (Fig. 13.3). Continue the roll to the right (in this case) until the index pointer is in the lower half of the instrument case and aligned with the lower left end of the 70° pitch line. Use the aileron toward the index pointer nearest the closest vertical indication. In this case, wrap in the left aileron. After you have noted the bank attitude and after rolling toward the index pointer and the closest vertical position, when the index pointer is either in or moving beyond the vertical indication and toward the upper half of the instrument case, determine whether you're climbing or diving. This will also be indicated in the pitch reference of the horizon. Now airspeed trends are also reliable indications, and a cross-check of the vertical speed indicator and altimeter will help.

If you are diving, (1) reduce power and continue to roll the wings to a level position, (2) extend the speed brake if you have one, and (3) smoothly apply the elevator to stop descent gently enough to prevent excessive G loads to a level flight attitude. Remember that back pressure in a steep turn or when the bank is more than vertical will result in excessive loss of altitude. Conversely, if the nose is well down, change the pitch attitude as rapidly as possible before the speed is excessive. Elevator use is related to speed in recovery from unusual attitude.

If you are climbing, (1) roll to bring the bank index pointer toward the nearest 90° bank index mark to establish a banked recovery attitude that will allow the nose of the aircraft to drop to the horizon smoothly from a loss of vertical lift; (2) add power if necessary; (3) maintain the bank and enough back pressure on the wheel to remain well seated; and (4) as the horizon bar and aircraft indicator come together as the nose approaches the horizon, roll the wings level with the nose slightly low and then stop your rate of descent when a safe speed is attained.

I hope you never inadvertently get into an extreme attitude. But if you do, just fly the wing, using the techniques we've just discussed and controlling attitude with the basic scan procedures, and you'll recover safely without imposing undue stress upon your aircraft.

14

Maneuvers at Minimum Speed

■ Slow flight, or maneuvers at minimum speed as it is now called, is no longer required in air carrier instrument checks and type rating rides. It may still be found in some flight training curriculums, though it is seldom used except when a student needs to improve instrument scan. I still use slow flight with most of my students early in the training program, because it is one of the quickest ways to determine actual instrument ability and scan capability. It also has a practical application when converting to new aircraft; it allows the student to feel the effects of flaps, gear, and power changes and to maneuver the aircraft at speeds and attitudes that will be common in the approach.

Basically, slow flight is flight at a speed of 1.3 (130%) over stall in all configurations—V_{ref} in a jet and 10–20 knots over clean stall in prop aircraft. But there are other maneuvers, which are not true slow flight, that may be included in the same category, such as the canyon approach or the vertical S series of maneuvers. Each of these maneuvers affords good practice in positive control of the aircraft and requires a fast, accurate, and correct instrument scan technique. They also give you a good chance to feel out the effect of the gear, flap extensions, and power requirements in different configurations.

Basic Slow Flight

Probably the best way to begin slow flight is to familiarize yourself with the stall speeds and characteristics of your particular aircraft as related to the effect of flaps. Chapter 12 fully covers the stall knowledge you should have for your aircraft.

Figure your gross weight and compute speeds that would be 1.3 over the stall for every flap configuration. These are the speeds you use in slow flight (the same as the flare speed on landing), and the best method of determining these speeds is to actually compute them for every configuration.

Another method, equally good for flight practice, is to determine a V_{ref} or flare speed 1.3 over the stall for landing flaps; then for every different flap configuration you add a figure that would be equal to the effect of the flaps in reducing stall speed for the slow flight speed.

For an example we'll use a DC-9-14 at a weight of 72,000 lb. and compute basic slow flight speeds; 72,000 lb. in this model of aircraft has a flare or V_{ref} speed of 125 knots with full landing flaps. Using 125 knots as a base, we can add 30 knots for our clean slow flight speed (since full flaps reduce stall speed approximately 29 knots, see Table 12.1) and reduce the 155 knots according to the effect of flaps in other configurations: 145 knots for flaps 10, 137 knots for flaps 20, 132 knots for flaps 30, and 125 knots for flaps 50. The speeds will not be exactly 1.3 over stall (they are only rule of thumb computations) but will be within 2 knots for every configuration except the clean, where it will be within 5 knots. This is because it is figured backward, beginning at V_{ref} and working back to the clean configuration, instead of computing slow flight speed and figuring from that point. However, it provides a quick way of computing slow flight speeds very close to those of actual 1.3 and just a little on the slow side and adequate for the maneuver.

Begin the slow flight by being in straight and level flight at your normal clean maneuver speed of 1.5 over stall or recommended minimum maneuver speed with 0 flaps (190 knots in the DC-9-14) and reduce to your clean slow flight speed (155 knots in our example). Note the attitude of the aircraft, its power requirements, and the amount of trim used in the stabilizer to maintain altitude. In this model the no-flap approach is made at a speed of basic V_{ref} plus 30 knots, exactly what we're using for slow flight, so we'll extend the gear and see its effects. There will be no change in the center of gravity with gear operation (there is in some aircraft), but carefully note the drag—let it slow the aircraft—and then the power required to overcome the gear drag.

Retract the gear and lower the flaps to 10°. Different aircraft will have different tendencies in flap extension. A Lockheed Jetstar, for example, pitches nose forward, requiring up stabilizer trim; the DC-9 balloons and requires down trim. Therefore, note the attitude change in flap extension. Hold your altitude and heading and slow to the desired speed of 145 knots. Make no power changes in the flap extension (to note the drag as well as resultant greater lift) and then make power changes as necessary to maintain the desired speed. We already know the effect of 10° flaps on stall speed; now we'll find the effect on flight characteristics and the power requirements to maintain altitude at minimum speed in the configuration.

Extend the flaps to 20°, reducing to the speed we've decided to use in this configuration—137 knots. A landing in this configuration uses a V_{ref} that may be computed by adding 12 knots to a basic V_{ref}. I led you in from the back door to show you the relationship of V_{ref} to stall speeds and how they may be safely and easily computed by rule of thumb with a little knowledge of the aircraft's stall speeds and the effect of flaps.

Use all of your flap configurations, maintaining altitude, and become familiar with the effects of flaps and gear in both attitude changes and power requirements.

An approach and landing with an engine failed is with 30° flaps and a flare speed of basic flare plus 8 knots. Compare this against the 132 knots we've computed by rule of thumb for slow flight, as well as the speeds in the other configurations, and you can see the practical applications of slow flight. You are familiarizing yourself with the handling characteristics of the aircraft at speeds that would be approach and flare speeds in all configurations.

Now we're going to repeat the maneuvers starting from the clean configuration and extend the flaps and gear while holding a constant altitude and reducing to V_{ref} speeds for each configuration. This time, instead of flying a constant heading, we're going to make 30° heading changes. We'll start on a cardinal heading, say 270°, and turn left with a 10° bank to 240°; we'll turn back and forth between 270 and 240°, extending flaps and gear each time we begin to reverse our turn to change heading.

After we've worked our way down to full flaps with gear down, retract the flaps to 30°, then 20°, then gear up, then flaps 10°, then flaps up, or just the reverse of the extensions we just used. Continue to turn back and forth between 240 and 270° with 10° banks, and accelerate to the proper speed for the configurations. Note the tendency to settle as the flaps are retracted; note also the trim and power requirements to maintain speed and altitude.

Keep your altitude constant with elevator control, correcting pitch changes with trim, and maintain your desired speed with power.

This exercise will improve your instrument scan immensely, since you have many things to watch and control simultaneously while your flight is being adversely affected by configuration changes at minimum speeds.

The next phase of basic slow flight is to add climb and descent to the turns while extending and retracting the flaps and gear. I use the following maneuver. Beginning at 1.3 speed with maximum takeoff flaps (137 knots and flaps 20° in the 72,000-lb. DC-9-14), turn for a heading change of no less than 45° or more than 90° to a preselected heading while descending exactly 500′ per minute at a constant speed of 137 knots to an altitude 1,000′ below our initial altitude. Use no more than a 10° bank and do not go below the desired altitude; stop on the exact heading and maintain a constant speed and exact rate of descent throughout the maneuver. If you come to the desired heading before reaching the desired altitude, stop the turn; if you reach the altitude first, stop the descent; but continue either descent or turn until you're on the preselected altitude and heading. This may sound difficult, but it may be performed with reasonable tolerances by using the instrument scan technique. Use pitch control for rate of descent and altitude control and power for speed control.

The next phase, of course, is to climb back to the original altitude and heading, using 137 knots indicated airspeed and 500′ per minute rate of climb.

The maneuver may now be repeated by extending the gear for descent and retracting it for climb.

Vertical S

Another good speed control and instrument scan practice is the old military vertical S series. They are not exactly slow flight, and many variations in speed and configuration are possible, but they are excellent maneuvers to develop scan.

VERTICAL S-1

One good practice configuration is with the aircraft clean and on the V_{NA} speed (220 knots in the DC-9). Beginning on a constant altitude, we want to descend for 2,000′ at a rate of descent of exactly 2,000′ per minute at the correct speed in straight flight. Just touch the lower altitude, not going below it, and then climb at 220 knots and 2,000' per minute back to the original altitude without going above it. Just touch the original altitude, and then descend as before.

With the aircraft trimmed on altitude, heading, and speed, extend the speed brake to begin the descent; control and maintain a constant rate of descent with the elevator, using power for speed control.

Lead your lower altitude by 10% of your rate of descent (200') by retracting the speed brake and beginning a smooth roundout of pitch attitude to just touch the lower altitude. Be careful not to go below the desired altitude.

As the altimeter reaches the desired lower altitude, smoothly establish a 2,000' per minute climb. As you apply back pressure to establish climb, simultaneously apply power to maintain constant speed and then control the rate of climb with the elevator and the speed with power.

VERTICAL S-2

Upon reaching the original altitude in climb in the vertical S-1, just touching the desired altitude but not going above it, repeat the climb and descent but this time with a constant 30° bank. It doesn't matter which direction of bank is used, and heading is no longer important. Concentrate on a constant speed, controlled with power; a constant angle of bank, using the horizon as primary instrument; and a constant rate of climb and descent, controlled with the elevator, with the vertical speed or rate of climb the primary instrument. Monitor the altimeter so as not to go above or below the desired altitudes.

VERTICAL S-3

This is an extension of the vertical S-2, using the same configuration, speed, rate of descent, altitudes, and bank angles but reversing the direction of bank at the top of the climb and at the bottom of the descent. In a perfectly performed maneuver, the climb or descent roundout should be started before reaching the desired altitude in an amount equal to 10% of the vertical speed, and the bank angle should be diminished at a rate that will result in wings-level flight just as the aircraft reaches the desired altitude perfectly level.

From that point, the vertical S series goes on into more complicated direction changes, speed changes, time patterns, etc.; but these three are more than adequate to give you a good workout and improve your scan. They may be done in any aircraft and with several variations or configurations other than at V_{na} and using the speed brake as described. Any speed, rate of descent and climb, altitude change, and configuration within the limitations of your aircraft's performance may be used.

The actual slow-flight maneuvers you may be re-

quired to do in flight training will vary with different airlines or in different aircraft training programs in civil jets. However, you can practice the maneuvers we've talked about in any aircraft with instruments; if you've become fairly proficient in their performance, you should have no difficulty doing slow flight in any training program. One objective of slow flight is to become familiar with the characteristics of a particular aircraft—its feel and control response—at various speeds and configurations. This can best be accomplished when the airplane is flown precisely on the desired numbers. It's a matter of good instrument scan and smoothness in use of the controls.

Prop Slow Flight

Slow flight can be modified to fit any aircraft. To compute a minimum safe speed, take the clean stall speed at maximum landing weight (very rarely will a training flight be heavier than this) or at an average weight for training and add 10–15 knots. Or compute a speed that would be equivalent to 1.2 V_s clean. This would be the minimum speed for flap retraction after second-segment climb. For an example, let's use 110 knots to outline the maneuver.

Trim the aircraft for 110 knots in clean configuration and maintain 110 knots throughout the following maneuver:
1. Extend takeoff flaps.
2. Extend gear.
3. Extend approach flaps.
4. Retract flaps to takeoff.
5. Retract gear.
6. Retract flaps to zero.

Maintain heading and altitude throughout. Repeat the maneuver while executing gentle turns (10° bank) 15° either side of a given heading.

Canyon Approach

This is no longer a required maneuver, but it was one of the best maneuvers in prop aircraft training. Even though it is not in the curriculum, it may be practiced in any aircraft and is an excellent exercise in slow flight. I'll describe it as a Convair 440 maneuver, but it may be modified to meet the performance of any aircraft.

Trim for level flight at a specified altitude and heading, using takeoff flaps, gear up, and proper takeoff flap holding speed. Upon passing an imaginary (or actual, if desired) fix, maintain heading, extend landing gear, and descend rapidly (at least 1,000' per minute) to an altitude 1,000' below that specified at the beginning of the maneuver. Maintain heading and altitude for 1 minute; then retract landing gear, apply

METO or max continuous power, and execute a 30° bank climbing turn of 180° back to the original altitude. Maintain holding speed throughout the maneuver.

When the desired altitude is reached on descent, do not go below that altitude. On the climb, do not go above the original altitude.

The maneuver is shown in profile in Figure 14.1 with applicable Convair 440 speeds.

Rapid Descent and Pull-up

This is no longer the canyon approach type of maneuver. In a jet training program you'll be shown and required to fly a high rate of descent demonstration and normal go-around. This will be done in the landing configuration, gear down and full flaps, by flying a descent to a selected altitude at V_{ref} with all power off. At the altitude selected as the bottom of descent, a normal go-around is executed. The whole purpose is to simulate a long unspooled approach—a final descent of about 1,500′ with power off—to a field elevation and then attempt to pull up and go around. The intent of the maneuver is to demonstrate the danger of making an unspooled approach.

Most instructors are content to demonstrate this high–sink rate maneuver only once to each student. They have their students initiate the pull-up at exactly

the selected altitude for simulated field level and thoroughly convince them that a full power-off, unspooled, approach descent is extremely dangerous. It takes 8–12 seconds for an unspooled engine to close the bleed valves and produce sufficient power to accelerate the aircraft safely through the rotation to a nose-up pitch attitude required to establish the climb for pull-up and go-around. A recovery started too late can be disastrous.

I have my students perform the maneuver three times, not to overimpress them with the danger of the unspooled approach, but to show them how to execute a safe recovery or a way of taking the danger out of the execution of such an approach.

The first approach I have my students do is the standard rapid descent and pull-up, descending at least 1,500′ to a selected altitude with full power off and gear and full flaps extended and then executing a normal go-around exactly on the selected altitude. However, prior to the execution of the maneuver, I thoroughly brief them on the things to look for and recognize, other than the fact that the aircraft would go below the selected altitude on go-around and would crash if there were not at least 4,000′ between us and the ground. I particularly want them to note the nose-down attitude of the aircraft, the angle of descent, the vertical speed for rate of descent, and the altitude required to stop the descent and begin the climb from the go-around. In every instance, when the go-around

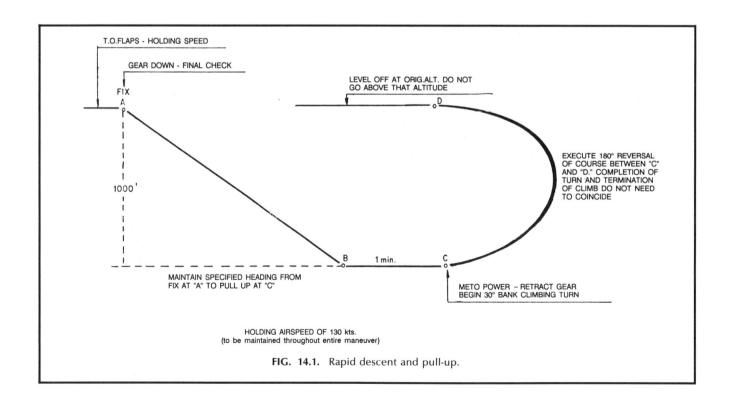

FIG. 14.1. Rapid descent and pull-up.

is initiated at the selected altitude, the altitude lost is almost exactly 10% of the vertical speed indicated in the rate of descent.

In Chapter 7 and in all references to stopping vertical speed, either climb or descent, I've recommended using 10% of your vertical speed in feet to lead your altitude for level-off. This was for smoothness and positive control in more normal climbs and descents, but this basic technique has a practical application here.

It is also important to know a power setting that produces little or no thrust and is just on the verge of the bleed valve closing. They teach you in ground school and flight training the RPM values at which the bleed valve is fully closed and the RPM at which the bleed valve will begin to open in power reduction, and they recommend that the engine not be operated at an RPM on the bleed valve setting because of possible vibration. But it is equally important to know the RPM at which the bleed valve is partially closed or just on the verge of closing. It normally takes a jet engine 8–12 seconds to spool up from full flight idle, but from a power setting just on the verge of closing the bleed valve it will take only 1–1½ seconds. This power setting is 40–42% N1 for the Pratt & Whitney JT-8D engine and produces 1.02–1.04 EPRs, virtually no power. I recommend that you determine a similar setting for the engine you are operating.

It would be wise to actually determine an average rate of descent for your aircraft in the full dirty landing configuration at V_{ref} flight idle at maximum landing weight and then use it as a rule of thumb on which to base your 10% lead. The DC-9 averages 2,000′ per minute, so the desired field elevation should be led by 200′ for the beginning of go-around pull-out.

In the second high-rate-of-descent approach, I have my students set 40–42% N1 and descend at exactly V_{ref} to 200′ above the selected altitude. This is done by applying power and simultaneously retracting the flaps to takeoff and raising the nose smoothly above the horizon the same number of degrees it was below the horizon in descent. If the gear is retracted when a positive rate of climb is indicated by both vertical speed and altimeter, the aircraft invariably just skims the selected altitude without going below it.

Another little trick that may be used in such a situation is to add 1% of your average vertical speed to your approach speed. By knowing the average rate of descent with full flaps and power off (which for safety should always be based on the rate of descent at max landing weight) and adding 1% to the normal V_{ref} speed for actual weight, a good pilot can safely execute a complete power-off approach and landing. Use an altitude equal to the average rate of descent of your aircraft at flight idle and in landing configuration as a high key altitude at a position downwind abeam the desired touchdown point at the takeoff flap configuration and speed At this point, extend the gear and maintain the same speed to allow the gear drag to initiate descent. Fly 10–12 seconds before turning base (depending on the wind) and extend approach flaps turning base leg. Your speed should now be maneuver speed for the approach flap configuration but never less than V_{ref} plus 20. The second key is on base leg, turning in early if low and delaying turn if high. If still low, drop the nose and build up speed to float; if high, drop the nose to keep the aircraft aimed at the landing target. You are still in the approach flap configuration, perfectly safe for landing, and full flaps can be extended after field is assured and speed of V_{ref} plus 20 maintained for flare capability upon landing.

To prove the technique and to show how it may be applied as insurance in an unspooled approach, I have my students do the high rate of descent and pull-up the third time in this manner. Descending on V_{ref} plus 20 and with the power set for immediate spool-up, descend to exactly the selected altitude and execute the go-around. The aircraft will not go below the selected altitude and the go-around will be safely executed, even from a full flight idle. This will be true for both executing the approach at V_{ref} and initiating go-around at a lead altitude of 10% of your vertical speed or by adding a speed of 1% of vertical speed to V_{ref}. The power setting just below the bleed valve closing is extra insurance.

The lessons learned in these maneuvers have a practical application in approach and help develop instrument scan. This isn't to recommend power-off approaches, but if you inadvertently get trapped into one, applying these techniques takes the danger out of it.

15

Landings: Approach Technique and Performance

The landing, unlike the takeoff, is given considerable attention in the flight training program, so you are sure to get a great deal of practice in whatever flight training program you enter.

The primary difference in traffic patterns for jets is that the approach pattern should normally be a little larger than for a propeller-driven airplane of comparable size. It takes a little more room for deceleration, flap management, and establishing a good final landing approach, and I'd recommend a pattern about 1½ times the size of the pattern of a similar sized prop aircraft. Landings are 90% the result of the approach, and approach is 90% the result of the initial approach or traffic pattern.

Proper entry into the landing pattern is the key to attaining the correct speed throughout the approach, since acceleration and deceleration are closely related to drag and thrust. Gear and flap management are of prime importance in establishing the correct speed and attitude for the landing approach; in any case, maneuvering speed and thrust settings should be established early in the approach. The entry into the final approach portion of the landing pattern is accomplished most easily and with a minimum of large thrust changes by flap management and gear extension in the initial approach at the proper time.

Normally the descent in approach is commenced as the increased flap and gear drag slows the aircraft to the desired approach speed. As recommended in Chapter 5, a glide path angle of 2.5–3° should be used

and will virtually eliminate the necessity of large thrust changes in the approach. Airspeed trends will indicate thrust requirements to maintain the desired speed schedule at a steady glide path angle. Recognition of and proper reaction to airspeed trends will limit corrections required in the approach to small attitude and thrust changes.

The point where you lower full flaps can vary; normally it coincides with a position at about the outer marker (or about a 5-mile final) of your final descent in the landing approach. As the wing flaps extend to the full down position, giving you more drag than additional lift, you'll probably have to lower the nose slightly to maintain a constant glide path profile. Some increase in thrust will probably be necessary to enable you to arrive at the runway threshold at the desired speed due to the high drag of full flaps. The control forces required, due to flap management, can and should be reduced if not entirely eliminated with the proper use of horizontal stabilizer trim control.

Your threshold of flare speed will be computed as 1.3 (130%) over stall for the landing configuration. I strongly recommend that you add 5 knots to this speed for an approach in calm or unknown wind conditions or add one-half the headwind and all of the gust to your threshold speed for an approach speed. However, in the latter case, the cumulative sum of all the above components should not exceed 15–20 knots except in extreme conditions.

To summarize the landing approach sequence:

1. Plan your entry into the pattern and approach to the landing to allow sufficient time and space to properly align the downwind, base, and final approach legs.

2. Enter the pattern on the downwind leg in the clean configuration and at the recommended maneuver speed for your aircraft in this configuration.

3. While still on the downwind, no later than at midfield, establish a maneuvering configuration by extending your flaps and slats to the maximum takeoff position.

4. After flap extension, reduce speed to 1.5 V_s (or the recommended average speed for this configuration in your aircraft) and stabilize your attitude and airspeed by retrimming the stabilizer as the flaps extend and the pitch attitude changes. There will be additional drag from the flaps, causing deceleration, but do not expect immediate speed reduction. Correct the pitch tendencies first by retrimming, and give the aircraft time to slow down before you make any power reductions. Most aircraft will usually slow to the correct speed, with no power changes at all from 1.5 V_s clean to 1.5 V_s with takeoff flaps, from drag alone in about 2 minutes or 5 miles. Altitude, attitude, and speed control are accomplished most easily at this point by trimming for pitch and altitude control first, letting the drag decelerate you to the desired speed, and then making any small thrust changes necessary after drag deceleration has eased.

5. If the landing is to be made under VFR conditions from a downwind position and circling back to the runway, the second stage of flaps—the approach flap configuration—should be established upon entry to the base leg. The maneuvering speed for this configuration now becomes 1.42 over the stall speed for the configuration. This is usually computed as an average speed or by a rule of thumb, such as V_{ref} + 20 or V_{ref} plus the flaps you don't have extended. As an example of the latter case, if your aircraft had 50° of flaps when full down and the approach flap setting was at 30°, then the maneuver speed for approach flaps could be V_{ref} + 20 knots, the difference between 30 and 50. These figures apply to one aircraft only, the DC-9, and are rough approximations of the correct speed. They are used solely as examples of one method of easily determining maneuver speeds. Something similar may be used for every aircraft; you will either be given the speed to add to V_{ref} in flight training or you may determine the effect of flap settings to stall speed for your aircraft and come up with your own rule of thumb.

6. Full flaps should be applied far enough out on the approach to stabilize trim changes and speed control "in the slot" prior to touchdown. The final approach should follow a linear descent path of approximately 2.5–3°, with speed decreasing to reference or flare sped (V_{ref} or 1.3 V_s) plus 5 knots for a calm or

unknown wind, or V_{ref} plus wind and gust factors, retrimming as necessary for near zero control forces. V_{ref} plus 5 knots may also be considered as maneuver speed in the full or landing flap configuration.

7. As you approach the end of the runway, begin a slow reduction to flare speed. You want to be at the threshold 50' high and exactly on V_{ref} if possible.

8. As the aircraft crosses the end of the runway, flare slightly to break the rate of descent and use a slow power reduction as the nose of the aircraft is rotated through the level flight to landing attitude.

Figure 15.1, illustrating a normal VFR traffic pattern, depicts the sequence very clearly.

Landing Performance

Performance and runway requirements vary with different aircraft, but, in general, the data used to compute landing performance are the same for all transport aircraft, FAA regulations specify certain requirements that must be considered when determining the runway length required at destination and alternate airports for all transport aircraft. The runway length must be such that a full-stop landing can be made at the destination or alternate airport within 60% of the effective length of the runway from a point 50' above the intersection of the obstruction clearance plane and the runway.

"Obstruction clearance plane" means a sloping upward from the runway at a slope of 1:20 to the horizontal and tangent to or clearing all obstructions within a specified area surrounding the runway (Fig. 15.2).

In addition to illustrating the obstruction clearance plane, Figure 15.2 shows the various portions of a landing: the 50' threshold where your speed should be V_{ref}, or V_{ref} plus wind and gust components; the air distance to touchdown at the 1,000' mark; the stopping distance, which combines with the air distance for 60% of the effective length; and required runway length.

Considerations in Landing Distance Chart Construction

An air carrier *Flight Crew Operating Manual* may contain only a chart for computing full or landing flap runway length requirements. This is about all that is required, since each aircraft also contains a *Gross Weights Manual* (prepared by the airlines engineering department or by the manufacturer specifically for the airline) showing the maximum weight allowable for takeoff or landing for every runway in the entire airline system. Included are corrections for wind, temperature, wet or dry runway conditions, and any me-

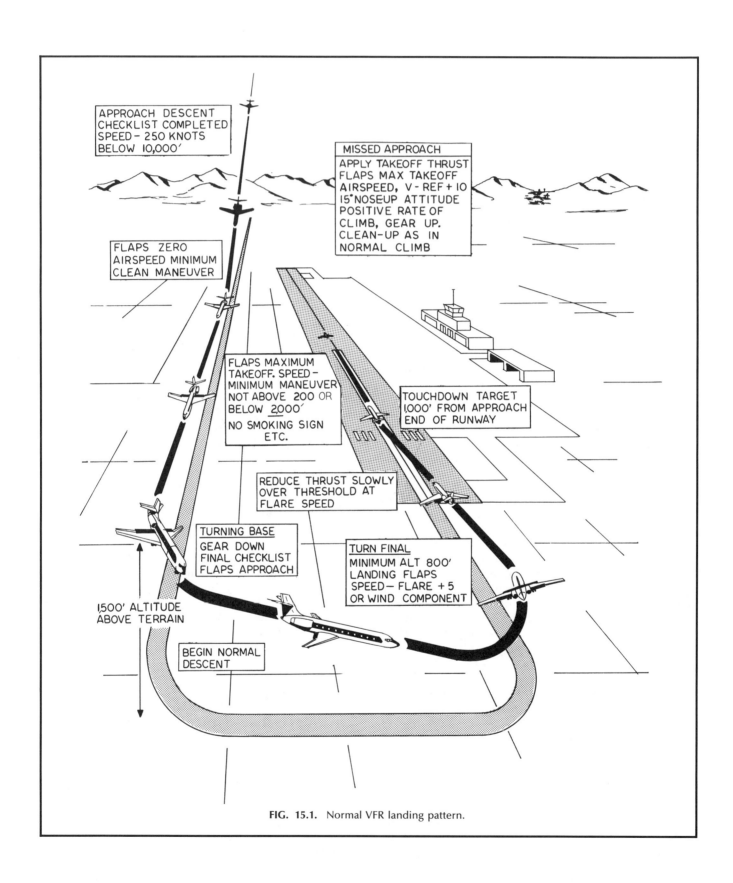

FIG. 15.1. Normal VFR landing pattern.

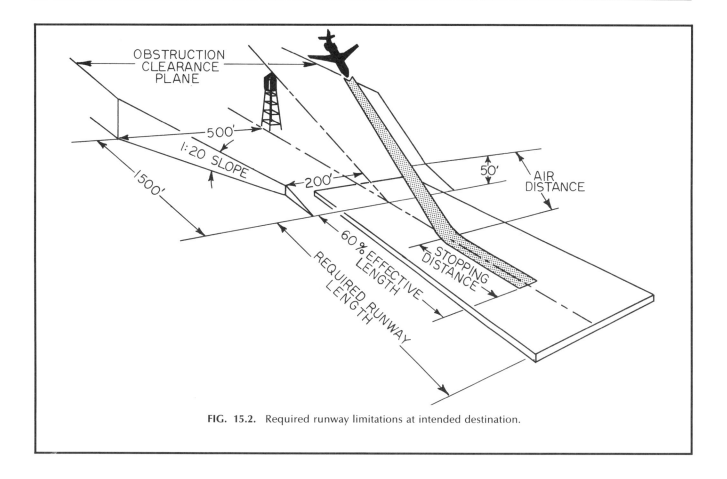

FIG. 15.2. Required runway limitations at intended destination.

chanical considerations such as failure of antiskid. All other considerations—runway slope, field elevation, and runway length—though not shown except for effective runway length, have already been taken into consideration in the preparation of the *Gross Weights Manual*. If the weight of the aircraft is at or below the maximum shown for the conditions prevailing, you know you are within the runway requirements for the aircraft.

However, if the weight is above maximum for the conditions, you are operating on bare minimum runway requirements; a thorough understanding of the performance techniques may be very helpful.

Figure 15.3 is a DC-9-30 chart but is similar in construction to all equivalent charts for other transport jet aircraft and may be considered as typical. It is based on the following considerations:

1. The aircraft being at 50′ at the threshold.

2. A threshold speed of 1.30 V_s, a normal descent to the 1,000′ point, and a touchdown speed of 1.26 V_s.

3. A 2-second time delay for distance traveled during transition from touchdown attitude to *maximum antiskid brake application* after reaching taxi attitude.

4. A dry asphalt or concrete runway with a conversion capability for wet runway requirements equal to a 15% increase.

It is equally important to note that Figure 15.3 does not take into account the following factors that also affect runway requirements for landing:

1. Reverse thrust.

2. Temperature.

3. Increased threshold speed corrections for wind and gust components.

4. Runway slope.

5. Runway conditions other than dry or wet.

The effect of additional speed above V_{ref} at the threshold is shown in Figure 15.4, where touchdown speed = 1.26 V_s. Therefore, this chart shows the effect of the additional air distance to touchdown required to decelerate from V_{ref}, plus whatever additional speed you are in excess of V_{ref}, to a touchdown at 1.26 V_s.

Landing Procedures and Techniques

Normally the point of touchdown should be approximately 1,000′ beyond the approach end of the

LANDING FIELD LENGTH AND SPEED (DC-9-30)
FLAPS FULL DOWN / SLATS EXTENDED

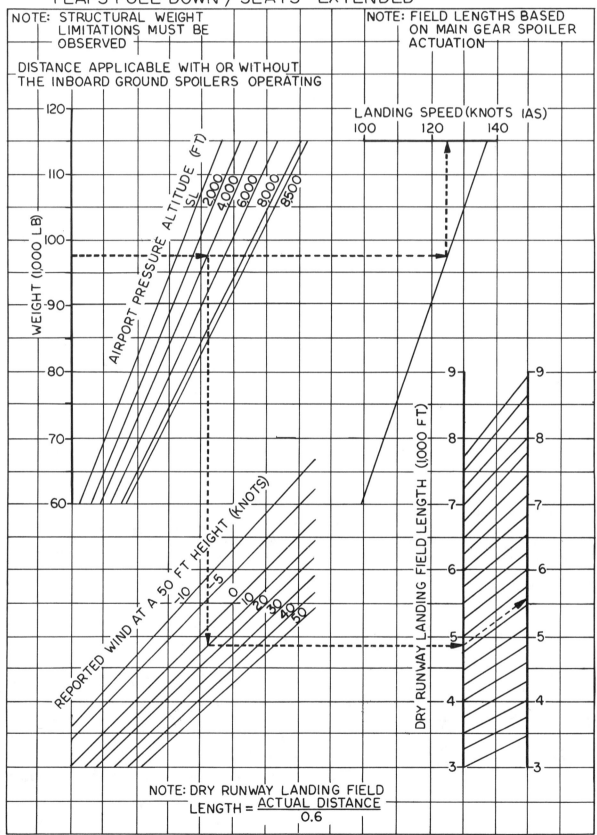

FIG. 15.3. Landing performance chart.

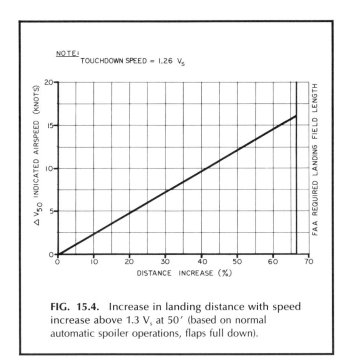

NOTE:
TOUCHDOWN SPEED = 1.26 V_s

FIG. 15.4. Increase in landing distance with speed increase above 1.3 V_s at 50' (based on normal automatic spoiler operations, flaps full down).

runway. Using the 1,000' runway markings (or an estimated position using runway lights) as a target for touchdown will ensure against undershooting the runway and, with a proper glide slope angle, will result in the correct air distance to touchdown and leave plenty of runway available for stopping.

Speed control in the approach, flare, and touchdown is important. If the airspeed is inadvertently high and a go-around is not executed, it is not only desirable but also important that the touchdown be made as close to the 1,000' target point as possible. You may have to push it on, shove it down through the ground effect cushion, and land a little faster and harder than the landing you normally try to make. The natural tendency when you are a little too fast is to hold it off, to ride the ground effect cushion while waiting for the aircraft to slow to get to a soft landing, but this may cause you to use more runway than is comfortable or available. Deceleration is much more effective on the runway using brakes than it is floating just off the pavement. I refer you again to Figure 15.4. You're safer every time to put it on the ground on your touchdown point and use brakes, reverse thrust, and spoilers to get stopped rather than holding it off.

Aircraft altitude, or more specifically height of the airplane at the runway threshold, has a very significant effect on total landing distance. A properly flown approach on the correct approach speed using a 2.5–3° glide path and a 500–700' per minute rate of descent with a landing touchdown target 1,000' from the approach end of the runway, should result in the

aircraft being on V_ref speed at an altitude of 50' at the runway threshold.

Assuming a normal 3° glide path, a height of 100' instead of 50' over the runway threshold will result in a touchdown 900' beyond the 1,000' mark. This is an increase in the air distance to touchdown and an increase in the resultant total landing distance of approximately 900'.

A proper normal glide path angle of 2.5–3° in the approach to the runway threshold is always important and may be considered as another factor affecting runway required. Simply stated, the total landing distance is also dependent upon the glide path angle. As the glide path angle is flattened, the total landing distance is *increased*. This is because there is no flare, or breaking of descent, prior to touchdown, resulting in a touchdown faster than normal. If you had a combination of excessive height at the runway threshold and a flat approach, you'd deprive yourself of the use of a considerable portion of the runway.

In any high-performance airplane, don't hesitate to go around if the approach isn't right. The decision to go around should be made as early as possible. However, the aircraft must have a capability of go-around at any time during the approach and landing. To continue the landing from a poor approach could, for obvious reasons, have adverse consequences, to say the least.

Another factor that greatly affects stopping distance is the coefficient of friction between the tires and the runway surface. There are many variables involved: tire and tread design, runway surface material, condition of the runway (dry, wet, slippery, icy, slush depth, water depth, snow covered), tire pressure (the square root of the tire pressure times 9 gives you hydroplaning speed on a wet runway), gross weight (it takes an X amount of kinetic energy to stop X number of pounds moving at X speed), and of course, the braking equipment of the aircraft and the technique used in employing it. The stopping distance for transport aircraft is determined without the use of reverse thrust. However, if it is available, normal procedures for stopping employ its use, and a significant reduction in landing roll is achieved with it.

The normal procedure for stopping after touchdown would be to set throttles to idle, apply reverse thrust to about 80% power, and use brakes as required. Power reduction from reverse thrust should be initiated at 70–80 knots in order to reach a minimum reverse thrust power setting by 60 knots. At speeds below 60 knots indicated airspeed, reverse thrust in excess of approximately 60% power will probably cause compressor stall due to hot gas ingestion. These are average figures; actual power for reverse thrust and speeds for setting minimum reverse will vary for different airplanes.

The secret of the whole technique is timing in the

use of all available means of stopping the aircraft. As soon as the main wheels are definitely rolling on the runway, reverse thrust should be initiated, since it is more effective at high speeds, and the nosewheel should be flown onto the ground smoothly with the elevators. Brakes should be applied as required only after the nosewheel is firmly on the ground. If you have spoilers, be sure they are extended. The spoilers reduce wing lift, thereby increasing the effectiveness of braking and adding drag. Hold the nosewheel down firmly with forward pressure on the wheel to improve directional stability, particularly in crosswinds; the negative angle of attack also reduces lift, further increasing brake effectiveness as well as increasing drag.

Skids and Skid Recovery

Even a dry runway surface loses some of its friction coefficient due to wear and residue left on the surface in the form of rubber, oil, and unburnt fuel from turbopowered aircraft. Precipitation in any form, even a light rain shower, mixes with this residue and leaves a slick mess. The aircraft landing on this slick runway surface has much less steering and braking ability. Also any reduced visibility, such as a rain shower, makes runway alignment on touchdown more difficult, and a crosswind compounds the problem of recognition of misalignment and makes the possibility of a skid imminent. All of the ingredients for a skid are present: a wet slick runway, poor braking action, poor visibility, a crosswind, and an aircraft landing that probably will have a slight amount of crab for crosswind correction.

With these facts in mind, let's examine the effects of nosewheel steering, reverse thrust, and brakes on a skidding aircraft.

NOSEWHEEL STEERING

Overcontrol of nosewheel steering can start a skid. You'll be less likely to get into a skid if you avoid the use of nosewheel steering until you have slowed to a speed at which rudder control is no longer effective. Fly the airplane aerodynamically with flight controls as long as possible. Use the rudder and ailerons, but once the skid has started, the nosewheel should be turned *toward the skid* until control is regained. Under normal crosswind conditions, sweptwing aircraft tend to turn *downwind* because the wing into the wind develops more lift. For this reason, the control wheel should be held forward for nosewheel traction and the wings kept level during all phases of the take-off and landing roll.

It's difficult to use rudder without nosewheel steering in aircraft that have nosewheel steering con-

nected with the rudder, such as the DC-9. But the pilot may override nose steering demands from the rudder by locking the nosewheel in the center with the left hand and using the right hand on the throttles and reverse, while the copilot holds the yoke forward with the aileron into the wind as the aircraft decelerates after landing.

REVERSE THRUST

Reverse thrust will compound your skid and can pull you off the runway. This has been verified by the fact that a high percentage of aircraft that have skidded off the runway were still in reverse when they went off the pavement. I strongly recommend that once a skid has developed and the aircraft is no longer headed straight down the runway, the use of reverse thrust should be discontinued until you have the airplane under control and headed straight down the runway.

Figure 15.5 shows the effect of reverse thrust during a skid. The aircraft has weathercocked to the left into the wind and is no longer aligned with the center line of the runway. The reverse thrust, being in line with the fuselage and pulling to the rear, will tend to pull the aircraft to the right and cause the aircraft to skid off the runway center to the right.

BRAKES

An unequal application of brakes can start a skid, but brakes are the last control to be used in skid recovery. They are usually mounted well behind the center of gravity (the spread between the main gear trucks is rather narrow) and therefore are not too effective in directional control with differential braking after the skid has started.

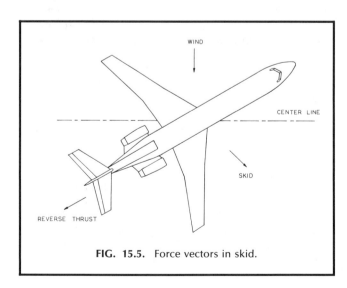

FIG. 15.5. Force vectors in skid.

Most skids develop on wet and slippery runways; if you do not have antiskid protection, your wheels are more likely to lock and your tires can start hydroplaning on a film of water. If this happens, the brakes should be released and then reapplied. In any event, remember that the coefficient of friction between the tires and the wet runway surface is greatly reduced, diminishing braking effectiveness.

You might also remember, with a long wheelbase aircraft or any of the transport jets with aft-mounted engines, that the main gear can be more than 50′ behind the cockpit. In a skid you may have the cockpit in the center of the runway with the nose pointed out toward the far side of the runway in relation to the direction of the skid, but your main gear may be out in the bushes!

On the theory that an ounce of prevention is worth a pound of cure, make *every* landing in the center of the runway; then you will have equal distance on either side if you do get into a skid. Keep your visual focus well ahead of the aircraft, as far ahead as visibility will permit. This enables you to recognize small heading variations and correct them before they become large variations.

SKID RECOVERY

The foregoing discussion may help you stay out of a skid. However, if you fly year round on a daily schedule, in and out of different airports in all kinds of weather, experiencing a skid is virtually inevitable. It's a disconcerting experience, but safe recovery can be accomplished with a little knowledge of what's going on and good pilot technique. Once a skid has developed, you need all the technique and control you can get to reestablish runway alignment.

Referring again to Figure 15.5, we find an airplane skidding off toward the right side of the runway. How does an airplane get into this situation? One of the basic laws of motion affecting flight is that an object in motion will continue in motion in the same direction unless affected by some outside force. This fairly well indicates the desirability of having the object in motion—in this case the airplane you're flying—in the *proper* direction initially. Assuming that this has been accomplished, we may then consider the possible outside forces: overcontrol of nosewheel steering, asymmetric reverse thrust, unequal brake application, improper crosswind technique after touchdown, or crosswind gusts.

Earlier I mentioned that sweptwing aircraft have a tendency to weathercock *downwind,* away from the wind rather than into it. This tendency reaches its maximum effect with the wind 30–40° off the nose. Of course, the actual maximum effect would depend on sweepback angle and the relative wind angle because the wing on the upwind side is developing more lift

than the one on the downwind side and tries to turn the airplane downwind. However, as the crosswind component becomes more than approximately 45°, the weathercock tendency develops *into* the wind. The tendency to turn downwind is negligible, but the weathercock tendency into the wind is much more pronounced. Turning the nosewheel too fast to overcome this tendency can start the skid as the nosewheel tires lose even more traction from the side forces imposed upon them. You may turn the nosewheel and find it is not effective, especially on wet pavement. Basically, recovery requires getting traction restored to the nosewheel tires. This is accomplished by turning the nosewheel into the skid. Looking at Figure 15.5 again, we find we have three forces working against bringing the nose back to alignment once the skid has developed: (1) inertia in the direction of the skid, (2) wind pushing the tail to the right, and (3) reverse thrust pulling the tail to the right. We can't turn the wind off, but the other two forces can be controlled. We can eliminate the reverse thrust until realignment is accomplished and have only wind and weathercock tendencies along with inertial forces to overcome with flight controls.

Directional control should be accomplished in this order: reject reverse thrust; use rudder, nosewheel steering, and brakes.

1. Reverse thrust forces a skidding aircraft *toward* the skid. On a long runway it's better to come out of reverse and get the airplane back into alignment, rather than to stay in reverse and skid off the first 3,000′. Once the aircraft is in control and headed in the right direction, reverse thrust can be used again. The illustration used as an example (Fig. 15.5) is typical of a DC-9 or Boeing 727 or any other aft-mounted engine aircraft that has very little asymmetrical thrust effect. An aircraft with wing-mounted engines (DC-8 or Boeing 707) or multiengine prop-driven aircraft might have better recovery results by reducing reverse thrust only on the engines on the side away from the skid. Generally speaking, however, it is better to reject reverse, particularly in jet aircraft.

2. If a skid develops, use rudder and nosewheel steering *into* the skid to regain nosewheel traction. The use of brakes will be as required according to their effectiveness and for speed reduction to reduce inertia.

Jammed Stabilizer Approach and Landing

The pitch trim of jet aircraft is accomplished by moving the position of the horizontal stabilizer. If the horizontal stabilizer becomes jammed, the ability to change its angle of incidence is lost, and special care and consideration must be taken, particularly at lower speeds in approach. The actual technique to be used will vary with airplane design and flight characteristics and, of course, the position the stabilizer is jammed in. However, the following comments will apply to virtually all aircraft.

The pitch change effectiveness of the stabilizer is about 2½ times that of the elevator; therefore, it is of paramount importance to avoid high rates of descent, especially on the final approach. Improper techniques could result in loss of elevator control if large changes in pitch were required at low speeds, and control of descent and pitch attitude without stabilizer trim must be accomplished by coordinated use of thrust and elevator control.

To compensate in part for this loss of trim capability and as insurance against the possible loss of elevator control, additional airspeed should be added to the computed threshold or V_{ref} speed for the normal approach, thereby providing more elevator effectiveness. The exact additional speed increment depends on gross weight, center of gravity, and the position the stabilizer is jammed in and varies for different aircraft—usually 5–25 knots.

It is logical that the more aft the position the stabilizer is jammed in, the less additional speed is required for flare; the more forward jammed position will require greater speed. For example, if the stabilizer jammed immediately after takeoff before the trim was changed from the takeoff setting, no more than 5 knots would need to be added to V_{ref} since the stabilizer is already set at a position relative to the takeoff center of gravity and gross weight for a speed of 1.38 over stall; this is very near the exact setting required for 1.30 for V_{ref}. If the stabilizer jammed in high-speed cruise or descent, a full 25 knots would be needed; if it jammed while the aircraft was trimmed for maneuvering speed in any configuration, 10–15 knots additional speed would be required.

Zero Flap Landing

Fortunately, this is a highly unlikely event, but a zero flap approach may be included in a rating ride.

Prior to a landing without flaps/slats, it is imperative that the pilot be fully aware of the total effect of total loss of the hydraulic system. This would be necessary only if hydraulic failure was, in fact, the cause of the situation rather than a mechanical failure in the flap/slat system. However, from a training standpoint it is best to assume the most critical situation.

The mechanical components that may be affected by hydraulic failure in addition to the flaps/slats include:

1. Landing gear—It may be necessary to "free fall" the landing gear by the alternate or emergency extension method. In some aircraft this may require

more time than in a normal extension to assure a safe condition of the gear. Therefore, extend the gear early enough to assure yourself of a safe extension prior to making the approach.

2. Ground spoilers—Some aircraft (DC-9) have spoilers that will automatically deploy upon landing (if the system is armed and functioning) when wheel spin-up reaches approximately 70 knots. Others (727) have spoilers that are deployed by the pilot after touchdown. In either case, if they are hydraulically actuated, they will be inoperative and add to the landing ground roll.

3. Nosewheel steering—In some aircraft it may be totally lost. In others, some nose steering capability may remain through an accumulator. In either case, consider it as limited or nonexistent.

4. Brakes—Will be dependent upon accumulator pressure only. Antiskid *should not be used.* The antiskid system modulates hydraulic pressure to the brakes by opening a valve in the return line and may deplete the accumulators if braking is heavy enough to cause antiskid wheel release.

5. Reverse thrust—Most hydraulically actuated thrust reversers have their own accumulators, so reverse thrust will probably be available. However, since reverse thrust is never considered in stopping distance to determine runway requirements, it should not be considered now.

In addition to the loss or reduced effectiveness of these components, you should also consider tire speed. The high landing speeds associated with landing at zero flaps and slats retracted will sometimes exceed the minimum tire limit speed. Do you know the limiting speed for your tires? A tire that is speed rated at 225 miles per hour is equivalent to a true ground speed at touchdown of 195 knots; a 200-MPH tire has an equivalent ground speed of 174 knots.

There will also be a maximum weight and temperature that limit brake energy. This will vary with field elevation, temperature, and surface wind. The higher the field elevation and temperature and the less the headwind component, the lower the maximum brake energy limiting weight. However, if recommended procedures are followed, it will seldom be limiting except when operating with maximum payload for your aircraft. But I'd suggest you determine what it may be from a performance chart; if your manual doesn't contain one, you might contact your maintenance department, tech service, or engineering department (airlines have different ways of naming departments from which you might get technical information concerning your aircraft).

From what we've discussed so far, it is apparent that the pilot faced with making a zero flap and slat approach must therefore consider the following:

1. Landing gross weight—It is desirable to land at the lowest weight practical for the lowest approach

speeds (less mass in motion and therefore less kinetic energy required from braking action to stop).

2. Weather—A zero flap approach and landing can be made from an instrument approach, but it will be more difficult. Also precipitation will have an adverse effect on runway conditions and stopping distance.

3. Runway available and surface conditions—A long, dry concrete runway (the longest within the range of your aircraft) is highly desirable.

I'm going to illustrate some stopping distances for a zero flap landing in a DC-9-30 (taken from a performance chart developed for the C-9A, the military version of a DC-9-30 with identical gross weights and performance). They are based on estimated data rather than actual flight test, but this doesn't mean they are inaccurate. It's not difficult to compute mathematically the foot-pounds of energy required to stop an object in motion, knowing its weight, speed, friction coefficient of the tires, and the kinetic energy capable of being developed by the brakes to perform the task; then accurately computing the distance required to stop is possible.

In the illustrations that follow, various speeds are shown that represent additions to V_{ref} for a normal full-flap landing configuration. In order to better understand the charts, it is necessary to be familiar with a V_{ref} chart for the aircraft (see Table 15.1).

TABLE 15.1. DC-9-30 — V_{ref} 50° Flaps/Slats Extended

Gross Wt. (in 1,000 lb.)	60	65	70	75	80	85	90	95	100	105
Knots IAS		99	103	107	111	115	118	122	125	128	131

The chart for the C-9A is almost the same except that it shows a V_{ref} beginning at a gross weight of 70,000 lb. (60,000 lb. is the average empty weight of the aircraft) and extends to 108,000 lb., the certified max gross.

These are easy reference charts for cockpit use and are used by taking the landing weight to the nearest weight shown on the chart and adding or subtracting 1 knot per 1,000 lb. variation. For example, with a landing weight of 98,000 lb. (98 is nearest 100), subtract 2 from 128 (V_{ref} for 100,000 lb.) for a V_{ref} of 126.

Figure 15.6 shows the estimated landing ground roll of a DC-9-30 for a zero flap landing. The conditions are: dry and level runway, slats retracted, spoilers inoperative, no reverse thrust, and antiskid inoperative. There are three examples shown on the chart (circled numbers). The first one illustrates the performance you might expect for the aircraft shown in the approach and landing profile of Figure 15.7. To use the chart, enter at temperature, go up to pressure altitude,

over to gross weight, down to speed at threshold in excess of desired threshold speed, to the left to wind, and down to landing ground roll. This is the landing ground roll you may expect after touchdown. The other two examples are at more realistic weights and are also shown at speeds that would be correct (according to some operators' recommended threshold speed) for a zero flap/slat landing in relation to that recommended by Douglas, as shown in Figure 15.7. Some operators recommend that the threshold speed be computed by taking the basic V_{ref} for the gross weight and adding 55 knots. This is done to provide the pilot with an easily remembered way of computing a threshold speed for no flaps/slats, slats and zero flaps, flaps and no slats, or split flaps or slats, by adding 25 knots to basic V_{ref} for loss (or split) of slats and 30 knots for loss (or split) of flaps.

The three examples in Figure 15.6 are given in Table 15.2.

The final figures are not the landing distance or the runway length required but solely the distance you may roll before stopping after touchdown. They are predicated on a dry runway and assume no slope. However, maximum use of reverse thrust may reduce these stopping distances by approximately 20% if reverse thrust is available.

Landings made in the clean configuration have required maximum flare distances of 2,000' from the 50' threshold height to touchdown. This was the worst case demonstrated and shows the importance of getting the aircraft to touch down on the proper target. To determine total landing distance, add air distance to touchdown to landing ground roll.

You probably won't find a similar chart for your airplane in any of your manuals or performance data, though I'm of the opinion there should be a chart based on flight test for runway requirements in every possible landing configuration. But you can determine the runway required as a minimum for a zero flap/slat landing under the conditions we've discussed by rule of thumb: Double the runway requirement for a wet runway. To illustrate this, work a runway requirement (wet) for any weight on Figure 15.3 and then a landing ground roll for the same conditions in Figure 15.6 and compare them. You will find the rule of thumb gives you an adequate margin of safety.

Figure 15.7 is a profile of the traffic pattern and approach for a zero flap/slat approach and landing. As it shows, a wider than normal approach traffic pattern should be flown for a landing without flaps, since you will have an increased turning radius because of the higher airspeeds you will have to use in consideration of the greater stall speeds and the resultant maneuvering speeds required.

The turn to base leg should be delayed to ensure a long, straight final approach. The landing gear should be extended long before beginning the actual ap-

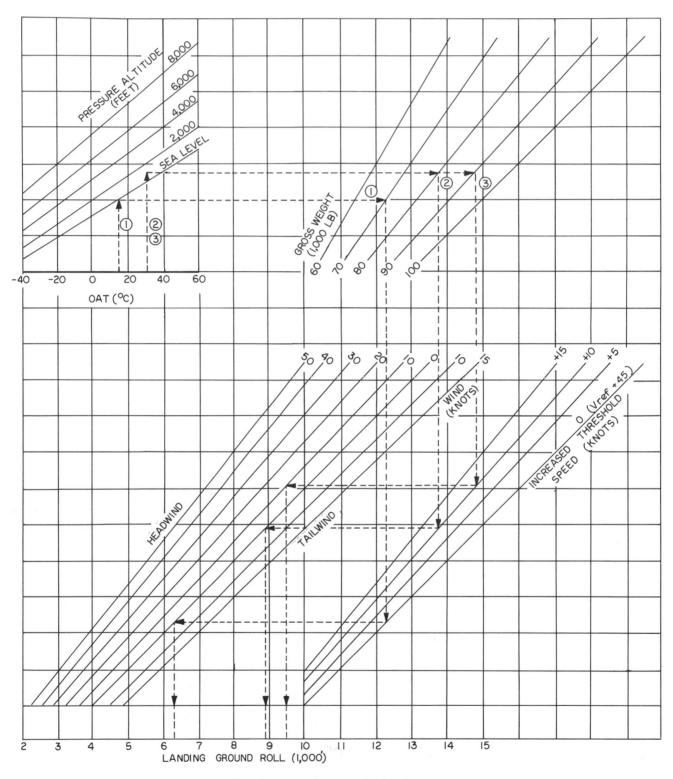

FIG. 15.6. Landing ground roll with no flaps.

110

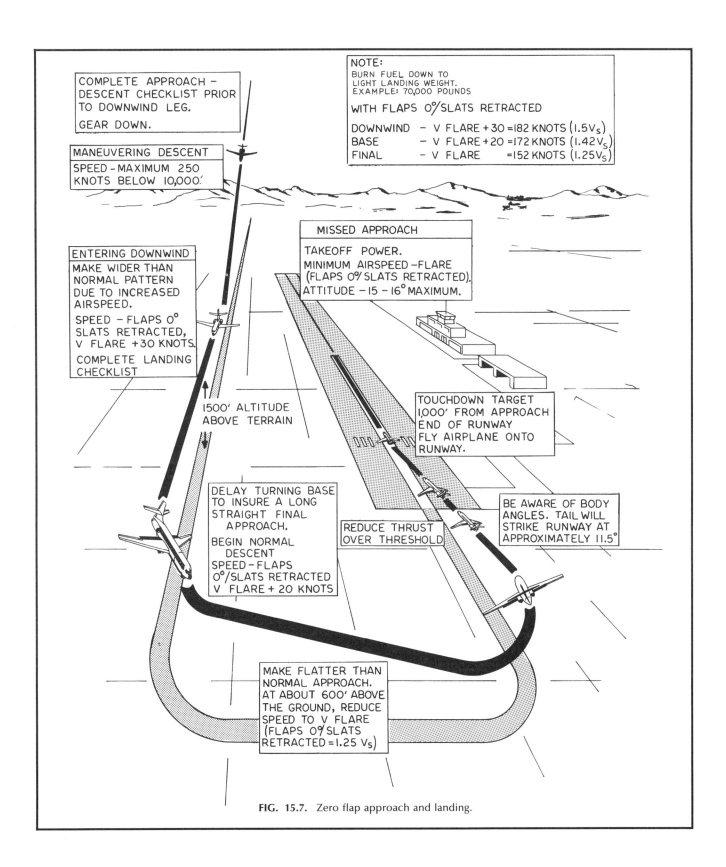

COMPLETE APPROACH –
DESCENT CHECKLIST PRIOR
TO DOWNWIND LEG.
GEAR DOWN.

MANEUVERING DESCENT
SPEED – MAXIMUM 250
KNOTS BELOW 10,000.'

NOTE:
BURN FUEL DOWN TO
LIGHT LANDING WEIGHT.
EXAMPLE: 70,000 POUNDS
WITH FLAPS 0%/SLATS RETRACTED
DOWNWIND – V FLARE + 30 = 182 KNOTS (1.5 V_s)
BASE   – V FLARE + 20 = 172 KNOTS (1.42 V_s)
FINAL   – V FLARE   = 152 KNOTS (1.25 V_s)

MISSED APPROACH
TAKEOFF POWER.
MINIMUM AIRSPEED – FLARE
(FLAPS 0%/SLATS RETRACTED).
ATTITUDE – 15 – 16° MAXIMUM.

ENTERING DOWNWIND
MAKE WIDER THAN
NORMAL PATTERN
DUE TO INCREASED
AIRSPEED.
SPEED – FLAPS 0°
SLATS RETRACTED,
V FLARE + 30 KNOTS.
COMPLETE LANDING
CHECKLIST

1500' ALTITUDE
ABOVE TERRAIN

TOUCHDOWN TARGET
1,000' FROM APPROACH
END OF RUNWAY
FLY AIRPLANE ONTO
RUNWAY.

DELAY TURNING BASE
TO INSURE A LONG
STRAIGHT FINAL
APPROACH.
BEGIN NORMAL
DESCENT
SPEED – FLAPS
0°/SLATS RETRACTED
V FLARE + 20 KNOTS

REDUCE THRUST
OVER THRESHOLD

BE AWARE OF BODY
ANGLES. TAIL WILL
STRIKE RUNWAY AT
APPROXIMATELY 11.5°

MAKE FLATTER THAN
NORMAL APPROACH.
AT ABOUT 600' ABOVE
THE GROUND, REDUCE
SPEED TO V FLARE
(FLAPS 0%/SLATS
RETRACTED = 1.25 V_s)

FIG. 15.7. Zero flap approach and landing.

TABLE 15.2. Examples for Figure 15.6

	Example 1	Example 2	Example 3
Temperature	15° C (59° F)	30° C (86° F)	30° C (86° F)
Field elev.	Sea level	1,000'	1,000'
Gross wt.	70,000 lb.	80,000 lb.	90,000 lb.
Threshold speed 1.25 V_s	152 knots	160 knots	167 knots
Speed in excess of 1.25 V_s	0	10 knots (170 IAS)	10 knots (177 IAS)
Wind	0	0	10 knots headwind
Landing ground roll	6,250'	8,875'	9,500'

proach to ensure that the gear is down and locked and to allow speed stabilization with the gear extended.

All flight while maneuvering banks in excess of 15° should be either at the minimum recommended maneuver speed of your aircraft in the clean configuration or at an actual computed speed of not less than 1.5 over clean stall speed for your gross weight. (As shown in Fig. 15.7, this is V flare + some speed component that would be appropriate for your aircraft. This illustrates a DC-9-30 making a clean approach but is typical; there is a similar speed that may be added to V flare for every aircraft.)

After turning on a long final approach, start a speed reduction to an approach speed of 1.25 over stall. This could also be a recommended speed addition to a basic V_{ref} that would result in an approach speed closer to 1.3 over stall. In any case, set the flare speed desired with the "bug" on your airspeed indicator.

Try to cross the outer marker (or an equivalent position on final of about 5 miles from the runway end) at 1,000' above the field elevation, and try to have the desired approach speed established when reaching 600' in the approach descent.

You may have to reduce power considerably to slow to the desired speed. Completely unspooled approaches in jets, sometimes called "formaldehyde glides," are not recommended. Therefore, if power control for airspeed slowdown is difficult, set 50% of your engines just under the bleed valve setting, where they may be spooled up quickly if necessary, and use the remaining engines for speed control. This has the effect of keeping at least half your engines fully spooled up and the others with power rapidly available for go-around if needed, making speed control at approach speed much easier.

The aiming point for touchdown and speed control are of paramount importance in this approach. The low drag resulting from the zero flap/slat configuration requires a rather high body angle, a nose-high attitude, and a flatter than normal approach. I recom-

mend putting your seat up so that you can partially overcome the visual difference of the unusual approach attitude and have good visibility throughout the approach. Deceleration will be slow. However, acceleration will be rapid, either by an increase in thrust or by lowering the nose. But you should have it "in the slot" at 600', on the correct attitude and speed, and flying a slightly flatter approach glide angle to the touchdown aiming point. If it appears that you cannot be on the correct speed at the threshold and the aircraft will touch down beyond the aiming point, don't hesitate to go around.

When your aircraft enters ground effect, there will be a noticeable increase in control effectiveness, especially in the longitudinal axis, so be careful not to overcontrol the elevator at this point. The airplane is already in the landing attitude, so the landing should be accomplished by a smooth bleed-off of thrust and with no change of attitude. To flare at this point or to make any attempt to make a smooth landing by holding it off is very poor technique. In this low drag configuration it would result in a long float, wasting runway needed for braking, and would increase the possibility of striking the runway tail first.

As soon as you touch down, apply reverse thrust if it is available and apply forward force to the elevator to lower the nose to a negative angle of attack, destroying lift and increasing brake effectiveness. Use brakes and reverse as necessary to effect a stop within the runway limits.

Landing with 50% Power Loss

Landings with all engines out on one side are required in obtaining a type rating in a T-category aircraft and are practiced in training. There are so many variables of gear and flap management, minimum control speeds in the event of a go-around, and procedures in sequence from aircraft to aircraft that approach patterns and final approach technique for all aircraft are impossible to discuss in a work such as this. The general factors to consider are:

1. Proper time to extend the gear—This depends on the drag effect of the gear and its effect on go-around capability. Most jet aircraft are not limited in gear extension or go-around capability, but most prop aircraft are.

2. Flap management—Most important is the effect of flaps on approach and how they might limit go-around capability. The final approach is usually flown with "approach flaps," an intermediate setting more than the maximum takeoff flap setting and less than full landing flaps. This results in a slightly higher approach speed but ensures go-around capability.

3. Minimum control speeds—Knowledge of minimum control speeds for your aircraft and operation at

a speed assuring positive control at all times in the approach is absolutely essential.

The actual landing is not difficult. The flap configuration varies with aircraft design and performance capabilities, and the approach speed may be 8–15 knots faster than normal, but the landing is at an attitude and speed in relation to stall for the configuration, as in a full flap landing. Any rudder trim for directional stability in the approach should be taken out once landing is assured, and normal techniques should be used for flare and touchdown.

It is after touchdown that things get a bit different. You have reverse thrust on one side of the aircraft only. Therefore, have the nosewheel firmly on the ground and use reverse thrust smoothly, avoiding a rapid application of power. I've found it highly effective to use a rather firm brake application because of additional stopping distance required by additional speed of approach and landing. Directional control may be maintained with rudder and nosewheel steering against the adverse yaw of asymmetric reverse thrust. It may even be advisable to lead the yaw a bit with opposite rudder; if you were going to reverse the engines on the right side, lead with a bit of left rudder a split second before you go into reverse. Then apply only the amount of reverse thrust that may be easily controlled in yaw with the rudder at maximum deflection. As your speed reduces, it will be necessary to diminish reverse thrust as the rudder becomes less effective and nosewheel steering is required. It's kind of a "feel" technique in relation to control effectiveness, yaw from asymmetric reverse thrust, and runway alignment. Considering all the factors that may affect you in approach, flying an approach with good speed control and proper gear and flap management and using brakes and reverse thrust judiciously on landing rollout makes the landing with 50% power loss on one side safe even if not simple to execute.

Antiskid Systems

An antiskid braking system modulates the brake pressures under all runway surface conditions. It is a self-adapting system that automatically modulates brake pressure through electronic devices that sense and determine wheel speed rate of change rather than actual wheel speed. Such a self-modulating system continuously controls brake pressure to the maximum that the aircraft tires and runway will tolerate, by relieving brake pressure into the return line before the wheel locks and skids.

Most antiskid systems offer individual wheel control that affords the maximum degree of braking power on wet or icy runways. This is considered optimum, since each wheel is regulated to its needs alone and not influenced by different runway surface conditions that may be encountered by the other wheels simultaneously.

When using antiskid, the pilot should hold a steady brake pressure on the pedals according to the particular stopping distance or braking action required. It is not necessary to pump the brakes. If you do not want full brake pressure, the system will adapt to whatever metered brake pressure you choose by brake pedal application. Therefore, antiskid provides the most efficient utilization of the amount of pressure applied, including full brake application.

Manual Braking: Antiskid Failure

When landing with the antiskid system inoperative, the antiskid switch should be off and brake pedals should be applied using normal braking techniques—holding a steady brake pedal force that provides moderate deceleration. Too much brake will blow the tires. If antiskid failure is indicated on only one wheel, leave the system operative with the antiskid switch on, but use braking as if all antiskid were inoperative. This will protect all tires as much as possible but will blow only one tire if too much brake is applied.

Braking action with antiskid inoperative and brakes providing moderate deceleration is approximately 60% of the effectiveness that may be obtained with maximum antiskid braking and brakes full on and is normally achieved with just slightly less than half of full brake pedal application. With the system inoperative, too much brake pressure will cause the tires to skid—losing braking efficiency, flat-spotting tires, and probably resulting in blowing the tires.

Landing Rules of Thumb

It still takes pilot skill, feel, and judgment to make good landings. It is important for every pilot to know the performance capabilities of the aircraft in the landing, how the various factors that affect landing performance affect the aircraft, and how to apply these factors to the landing and stopping distances in a manner that will assure adequate runway.

Every aircraft has performance charts (in the performance section of the pilot's operating manual) that give landing field lengths similar to the chart in Figure 15.3. This will be the type of chart you might have to use to work a problem in your oral and provides sufficient information for most purposes. However, the chart in Figure 15.8 is based on the same flight test data and considerations. It is far superior in that it also considers temperature and any increased speed you may be carrying above V_{ref} as a wind and gust component and gives you a reading of the landing distance

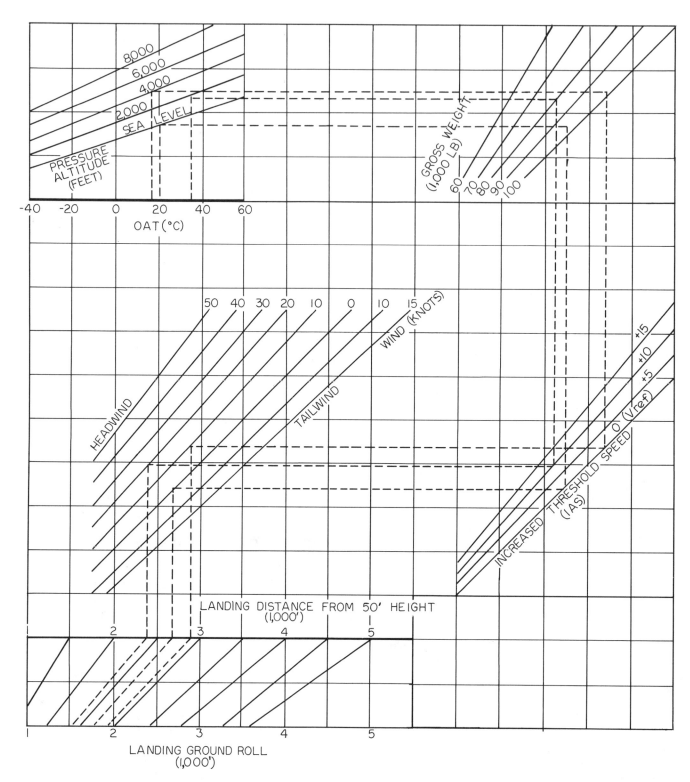

FIG. 15.8. Landing distance and ground roll with full flaps.

from the threshold as well as actual landing ground roll. This leaves only two factors to landing performance not accounted for—runway slope and surface conditions. In air carrier operation and nearly all civil operations, surface conditions of the runway are considered as an adverse condition to braking and stopping distance by adding 15% to the total runway requirements.

Every airline has gross weight manuals in all their aircraft from which maximum weight requirements may be found for both takeoff and landing on every runway and airport the airline serves. These are prepared by the airline's engineering department and make the job relatively simple for the pilot. The runway length, slope, and elevation are fixed; the only variables are temperature, runway condition (wet, dry, or slush depth), and wind. These are factors over which the operator has no control; the gross weight of the aircraft is the only variable that may be adjusted to fit into the framework these factors build. In short-range operation, many times the fuel burn-off to the next stop may be a control factor in computing takeoff weight in order to be down to the maximum landing weight or less for the next stop.

But how about pilots that don't have all this prefigured information available? They have to do all the computations themselves using the performance charts for every takeoff and landing. They may have occasion to land at a field for which they just don't have time to go into the performance charts and figure the runway requirements. Or they may be faced with a zero flap landing and have no performance chart for runway requirements. Or perhaps they have some other malfunction—inoperative antiskid or inoperative spoilers—and don't remember the exact effect of such malfunction on performance. Or the runway is icy. How much runway do you need for safety? What will your landing ground roll be?

That's where some rule-of-thumb applications of the factors that affect landing performance come in handy. And there are reliable rules of thumb that apply to landing in every airplane, including jet aircraft. An aircraft is mass in motion, and all are subject to the same laws of motion and inertia. The only variations are in mechanical malfunctions that differ for different aircraft.

You can work out your own figures for landing performance. You start by computing the runway requirement for landing, on a chart similar to Figure 15.3. Determine the dry-field length for maximum landing weight, sea-level pressure, standard temperature, and zero wind for a normal landing with your aircraft. You'll come up with a runway length requirement based on the air distance to touchdown from the threshold and the stopping distance or landing ground roll equal to 60% of the total field length required. As an example, if you required a 5,000′ runway under

these conditions, it is based on your being at exactly 50′ and on V_{ref} at the threshold, touching down at a speed of 1.26 V_s at the 1,000′ point and then stopping with the use of maximum braking and no reverse within 60% of 5,000′, which is 3,000′. Since you have an air distance to touchdown of 1,000′, it follows that you have a landing ground roll of 2,000′. Of course, you can determine the same thing from a chart as shown in Figure 15.8 much more simply, and such a chart goes a long way toward making landing rules of thumb almost unnecessary, except for runway slope and runway condition.

Then you can figure out runway requirements for every weight from empty plus one hour's fuel to maximum gross—at different speeds, at every field elevation the aircraft is certified for landing, for different temperatures and wind components, and for different inoperative components—to determine the effect of each of these factors to the base landing ground roll determined for sea-level dry standard. Or you can accept the rules of thumb I've developed from similar research and actual experience in over 50 years of flying many different aircraft.

The air distance to touchdown is affected mostly by altitude at threshold, though the speed of the aircraft is determined by the speed necessary for its weight, temperature, field elevation, and actual indicated airspeed. The actual distance to touchdown will be 900′ greater if 100′ high at the threshold rather than 50′, and will also be greater for additional speed. However, the landing distance is a total of air distance to touchdown and landing ground roll, so all factors will be considered as affecting landing ground roll only.

The aircraft as a mass in motion, moving at a true airspeed velocity at touchdown, must then be considered for its momentum (its mass or gross weight multiplied by its velocity). Then its inertia must be overcome by braking action and other available outside forces such as wind (which also is a factor in determining its ground speed and actual velocity), reverse thrust, spoilers, and other factors that may add drag in reducing its velocity.

The effect of braking action is dependent on the momentum it must overcome and the coefficient of friction between the tires and runway surface. This was one of the considerations we discussed for zero flap landing as a maximum weight limiting available brake energy and was one of the reasons for getting the weight down prior to landing. The lower the weight of the aircraft, the lower the necessary velocity; the result is less momentum, allowing the brakes to reduce inertia more efficiently. However, we are now concerned with runway condition, since maximum brake energy limiting weight is not a factor for any but zero flap landings. There is sufficient break energy available, and the coefficient of friction affected by runway surface changes landing ground roll.

The FAA requires that 15% be added to the dry runway requirement for a wet runway. It doesn't take too many years of flight experience to find that there are also other runway conditions for which no official additions to runway requirements are stated. You will find that runways are not just dry or wet; they are also often wet and slippery or icy. Some additional runway is required for stopping distance on such runway conditions. It would be easy for the manufacturer to actually test each model of aircraft for stopping distance after landing for *all* runway conditions, but this has not been done. Or a chart of estimated performance could be made by an aeronautical engineer, but this has not been done either. The Air Force uses landing ground roll charts, as shown in Figure 15.8, corrected for slope and runway conditions. In the interest of safety for airline operation, the FAA should adopt something similar.

To show what such an extension of stopping distance would look like, I have drawn the chart shown in Figure 15.9. If you can visualize it as being an extension to the landing ground roll result of the chart in Figure 15.8, that would be one manner in which such a graph could be used. It would be even better as a separate chart that would correct the results of base landing ground roll for both slope and runway conditions. You would then have a more accurate landing distance—60% of the effective runway length—to work with.

landing distance for a DC-9-30 (or C-9A) of about 2,700' and a landing ground roll of 1,750'. It is this 1,750' to which the following would apply:

100' high instead of 50' at threshold	additional 900' to touch-down
1% increase in speed over correct V_{ref}	2% increase in landing ground roll
1% change from max landing weight.	1% change in landing ground roll
500' change from 29.92 field elev	2% change in landing ground roll
15° change from standard (F). .	4% change in landing ground roll
10 knots wind, as increment of change	15% change in landing ground roll
1% uphill runway slope	4% decrease in landing ground roll
1° downhill slope	6% increase in landing ground roll

These are rules of thumb based on the factors affecting your mass in motion. Let's figure an example of adjusting the landing ground roll for the following conditions: 80,000 lb., pressure altitude of 28.82, tem-

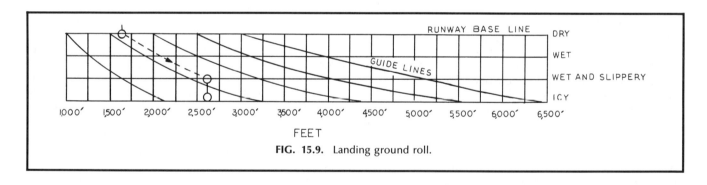

FIG. 15.9. Landing ground roll.

The following rules of thumb, however, are very accurate and may be used in lieu of the more sophisticated charts until we actually have them. They are stated, for the most part, as percentage changes to a landing ground roll that would be a base stopping distance found for your aircraft at maximum landing weight at sea-level standard, 29.92 pressure altitude, and 15° C, with zero wind on a dry runway and using maximum braking and no reverse. An example is shown on Figure 15.8—starting at 15° C, going up to sea-level pressure, for a weight of 98,000 lb., with no increased threshold speed, and zero wind showing a

perature 95°F, a 20-knot wind for which we'll carry an additional 10 knots over V_{ref} (the correct V_{ref} for a DC-9-30 or C-9A would be 114 at 80,000 lb., or 11.5% increase). Therefore, 80,000 lb. weight = -18% change from max landing and -18% change in landing ground roll; a 10-knot wind component would be 124 indicated airspeed, 11.5% increase, or $+23\%$ increase in landing ground roll; a pressure altitude of 28.82 would be approximately 1,000' of change from 29.92, or $+4\%$ change; 95°F is a 36° change from 59°, or about a $+10\%$ change; and a 20-knot headwind is a -30% change.

The algebraic sum of all the above: -18, $+23$, $+4$, $+10$, $-30 = -11$; 11% of a base landing ground roll of 1,750' $= 192.5'$, or a landing ground roll of 1,750 minus 193 for a total of 1,557'. Being a rule of thumb computation, this will not be an exact figure but will be very close to what the chart would show and extremely close to actual performance. I didn't show any correction for threshold altitude (assuming it to be correct) or any slope correction, since I have no chart to prove it. However, I can prove the above computations on Figure 15.8: 95°F $= 35$°C, so see the illustration of landing distance and landing ground roll starting at this point on Figure 15.8. It comes out at about 1,600' for landing ground roll.

This rule of thumb computation will give you a landing ground roll corrected for weight, speed, temperature, pressure altitude, and wind. Now we need to apply the corrections for slope and then have some manner of considering the effect of runway condition. I've already given you the percentage figures of the effect of runway slope. Before I give you the rules of thumb for runway conditions, let's define wet, wet and slippery, and icy:

1. Wet—Any concrete runway wet with braking action reported *good*.

2. Wet and slippery—Any asphalt or blacktop runway wet; *any* heavily trafficked runway; having fuel from jet spillage, exhaust smoke, rubber residue, etc., mixed with water on the runway; snow-covered runways; any runway wet with braking action reported *fair*.

3. Icy—Any runway covered with ice; any runway with braking action reported *poor* to *nil*.

All of these are presently lumped together and 15% added to runway requirement. However, runway conditions affect landing ground roll by the following approximate percentages:

1. Wet—Adds 25% to landing ground roll. This very nearly allows you to land and stop within 60% of the runway required.

2. Wet and slippery—Adds 60% to stopping distance! Runway requirement?

3. Icy—Adds an average of 120% to landing ground roll.

As an example, suppose we had a landing ground roll of 2,000', or a total landing distance of 3,000'; 3,000 is 60% of 5,000, so we'd need a 5,000' runway for dry runway requirements. Runway conditions

would affect this base 2,000' landing ground roll and total landing distance as shown in Table 15.3.

TABLE 15.3. Affect of Runway Conditions on Landing

Runway Condition	Landing Ground Roll	Landing Distance	Runway Required
Dry	2,000'	3,000'	5,000'
Wet	2,500'	3,500'	5,750'
Wet and slippery	3,200'	4,200'	5,750'
Icy	4,400'	5,400'	5,750'

I showed the runway presently required by the 15% addition to runway for wet requirements. The aircraft will fit into the airport, without a large margin of safety, if you're not too fast or too high at threshold (900' added would be a disaster if you were only 50' too high) and the touchdown was made at a point no further down the runway than the 1,000' point followed by maximum braking to a complete stop! A skid would also be bad news.

You can see, even though reverse thrust isn't taken into account in stopping distance, that you are depending heavily on reverse when landing on a wet and slippery or icy runway of minimum length for wet conditions.

RULES OF THUMB FOR RUNWAY CONDITIONS

You may expect reverse to affect stopping distance by 10-20%, varying from normal reverse thrust application to maximum continuous applied within 2½ seconds after touchdown to full stop.

When landing with normal use of brakes, you may expect a 60% increase above the maximum braking ground roll.

If your antiskid is inoperative, you may expect a 75% increase above the maximum antiskid braking landing ground roll.

Aircraft equipped with ground spoilers may add an average of 25% to landing ground roll if they are inoperative.

Aircraft equipped with automatic ground spoilers, only if the automatic feature is inoperative and the spoilers may be manually deployed after touchdown, may add 12-15% to landing ground roll.

16

ILS Approaches

■ This has been the hardest chapter of all to write. It's difficult, to say the least, to tell a pilot how to make a good ILS approach. But a good ILS approach, like a good landing, is not all due to luck. It takes pilot skill, good instrument scan, good technique in approach speed control, and an understanding of your instrumentation and the characteristics of the ILS; real proficiency comes only with experience and practice. I've seen all kinds of approaches and approach techniques over the years and have attempted to standardize a method of making the approach that has proved most effective. This method makes up this chapter.

Before making an ILS approach, or any type of instrument approach, the pilot *must* review the approach plate. There are several important items to look for in addition to frequencies, inbound course, glide slope interception altitude at the outer marker, etc. The missed approach procedure should also be reviewed. Both NAV receivers should be tuned to the ILS frequency, the marker beacon should be on low, and both ADF receivers should be tuned to the outer marker. Then the preplanning and methodical approach considerations come into the picture.

I'll break down the entire ILS approach, including gear and flap management schedule, for a jet aircraft. The same procedure would apply generally to a propeller-driven aircraft except that you would normally be on speed with flaps takeoff and would extend the gear upon glide path intercept (at the outer marker) to begin descent and descend in this configuration without placing the flaps full down until visual contact is made.

Wind

Wind should always be the first factor considered in making an instrument approach. You should have some knowledge of the wind at approximately 2,000′ above field elevation as well as the reported surface wind, which you might remember from your weather and wind reports. Many pilots never take the wind, other than surface reported, into consideration at all. There's an old rule of thumb, reasonable accurate, which estimates wind 1,000–2,000′ above the surface from the direction and velocity of the surface reported wind: Add 90° to the reported direction and double the velocity. As an example, with a reported surface wind of 150° at 10 knots, you would probably have a wind from 240° at about 20 knots at the outer marker.

No matter how you determine the wind, be conscious of the effect of wind on your approach and estimate the effect of wind on:

1. Approach ground speed—This takes into consideration the planned approach speed including wind and gust factors discussed in Chapter 5.

2. Resultant rate of descent to maintain the glide slope—There is a box on every ILS approach plate that indicates a rate of descent to maintain the glide slope, considering the angle of the glide slope and the ground speed of the aircraft. This can be very useful information.

3. Procedure turn—Wind affects the pattern of the procedure turn; a little preplanning will be helpful in knowing when to make the turn after passing outbound from the marker.

4. Localizer needle movement on intercept— When you have a crosswind on the ILS, you'll have either a headwind or tailwind component while flying the intercept, which will affect the rate of needle movement on interception of the localizer.

5. Inbound heading and anticipated drift—This will be helpful later in making corrections to maintain the localizer track.

Gear and Flap Management

Be on 1.5 V_{so} with flaps at maximum takeoff to the ILS intercept or during the procedure turn, or on your minimum average recommended maneuver speed for the takeoff flap configuration, and be at your glide slope intercept altitude within a safe distance of the outer marker. The prelanding final check should be completed at this point except for those items associated with landing gear, flaps, hydraulic pressure and quantity, and monitoring for malfunctions inside the marker.

1. Extend the landing gear at the first indication of the glide slope indicator moving off full-scale "fly-up." Maintain your speed in the gear extension and have everybody in the cockpit check three green lights for a safe gear. If your aircraft is so equipped, you would check for antiskid and autospoiler malfunctions before arming the spoilers. Accomplish the remainder of the final check.

2. After three green lights of gear down and at ½ dot above the middle dot or first dot above the center of on-course of glide slope, extend flaps to approach. You may accept the speed bleed-off of the flap extension, usually without any power changes, but maintain your altitude with *stabilizer trim*.

3. Extend landing flaps ½ dot before the glide slope on-course and glide path capture. Once more you may accept speed bleed-off, since it is desired to bleed off to approach speed, and use stabilizer trim to maintain altitude.

4. If the above has been properly done, check that you are on speed and configuration at glide slope intercept. You should be very close.

Glide Slope Intercept

1. On glide slope intercept, pitch the nose down to your estimated descent rate. This is usually somewhere between 500 and 700' per minute on the vertical speed indicator; about 600' is a good average at this point. This should be accomplished with the elevators and speed adjusted with the throttle.

2. Trim out the elevator with stabilizer trim. Now the aircraft is trimmed for rate of descent at a given ground speed.

3. Tie down the glide slope in the first mile and attempt to fly slightly above the center line at all times. About ½ dot fly-down has the aircraft at a slightly higher angle of attack and makes most of your corrections for glide slope a mere releasing of back pressure as the glide slope indicator wants to drop below ½ dot fly-down. If using the command bars, put your nose right into the vee as if in real tight formation, with the command bars just touching the wings of the little delta airplane in the instrument. This makes flying the glide slope and maintaining speed control much easier.

4. After capturing the glide slope, increase your scan rate. Continue to track the glide slope by main-

taining ½ dot fly-down, or ½ dot high, for proper aircraft attitude. Use *elevators*, not stabilizer trim, to adjust and alter pitch or to fly up or down for glide slope and vertical speed; maintain speed with the throttle. When off the glide slope, if not using the command bars, use no more than ±200' per minute corrections to vertical speed and avoid high sink or descent rates.

Localizer Tracking

When tracking inbound on a known track and considering the drift effect of the wind, corrections to course are most easily made according to *rate* of off-course displacement of the course indicator needle:

1. Slow needle movement—5° heading change.
2. Fast needle movement—10° heading change.
3. Keep bank angles small, 5–10°, and never bank more than the number of degrees to be turned in the heading change.
4. If the needle stops, stop the turn immediately.
5. When back on the localizer, take out half of the heading change or correction. Make all heading changes with a coordinated turn, or aileron only if you have a yaw damper, and remember that use of the rudder could excite Dutch roll.

About halfway down the localizer (about 700' and at least 500'), double your scan rate and make sure that all your corrections are small and smooth.

Expect your wind to begin to change at the base of the clouds (or at about 300' if still in the clouds) and to become nearer the direction reported on the surface though perhaps changing in velocity. You might look again at the ratio of wind velocities between the surface and 300' according to lapse rates in Figure 5.2.

By the time you reach 200' (or whatever your approach minimums may be), your deviation from the localizer and glide slope should not be more than 1 dot from the center line (preferably less), and you definitely should not be below the glide path.

At Minimums

We just mentioned deviation limits for 200' and ½ mile or greater; your speed, rate of descent, and aircraft attitude should be such as would allow normal flare and landing. Visual contact with the runway must be made, and large corrections for runway alignment and landing should not be required.

That's about it for a method of flying the ILS. It's not difficult for the pilot who thinks and plans ahead of the aircraft. I hope this breakdown of flying the ILS will prove helpful to you. It has been very useful to me in instructing, and application of the same principles in line operation makes actual approaches easier.

17

Missed Approaches and Rejected Landings

■ A pilot's mind is always geared to completing the approach and landing if the weather is reported to be minimal or better. However, a missed approach is always a possibility from any instrument approach, and an aborted landing may be necessary for a number of reasons, such as an obstruction on the runway or an unsafe gear warning when the throttles are closed. A good pilot even though fully expecting to establish visual reference and to land, is always prepared to go around. Both maneuvers, the missed approach and aborted landing, are required in the rating ride or instrument check.

I've always been of the opinion that either the missed approach or rejected landing, if executed with a technique based on common sense, is both easy and safe to perform. There may be some difference of opinion, however, and there have been some fatal accidents attempting a go-around.

The decision height of an instrument approach is the point where a decision is made to land or go around, based on visual contact with the runway and aircraft position and alignment with the runway to effect a safe landing; rarely would this decision be reversed. It is much safer to roll out in a heavy fog bank after landing than to attempt an aborted landing. Conversely, it is safer to go around from the decision height than to continue for a landing if minimal landing conditions do not exist. A good decision is of the essence and should be followed if at all possible.

Either the missed approach or aborted landing is, in reality, simply a takeoff begun at a higher altitude and without the necessity of accelerating on the ground. The 1.3 V_{ref} speed in a jet will be very close to

V_2 or takeoff safety speed; the flare speed of a propeller airplane will be closer to V_1, but it will accelerate faster than a jet.

An aborted landing is a go-around executed anywhere below the 50′ flare point and before actual touchdown on the runway—in effect, a takeoff from a point higher than field level and from an intial speed at least in excess of V_1. It may be best executed by (1) applying takeoff power, (2) retracting the flaps to maximum takeoff position (a command by the pilot flying the aircraft, calling for the degrees of flaps required), (3) holding the aircraft's altitude (which may require an attitude change as the flaps retract), and (4) retracting the gear when a positive rate of climb is indicated. In other words, apply takeoff power and get into a takeoff configuration, retracting the gear after you have begun to climb (just as in a normal takeoff) and there is no longer a possibility of touching down on the runway.

Prior to executing an instrument approach, the pilot should always review the approach plate and be familiar with the go-around or missed approach procedure. Then, if a missed approach procedure is required, it should be executed exactly like an aborted landing—a takeoff from minimum approach altitude and beginning at a speed at least 30% above stall in the landing configuration. Apply takeoff power, go to the go-around mode on your flight director for pitch attitude speed command if you have the equipment, retract the flaps to maximum takeoff setting, and retract the landing gear upon a positive rate of climb. Without a flight director and speed command, altitude and airspeed together are the primary instruments for

120

pitch control, and heading is primary for bank. The horizon becomes primary for pitch attitude after a positive rate of climb at a speed in excess of V_2 is attained.

After executing either the aborted landing or missed approach and retracting the gear, follow a normal flap retraction and climb schedule, and you will have accomplished the maneuver safely and correctly. Of course, don't forget to advise approach control or the tower of the missed approach on instruments.

For certification for Category II approaches you are required to make an approach with an engine out, flying manually instead of on autopilot, and execute a missed approach at the lowest altitude presently used as a minimum in Category II decision height (100′). The three-engine go-around in a four-engine aircraft is not very different from a four-engine go-around in performance capability, but a 50% power loss in a two-engine airplane is more critical. The reasons for not using full flaps in approach are the limited or nonexistent go-around capability with 50% power and the drag of full flaps. However, the speed in approach for a 50% power loss approach flap setting will be faster than the 1.3 over stall minimum approach speed with full flaps and at least 1.3 over stall for the appropriate approach flap setting. This should be considerably in excess of the takeoff safety speed of V_2, and acceleration will not be a problem.

Relate this maneuver to the takeoff with an engine failure at V_1 and a V_2 climb to the end of second-segment altitude. For execution you will, of course, fly the missed approach procedure. But the first consideration is the go-around. Do it like the normal missed approach; apply takeoff power to the good engine or engines, retract the flaps to takeoff, and retract the gear upon a positive rate of climb. Then climb at a speed no lower than V_2 (just like an engine failure on takeoff) to an altitude that is the same as the altitude for obstruction clearance or the end of second-segment climb, level off at that altitude and retract the flaps on a normal flap retraction schedule for your aircraft while holding altitude, continue to hold your level altitude until engine-out climb speed is reached, reduce power after beginning engine-out climb, and then call for the climb check and notify ATC of the missed approach.

The biggest difficulty in the engine-out miss is directional control. Don't forget to hold the wings level and use enough rudder to smoothly put the ball right in the center, and you'll fly straight out. It is almost impossible to fly the back course of the ILS during this phase of the go-around; a heading is recommended until you are far enough from the transmitter, at which time a correction of heading may be made to intercept the localizer if this is part of the missed approach procedure.

This method of relating the aborted landing and missed approach to a takeoff begun at a higher altitude than field level and at a higher speed than a takeoff from the runway makes their execution much easier to understand and perform. Just get takeoff power on the aircraft, get it into a normal takeoff configuration, and then fly as normal a takeoff path as possible; you'll accomplish the aborted landing or missed approach in the safest manner possible every time.

18

Category II and III Approaches

The ultimate goal of the air transport industry, all-weather capability of safe instrument approaches and landings under zero conditions, is a virtual reality. With the proper equipment it is possible to land with no ceiling and the visibility absolutely zero. Taxiing to the ramp, however, would be extremely difficult, and visibility limits are required.

The logical extension of the Category II approach to the Category III approach began in 1972 when the Lockheed L-1011 TriStar became the first aircraft to be certified for Category IIIA operation. On May 25, 1972, Tony LeVier flew an L-1011 from Palmdale, Calif., to Dulles Airport at Washington, D.C., with the autopilot flying the airplane from brake release on takeoff roll to the end of rollout on landing, and the 1011 become known as "the most intelligent jetliner ever to fly." It was the last airplane I was rated in, flying it for three years until retiring in 1983. It is the only aircraft I have ever made Category III approaches in, and it is quite an experience to monitor the systems in approach, watch the aircraft correct for runway alignment by going to runway heading with a slight slip with a wing down into the wind, flare, close throttles, touch down as soft as a feather, lower the nose gently, and see the runway for the first time just before the nosewheel touches the runway in the touchdown zone and right on the center line. It's something to look forward to.

As you can see, the limited visual cues associated with Category III make it necessary for the flight crew to have complete confidence in the autoland system. The crew must also have a thorough understanding of the system and monitor it from localizer capture through rollout.

Category II approaches are an extension of Category I, requiring a coupled approach using the autopilot to a decision height; Category III is an extension of Category II, requiring a coupled approach with an alert height with autoland capability. The decision height is the height above the airport where the pilot must establish visual reference with sufficient visibility to execute the landing. The alert height is an established height 100′ feet or less above the highest elevation in the touchdown zone and is based on the characteristics of the airplane and its particular airborne Category III system.

The procedures used in Category II and III approaches have such similarities that they may be combined in training and certification. Certain requirements must be met, and oral examination for rating or qualification in the the airplane will include questions related to both categories.

Oral Requirements

Federal Air Regulations require that an oral operational test be given a pilot qualifying for Category II or III operations. Without quoting the regulation verbatim, I will break it down to its essence. It states that the pilot applicant must demonstrate knowledge of:

1. Required landing distance.

2. Recognition of decision or alert height.

3. Missed approach procedures and techniques utilizing computed or fixed attitude guidance displays.

4. Runway visual range (RVR), its use and limitations.

5. Use of visual cues, their availability or limitations, and altitudes at which they are normally discernible at reduced RVR readings.

6. Procedures and techniques related to transi-

tion from nonvisual to visual flight during the final approach under reduced RVR.

7. Effects of vertical and horizontal wind shear.

8. Characteristics of the aircraft flight director system, automatic flight control system, autopilot(s), autoapproach coupler, autothrottle (if so equipped), and other required Category II or III equipment.

9. Characteristics and limitations of the ILS and runway lighting systems.

10. Assigned duties of the copilot or second-in-command during Category II or III approaches.

11. Instrument and equipment failure warning system and actions required in relation to where they may occur in the approach.

Due to the great variety of flight instruments, autopilots, automatic flight control systems, etc., it is impossible to provide answers other than in a general form based on one system. I will try to make the following comprehensive enough that it will provide information that may be related to any system approved for the approaches.

Pilot Qualification

Category II and III operation is essentially an airline operation for which the airline and its training program, the aircraft, and the flight crew must be certified and approved. This is not to say that the sophisticated business aircraft of today may not also be certified; they may. But to certify the crew requires a certified training program as well and, if it is provided, pilots other than airline pilots may be qualified.

The minimums for the approach depend on the particular airport as well as the aircraft. These vary, ranging from 1,600′ RVR with a decision height of 150′ to 1,200′ RVR and a decision height of 100′ for Category II. Category III limits are for approaches below 700′ RVR and restricted to 600′ RVR at present by transmissometer limitations in the United States. When transmissometers capable of reading in 100′ increments are installed, minimums will become 300′ RVR or less. Category II decision height and RVR minimums are denoted as RA (radio altimeter) for decision height and RVR for visibility limits on the appropriate approach plate.

Only the pilot in command is authorized to make any approach below basic minimums and, therefore, must execute the Category II or III approach. To do so, the pilot in command must have:

1. Completed the approved Category II and III training programs. Pilots in a particular air carrier aircraft meeting initial qualifications will be "high minimum" captains, restricted to 300′ and ¾-mile visibility minimums, 100′ and ¼ mile over the basic minimum of 200′ and ⅓ mile. The Category II and III requirements will qualify pilots for the use of these

minimums but they may not use them until becoming "low minimum" captains.

2. To become a low, or basic, minimum captain, requires a minimum of 300 hours as pilot in command of turbojet aircraft if the certification is for jets and a minimum of 100 hours as pilot in command in the aircraft used for the Category II and III approaches.

3. The copilot or first officer (I prefer first officer, but the second in command regardless of what the title may be) must also have completed the Category II and III training programs.

Required Equipment

Certain airborne flight guidance equipment is required for various categories of approaches. Their number and status of operation limit the operation of the aircraft differently for each category. It's impossible to describe all aircraft, and I will use the required equipment list for the Boeing 757. Other aircraft will require the same type and number of required equipment.

I will list equipment for Category I through Category IIIB, give the manner or autopilot mode in which they may be flown, and use abbreviations for certain items and notes for explanatory remarks (Table 18.1). The abbreviations are: ADI = attitude director indicator, AFDS = autopilot flight director systems, AH = alert height, APU = auxiliary power unit, ASA = autoland status annunciator, ATS = autothrottle system, DH = decision height, FD = flight director, IRU = inertial reference unit.

I have never flown the B-757. Some of my students have become flight instructor/check pilots in the aircraft and have given me information regarding its performance and capabilities. I have had the opportunity to ride the jump seat numerous times and have observed the accuracy of the navigation system and the autopilot's ability to fly the aircraft with great precision in every phase of flight. I have been told that its inertial reference system of navigation, flight management system, flight data computers, and control display units are identical to those in the space shuttle. At any rate, in my opinion the B-757 and B-767 are at least as "intelligent" as any airplane in the sky today, explaining their selection as models for Category II and III approaches. The 757 has three autopilots; one is required for Category II operation, the second (Land 2) for Category IIIA, and the third (Land 3) for Category IIIB approach and landing.

The ground navigation system (ILS) must be approved for Category II and III operation, which would be indicated by a Category II and IIIA approach plate, and the following elements of the ground navigation system should be operating normally:

TABLE 18.1. Automatic Flight Control System (AFDS) Flight Guidance Required Equipment

Required Equipment	Category I	Category II	Category IIIA	Category IIIB[a]
	Coupled/ Manual	Coupled	ASA land 2 700 RVR; 50′ DH	ASA land 3, 600– 300 RVR; 50′ AH
Autopilot	1/0	1	2	3
FD display	Capt.'s/2	2	2	2
Electronic ADI	2[b]	2[b]	2[b]	2[b]
ILS deviation	1	2[b]	2[b]	2[b]
Radio altimeter readout	0	1	2[b]	2[b]
AFDS mode annunciation	1	2	2[b]	2[b]
Missed approach guidance	Electronic ADI, FD pitch command, or automatic go-around			
Autoland status annunciator	0	0	2	2
IRU (in nav mode)	1	2	2[c]	3
ATS	w/wo	w/wo	w/wo	With
Rollout guidance	w/wo	w/wo	w/wo	With
Hydraulic system	C or R[d]	C or R[d]	3	3
Electrical power source	1	2[e]	2[e]	2[e]
Engines	1 or 2	2	2	2[f]
Windshield wipers	w/wo	With	With	With

Note: w/wo = with or without.

[a]If autobrakes are inoperative, antiskid and thrust reversers are required when RVR is below 600′.

[b]Must be supplied by separate symbol generators.

[c]Associated with engaged autopilot.

[d]The two pitch trim motors are hydraulically powered, one by the center (C) system and one by the right (R) system.

[e]The APU may be used as an independent power source.

[f]Two engines must be operative at AH; if an engine fails after AH, the approach may be continued.

1. All components of the ILS system.
2. High-intensity runway lights.
3. Standard approach lighting system. This system is 3,000′ in length, and the strobe light ("rabbit") should be (though not required) visible to the first officer at least 50′ prior to decision height (or at least prior to the actual decision height), and sufficient forward visibility is required at decision height to make a landing.
4. Sequenced flashing lights.
5. Touchdown zone and runway center-line lights.

There is another factor that may be restricting in the approach: a crosswind component varying from 10 to 25 knots depending on aircraft limitations, type of approach made, and the category of the approach facility used. The 757, for example, has the crosswind limitations for autoland as shown in Table 18.2. For all conditions, maximum allowable headwind is 25 knots and tailwind 15 knots. Other aircraft may be limited to 10 knots of tailwind and crosswind. You must know the limits for your particular aircraft.

The runway length for landing, at least 15% greater than that required by FARs, is the same as the requirement for a Category I or wet runway landing. This doesn't present a problem.

Category II

The Category II approach is approved *only* for the use of the automatic approach coupler, using one or more autopilots controlled and monitored by the pilot in command, except that the approach may be continued and flown manually using the flight director if the automatic approach coupler or autopilot malfunctions and is disengaged below 400′ above the elevation of the touchdown zone. However, unless stabilized on the ILS and glide path, I recommend that a missed approach be executed. It's a judgment item. In your rating ride you may be required to hand-fly a Category II approach with an engine out even though it may be prohibited in actual flight.

Multiple autopilot approach and autoland is rec-

TABLE 18.2. Maximum Allowable Crosswind for Autoland

	Crosswind Limits		
Weather Conditions	ASA Reads	Approach Facility	Crosswind Limit
Category I or better	Land 2 or 3	Category I–Category III	25 knots
Category I–Category III	Land 3	Category III	15 knots
Category I–Category III	Land 2	Category III	10 knots
Category I–Category III	Land 2 or 3	Category II	10 knots

ommended, provided that all required equipment for Category II autoland is available and visual reference is established at the decision height. However, the landing may be accomplished manually.

At the decision height, the aircraft should be in a position to continue for a normal approach and landing without requiring any excessive changes in heading, airspeed, or pitch attitude.

Sufficient training will be given covering all situations, and the certification for both Category II and III operations will be renewed by the pilot during required proficiency checks, semiannually for the pilot in command and annually for the first officer.

TRANSITION FROM NONVISUAL TO VISUAL APPROACH

With a decision height of 100′ in a Category II approach (measured by radio altimeter as an actual height above the touchdown zone) and a rate of descent varying according to the ground speed of the aircraft on the 2.5–3° glide slope from 500 to 700′ per minute, the time from visual contact with the runway to landing is rather short. There is a transition time of visual adjustment (refocusing of the eyes) from the instrument panel to the forward visibility required for depth perception for the landing. This visual adjustment time varies but averages about 4 seconds. An aircraft with a rate of descent of 500′ per minute at 100′ above the runway requires 12 seconds to touchdown, a 600′ per minute rate of descent requires 10 seconds, and a 700′ per minute descent requires 8.3 seconds. At the decision height, where the autopilot is normally disengaged in the Category II approach, the aircraft must be on the proper speed, attitude, and rate of descent; in trim when the autopilot is disengaged; and properly aligned with the runway to execute the landing.

The aircraft will continue to descend (if it is in proper trim and you make no large corrections, which you should not do) one-third to one-half its actual height above the runway before your eyes are properly readjusted to the outside world for reliable visual reference and depth perception to make the landing. You will descend 30–50′, or very nearly to the flare point of the beginning of the landing, before you can rely on what your eyes tell you.

It is important, therefore, that you be aware of the maximum deviation tolerances allowable from both the localizer and glide slope and from the scheduled approach speed upon reaching the minimum decision height. You must also depend greatly on the report of the first officer, who should be looking for visual cues (beginning at least 100′ before decision height) and approach lights in relation to runway alignment and when you will first establish ground contact. You should know from your instrument readings when to execute the missed approach upon reaching the deci-

sion height. If the readings are not within the limits of allowable deviation, you should initiate a go-around without even looking out. However, if the instrument readings are within tolerance, you should not make any large changes in power, attitude, or direction of the aircraft until you have had sufficient time to refocus your vision to execute the landing. The most common mistake made is correcting for crosswind in aircraft that are not certified for Category III and do not have runway alignment capability. The aircraft will be crabbing into the wind, and the pilot has a tendency to turn to runway heading; the aircraft, without drift correction, will begin to drift away from the runway center line. Hold your heading, it will keep you on track; do not make an adjustment until visual accommodation has been made.

APPROACH PROCEDURE

It is very difficult to recommend a procedure for the flight portion of the Category II approach without using a specific aircraft as a basis. However, in general, the following would apply to all aircraft:

1. The descent and in-range checklist should be complete.

2. The approach plate should be reviewed and an approach briefing accomplished.

3. Test all radios to be used by selecting and identifying all navigational facilities to be used in the approach.

4. Test your instrument warning system.

5. Test the radio altimeter and note that it indicates 100′ ± 5′, the proper warning flags appear in the radio altimeter, and the decision height lights on the radio altimeter and flight director instrument come on. Then set the bug on the radio altimeter, as indicated on the Category II approach chart, for the proper decision height.

6. When descending below 2,500′ above the terrain, note that the radio altimeter pointer begins to function.

7. When being vectored for the ILS approach, the flight director control panel selector should be in the heading mode.

8. You should fly a normal ILS procedure, using normal recommended airspeed. The gear extension and flap management schedule should be normal for an ILS approach.

9. If not on autopilot throughout the descent, vectoring and maneuvering to the localizer, I recommend that the autopilot be engaged and in use for the approach prior to the procedure turn or an equivalent position if being vectored.

10. The autopilot should be flown in the heading mode until after establishing a position inbound from the procedure turn or equivalent position, and the autopilot mode selector should not be changed to the

automatic approach mode until cleared for the approach. After selecting the automatic approach mode (Auto/GS mode), note that the autopilot mode indicator (an annunciator panel with lights) indicates both the localizer and glide slope as armed.

11. After selecting the automatic approach mode on the autopilot, the flight director control selector should be switched to Approach/Auto. Then check the flight director mode indicator for an indication that the heading mode is still in use, the command bars are still referenced from the heading bug, and the NAV/LOC and GS (glide slope) modes are armed by noting the proper indicating lights on the panel.

12. By this time both pilots' flight directors should be in the same mode and both navigation sets tuned and set up for the ILS.

13. The approach should be planned and executed so that localizer intercept will be 3–5 miles outside the outer marker to allow the approach coupler to compute crosswind correction prior to intercepting the glide slope.

14. Upon localizer interception, check the autopilot mode indicator readout to verify that you have a localizer capture by the autopilot; verify that you have a capture by the flight director. At this point the heading bug of the flight director should be changed to "missed approach" heading. Heading information would then be available for the flight director in the event of a missed approach and after attaining a safe altitude in the go-around mode.

15. Upon glide slope interception, check first the intercept by the autopilot (after all, it's flying the airplane), and second the GS/cap (glide slope capture) should be checked by both flight directors.

16. By using a normal flap management schedule prior to glide slope intercept, the aircraft should be at or very near the programmed approach speed. Try to keep throttle adjustments to the minimum and in small increments when on the glide slope to minimize pitch compensation required of the autopilot.

17. It is desirable to intercept the glide slope at a higher altitude than the crossing altitude at the final approach fix to have the landing configuration and "before landing final check" accomplished prior to reaching the final approach fix. If approaching the final approach fix at glide slope intercept at that point, use the gear and flap management recommended in Chapter 16.

18. When crossing the outer marker, check altitude, airspeed, and rate of descent on the glide slope and make a "flag scan" of the instrument failure warning system and the autopilot and flight director annunciator panels. Complete the final check if necessary. From this point on, any improper annunciator light in either the flight director system or autopilot modes should receive the same consideration as a red warning flag.

19. After passing the outer marker for the coupled approach, the pilot should keep the left hand on the control wheel and thumb in position to hit the disconnect button to disengage the autopilot in the event of malfunction.

20. At 700', the average halfway point (see Chapter 16), the aircraft must be tracking within ½ dot of the localizer and within ¾ dot of the glide slope. This is a reference position taken from raw data and not computed intercepts taken from the flight director.

21. After noting the 700' deviations, there should be a 500' check consisting of altitude, airspeed, and rate of descent (in that order) and an instrument warning flag and autopilot and flight director mode scan.

22. The middle marker should trigger the glide slope extend mode, so check that both the autopilot and flight director systems are operating in the proper mode.

23. When reading decision height (indicated by lights on the radio altimeter and flight director), cross-check altitude by the barometric altimeter and check deviation from the localizer and glide slope. (The flight director may now be considered as depicting actual position on the localizer and glide slope rather than computed intercept commands).

a. If the decision is to land, disconnect the autopilot with the button on the control wheel and complete a normal landing.

b. If the decision is to go around, disconnect the autopilot and execute the published missed approach procedure. If you have a flight director with go-around attitude guidance (such as the FD-109), hit the go-around mode (GA light on the flight director should appear) either with the throttle palm switch, if so equipped, or by selecting the go-around mode with the flight director control panel. With this system the go-around annunciator illuminates, and the V-bars (command bars) command a pitch-up for the optimum climb-out speed. To ensure maximum lift, the command bars will always call for a "wings level" attitude. This mode should be used only until a safe climbing attitude, configuration, and speed are attained; then the flight director should be switched to heading mode so that heading information will be available in the missed approach as required.

24. A missed approach shall be initiated when (a) the final check of the localizer and glide slope at reaching the decision height indicates a deviation of more than ⅓ dot on the localizer or 1 dot on the glide slope; (b) airspeed is within 5 knots of programmed approach speed but never less than V_{ref} or flare speed; (c) adequate visual reference has not been established upon reaching the authorized decision height; (d) the aircraft has cross-tracking velocity or is tracking in such a manner as not to remain within the confines of the runway as extended; (e) a touchdown cannot be

accomplished within the touchdown zone of the runway (first 3,000'); (f) the aircraft is not in trim at auto-pilot-disconnect to permit normal descent and approach to flare and landing; (g) there is any unusual roughness or excessive attitude change (due to air-borne or ground equipment) occurring after passing the middle marker; (h) any of the elements of the ground navigation system become inoperative during the approach; (i) any of the required airborne equipment becomes inoperative during the approach, except that the approach may be continued manually using the flight director system if the autopilot (or approach coupler) malfunctions and is disengaged anywhere below 400' above the touchdown zone.

These procedures are rather general and may be used for virtually any flight director system and autopilot with coupled approach capability, but I recommend that you be familiar with your particular instrumentation and autopilot's operational capabilities and limitations.

The duties of the first officer are normal duties for any ILS approach, but particular emphasis should be given to immediate compliance with the captain's directives. These duties include:

1. Handling all necessary radio communications with ATC, approach control, and the tower.

2. Tuning and identifying navigational radio as directed and requested.

3. Properly executing all appropriate checklists.

4. At the outer marker, performing a flag scan to include autopilot and flight director annunciator lights, calling out, "Outer marker, no flags," to verify that all mode lights are correct and there are no warning flags. Or, conversely, calling out any warning flags that appear.

5. Monitoring the flight director and approach coupler with the captain and cross-checking both flight directors on approach.

6. Monitoring the instrument failure warning system.

7. Calling out significant departures from the localizer and/or glide slope.

8. At 500' above field elevation, calling out the flag scan, altitude, airspeed, and rate of descent in that order to preclude misunderstanding. (For example, "No flags, 500', 135 knots, 700' per minute.") Thereafter, call out significant deviations from airspeed and desired descent rate.

9. At the middle marker, calling out, "Extend mode." If it does not trigger, manually select and then start looking for the runway, approach lights, etc.

10. Advising the captain when approach lights and/or runway are in sight (straight ahead, left, or right).

11. If it is a missed approach, handling flaps, gear, and power as the captain may direct. Monitor the flight director system for the go-around mode, tune radios, and monitor the flight director as required for a missed approach and handle appropriate radio communications.

In general, these duties also apply to Category III approaches.

Category III

The limited visual cues associated with Category III make it necessary for the flight crew to have complete confidence in the autoland system. The autoland system must be monitored from localizer capture through rollout.

During Category III approaches the captain and first officer will continually monitor the flight instruments. The captain will be head down and rely primarily on flight instruments through touchdown until rollout.

The captain's decision to land or go around will be based on continual monitoring of the approach through flight instruments. Misalignment or excursions from the final approach to flare will alert the captain to the possibility of having to execute a missed approach.

Category III does more than reduce landing minimums below 1,200 RVR; it requires the flight crew to depend on the autoland system. Both pilots become computer monitors. They observe the various changes in annunciations while monitoring the progress of the airplane throughout the approach landing and rollout by use of flight instruments.

Airplane position, attitude, altitude, speed, etc., can be determined more accurately by referring to the associated instruments than by limited visual cues available with Category III RVR. The airplane is automatically controlled in yaw, roll, and pitch axes. The cues indicating change in these controls and flight parameters will be recognized and evaluated more accurately by flight instruments.

Category III below 700' RVR requires autoland with Land 3 capability—all three autopilots functioning. The pilots are required to monitor the cockpit displays once they have programmed the autoland system. Visual cues at alert height are not adequate to manually control the airplane. The captain should monitor the flight instruments from the alert height though the rollout until the speed has decreased to where the airplane may be controlled visually. During rollout, the expanded localizer is of primary importance in assessing (it is controlling heading) the performance of the autorollout system in combination with the visual cues that are available. (Transmissometer limitations in the United States restrict operation to 600'. When transmissometers capable of reading in 100' increments are installed, minimums will be 300' RVR.)

Category III approach and landing may be accomplished with Land 2 capability if the RVR is 700′ or above. In this case, the autoland system is in the fail-passive mode; therefore, a decision height of 50′ is required. At 50′ the first officer must call out, "Runway," or a missed approach must be executed.

During an approach the pitch attitude of the B-757 will be approximately 2.2° nose up. The slant visual range at 50′ may be about the same as the reported RVR. Therefore, the threshold and possibly the touchdown zone lights may be visible. However, in dense fog, the slant visual range (SVR) and the resultant visual segment are significantly less. For example, the SVR with 600′ RVR at 50′ will be approximately 450′. (See Fig. 18.1.)

Visual cues improve as the airplane reaches the three-point attitude after touchdown. The B-757, for example, has a windshield cutoff angle of 23° below the optimum eye level position. Other aircraft may have different angles. (See Fig. 18.2.)

The touchdown zone lights extend 3,000′ past the threshold. At touchdown speeds those lights will be visible only for a few seconds, then single center-line lights will continue with 50′ spacing. At 3,000′ from the far end, the center-line lights become alternately red and white. During the last 1,000′ from the runway end the lights will be all red. If the touchdown zone and/or the center-line lights are inoperative or obscured, a Category III landing may not be accomplished.

Automatic brakes, if available, will be used at a minimum setting. In the unlikely event that the airplane enters the last 3,000′ of runway (identified by the alternate red and white center-line lights) above 80 knots, the pilot should revert to manual maximum braking.

Category III approaches must be accomplished with Land 3 or Land 2. Category III landings require: Land 3 capability at alert height of 50′ radio altitude when the RVR is reported below 700′ or Land 2 capability at 50′ radio decision height when the RVR is reported at 700′ or above.

During an autoland approach in the B-757 the stabilizer is trimmed approximately three units nose-up trim in addition to that normally required. This extra trim is easily compensated for by the autopilots and is to protect against a hard landing if they should disconnect during the flare maneuver. The trim is added below 330′ radio altitude. If a manual missed approach is initiated below this point, the pilot must be prepared to counteract the nose-up trim forces, which requires about 30 lb. of stick force when go-around thrust is applied. This may not be true in other aircraft.

INSTRUMENT APPROACHES

The captain will make all approaches and landings in weather below basic minimums. Autoland will be used for all Category III approaches and landings.

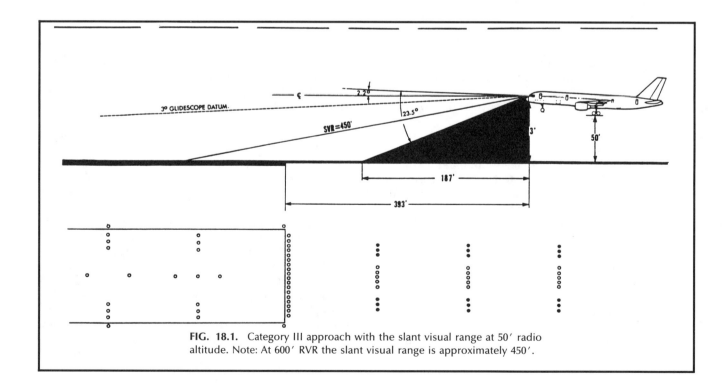

FIG. 18.1.   Category III approach with the slant visual range at 50′ radio altitude. Note: At 600′ RVR the slant visual range is approximately 450′.

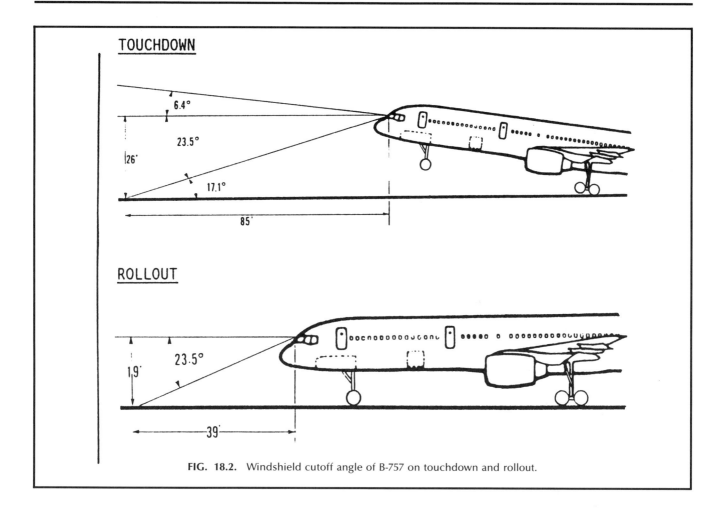

FIG. 18.2. Windshield cutoff angle of B-757 on touchdown and rollout.

All autopilot approaches should be flown so as to intercept the localizer at least 5 miles outside the outer marker with an intercept angle of less than 45°.

When the autopilot is controlling the airplane, particular attention should be used in monitoring the progress of the approach through the display of the attitude director indicator or flight director and the horizontal situation indicator.

PILOT DUTIES

The captain will program the approach or give the appropriate commands to the first officer to select the proper frequencies, speeds, headings, etc., required to set up and monitor the approach. The radio altimeters will be set to the appropriate decision height for Category II or 50′ to identify the alert height and/or decision height.

While monitoring the approach, both pilots will constantly compare attitude direction indicator commands, localizer and glide slope positions, and the rate of descent required to maintain the glide slope.

This awareness will ensure early recognition of wind shear conditions during the approach. The captain will keep the left hand on the wheel and right hand on the throttles and be ready to disconnect and take over manual control of the airplane if it becomes necessary.

The captain will call out the following information: localizer capture, glide slope capture, altitude crossing the outer marker, Land 2 or Land 3 as appropriate on the autoland status annunciator, flare and rollout arm, 1,000′ above touchdown, 100′ above touchdown, and rollout. (At 500′ radio altitude the align mode is engaged but not annunciated in the B-757. Other aircraft, such as the L-1011, with only Land 2 capability may not align until flare.)

FIRST OFFICER

The first officer will compare annunciations with the captain, respond to callouts, and remain alert for warnings. During the final stages of the approach the first officer will advise the captain of cues observed.

At 150′ and 100′ above the alert height/decision

height the first officer will adjust the scan to include the runway environment and compare cues with instruments required to monitor the approach. As prime monitor of visual cues during touchdown and rollout, the first officer will call out any deviations from normal. The color-coded runway center-line lights (last 3,000') should be visible with runway visual range as low as 300'.

The first officer calls out 100' and 50' radio altitudes during the final descent.

MISSED APPROACH

During the critical missed approach period the airplane must be rotated from a descending to a climbing attitude. The B-757 (and most other airplanes approved for Category III) has the capability for autogo-around with the autopilot controlling the airplane and flight director visual cues for monitoring or the capability for pilot-controlled manual go-around following flight director pitch commands, which are controlled by existing or selected airspeed, whichever is higher. In either case, the go-around mode is initiated by pushing the GO-AROUND switch.

The following are causes for executing a missed approach during autoland operations at Category III minimums:

1. Loss of Land 3 capability as indicated by the autoland status annunciator and runway visual range below 700'.

2. Loss of Land 2 capability indicated by the autoland status annunciator with RVR below 1,200 down to 700' RVR.

3. Loss of the ground station signal.

4. No visual contact with the runway at 50' radio altitude when conducting a Land 2 (700' RVR and above) Category III approach.

ALERT HEIGHT

Item 4 for a missed approach as stated above is based on an alert height, which is defined as an established height 100' or less above the highest elevation of the touchdown zone. It is based on the characteristics of the airplane and the particular airborne category system. If a failure occurs in one of the required redundant operational systems (in airplane or ground equipment) above the alert height, the Category III approach will be discontinued and a missed approach executed.

The alert height for B-757 Category III operations is established at 50' so as to coincide with 50' decision height for the fail-passive case. If Land 3 exists at 50', the approach may be continued with runway visual range below 700'. If Land 2 exists, the approach will be discontinued unless the runway environment is in sight.

AUTOMATIC FLIGHT CONTROL SYSTEM (AFCS)

In the new aircraft of today it is recommended that the AFCS should be used throughout the flight, either in manual (FD, or flight director) or in the automatic (CMD, or command) mode. And since Category II and III approaches are "coupled" approaches requiring the use of from one to three autopilots, a brief review of their capability and use is appropriate.

The pilot flying should use the system associated with the particular crew member position; the captain uses the left autopilot and flight director; the first officer uses the right autopilot and flight director.

When making any selection on the AFCS mode control panel, update the following as required: heading/bug/window to the desired heading, command airspeed bug/IAS-MACH window to the desired airspeed/Mach, and altitude window to the desired altitude.

The AFCS modes can be engaged by pressing the respective switch. A light bar in the lower half of each switch illuminates to indicate that the mode has been requested. Mode engagement is indicated by flight mode displays on the attitude director indicator. The AFCS will operate in only one pitch mode and one roll mode at a time.

When operating with the autopilot in the CMD mode, the pilot flying will program the AFCS as necessary to achieve the desired task or performance. When flying the airplane manually with the flight director engaged, the pilot not flying will select the necessary modes when called for by the pilot flying.

AUTOTHROTTLE SYSTEM

The autothrottle system (ATS) reduces pilot workload and improves the accuracy of the parameters to be maintained. The ATS operates in one of six modes: EPR, SPD, FL CH, GA, IDLE, or THR HOLD.

1. EPR mode—When used during takeoff, the ATS will adjust throttles to the computed maximum engine pressure ratio for the takeoff mode selected (normal or derated) prior to 80 knots. When used in climb, cruise, or for maximum continuous thrust operation, the ATS will acquire and maintain the EPR for the mode selected.

2. SPD mode—The speed mode is used to maintain a desired airspeed in cruise and holding patterns, reduce airspeed and maintain desired maneuvering speed on approach, and maintain an accurate target speed.

3. FL CH mode—The flight level change mode is used to climb or descend to a new altitude. It holds climb or descent thrust, and ATS SPD mode automatically engages when altitude is automatically captured.

4. GA mode—The ATS advances the throttles to-

ward the go-around thrust limit, not to exceed an air-plane climb rate of 2,000′ per minute.

5. IDLE mode—The ATS is reducing or has reduced the thrust to idle. It automatically engages during FL CH and certain VNAV (vertical nav) descents and in approach at 45′ radio altitude in preparation for autolanding.

6. THR HOLD mode—Throttles will remain where manually repositioned with the ATS engaged. It automatically engages during takeoff with speed of more that 80 knots or with FL CH and VNAV descents after the throttles have been manually moved.

FLIGHT WARNING GUIDANCE

The use of AFCS and ATS in approach requires constant monitoring for failure warning and certain corrective action to be taken in the event of malfunction. The following corrective actions are recommended in the event of certain AFCS malfunctions. However, if doubt exists as to the system status or capability, execute a missed approach and investigate.

Failure Warning, ASA (autoland status annunciation)

Category II—If Land 2 loss (NO AUTOLND appears) occurs during autoland approach, attempt to restore Land 2 capability. If unable to restore Land 2, continue coupled approach to Category II minimums. At decision height, if runway environment is in sight, land manually. If runway environment is not in sight, execute a missed approach.

Failure Warning, A/P DISC

When loss of the autopilot occurs during a coupled approach or loss of all autopilots occurs during autoland (the red A/P DISC light comes on), attempt to restore the system to original capability if *more than* 400′ above the touchdown zone. If no autopilots can be restored, execute a missed approach. If an autopilot is restored, continue the coupled approach to decision height. If *less than* 400′ above the touchdown zone, execute a missed approach.

Category III—No autoland: execute a missed approach. No Land 3 with Land 2: Attempt to restore Land 3 capability. If unable to restore Land 3, continue Land 2 approach to 50′ decision height. At decision height, if runway environment is in sight, complete the landing. If runway environment is not in sight, execute a missed approach.

Failure Warning, A/T DISC

Categories I and II—If unable to reengage the autothrottles, continue approach to decision height, manually controlling the throttles.

Category III—Attempt to reengage the autothrottles. If unable to do so, continue the approach to 50′ decision height, manually controlling the throttles. At decision height, if the runway is in sight, complete the landing. If the runway is not is sight, execute a missed approach.

Failure Warning, ADC (air data computers) Failure

Categories I and II—With the loss of both ADCs the autothrottle will be lost. However, with the loss of one or both ADCs continue the approach to decision height.

Category III—With the loss of one or both ADCs, continue the approach to 50′ decision height. At decision height, if you can see the runway, land; if not, execute a missed approach.

19

Nonprecision and Circling Approaches

Tools, skills, and talents that are seldom used grow rusty. Professionals in every field (the performing arts, education, medicine, music, law, etc.) constantly practice to maintain their skills and study to retain and increase their knowledge. The pilot flying air transport aircraft, the airline pilot, and the corporate pilot flying complex and sophisticated business aircraft all practice a profession requiring a high level of skill and proficiency. They undergo a great deal of training to obtain their licenses and ratings. Unlike the other professions, in this one they are also examined and checked frequently to prove that their proficiency, skills, and knowledge are of such a caliber that people's lives as well as multimillion dollar equipment may be safely placed in their hands. But there is one phase of their skill that they rarely get to practice—the nonprecision approach.

Due to infrequent use of nonprecision approaches in actual flight operation, pilots get a bit rusty in their execution. Nonprecision approaches are a part of flight training, proficiency, and rating checks; this is now all usually done in simulators. The simulator is an excellent tool for this purpose, but it has one characteristic that does not exist in actual flight. Its ceiling, visibility, and wind, when set for an approach, remain constant throughout. It is appropriate, therefore, to discuss all the various approaches to permit the pilot to review the problems, techniques, and factors to be considered in execution of nonprecision approaches, both in actual flight conditions and simulator training.

The newer aircraft navigation systems and autopilots make nonprecision approaches a simple matter, enable the pilot to use the autopilot and autothrottles

to reduce crew workload, and allow more time for management and monitoring of the approach. If the autopilot is to conduct the nonprecision approach, it must be disconnected prior to descending below minimum descent altitude (MDA), where the pilot takes over. Sufficient training, however, will be given to ensure that the pilot can successfully conduct such an approach if required. You may expect to be required to hand-fly one (or more) nonprecision approach in any check.

The inertial reference system and the flight management system allow selection of approaches through the control display unit, giving the pilot a complete picture of the procedure on the attitude director indicator and horizontal situation indicator. Any approach *not* stored in the data base is to be flown using raw data. Still available in these systems, at the pilot's fingertips, are wind direction and velocity, true airspeed, ground speed, drift angle, etc., when flying raw data. It is the use of raw data we will consider in this discussion, since most aircraft do not have the above systems.

Wind and Weather

In every approach, in the simulator or actual flight, the pilot should always have the existing weather information to ensure that minimums in fact are present. The pilot should also pay particular attention to the wind. The reported surface wind is the base upon which to make an estimate of ground speed in the approach.

"Wind is an important factor in flight. The direction, velocity, and characteristics of air moving horizontally determine takeoff and landing procedures and are an *indication* of winds the pilot may expect to encounter aloft." Surely you remember reading that statement in the study of basic meteorology.

Without going into a lengthy discussion of high- and low-pressure systems, I will simply quote from the meteorology study you did. Low-pressure areas: "Because of Coriolis force, the inward movement of the air is *deflected to its right*." High-pressure areas: "Air starts to move outward and is *deflected to its right*." Then, in discussion of surface weather map analysis, "Isobars on the surface weather map are *usually* representative of the wind flow up to 2,000' above the surface." At the surface, the wind flows generally at a small angle (23° approximately) to the isobars, due to surface frictional influences.

These changes vary, of course, in relation to the station, to the center of the pressure system, and the distance between the isobars. I merely want to prove that the wind at the final approach fix will virtually always be different from the wind on the ground and will constantly change in the approach. The reported wind, affected by surface friction, may be valid to 300' above the airport, but the wind in the approach will be different, both in direction and velocity.

With airborne equipment that provides wind information aloft, I found that I could take the surface reported wind and the wind 2,000' above the airport, average their direction and velocity, and have a reasonably accurate analysis of the wind to be encountered in the approach. Or, if the wind at 2,000' feet wasn't available, I could, as a rule of thumb, double the velocity of the surface wind, move its direction 23° to the right, and have almost equally good results. In both cases this was much better than just estimating my ground speed and drift in the approach based on surface winds alone. The practical application of this is particularly useful in the nondirectional beacon and back-course approach.

Definition and Categories

A nonprecision approach is any instrument approach made without vertical guidance, that is, a glide slope. It requires more preplanning. The pilot must plan the vertical flight path and, in some instances, also consider the effect of wind on the approach flight path, particularly in the automatic direction finder and back-course approach. The pilot that does so has little difficulty in approaches and executes them with a high degree of accuracy and safety.

Approach minimums are based on aircraft category, and the pilot should know the category the aircraft falls into. Aircraft categories are (in the United States) a grouping of aircraft based on their approach

speed at maximum certificated landing weight. The categories are: (1) 91 knots or less, (2) 91–120 knots, (3) 121–140 knots, (4) 141–165 knots, and (5) 166 knots or more. Category 5 contains only certain military aircraft and is not shown on civil approach plates.

General Recommendations

When a nonprecision approach is to be made, the pilot should review the approach plate, study the approach, and plan how it is to be accomplished. I recommend that the following should be decided upon well in advance of arrival over the final fix:

1. The landing gear will be extended and the final checklist completed *prior* to the start of the final descent to MDA or to landing.

2. Select flap configuration and estimate ground speed and drift. If the missed approach point (MAP) is determined only by time (that is, no distance measuring equipment or other radio facility for MAP resolution), landing flaps should be extended just before reaching final approach fix (FAF). This will enable the final approach speed to be stabilized when the descent is started and thereby result in a more accurate MAP.

3. Some nonprecision approaches utilize a visual descent point (VDP) reference. VDP is defined as a point on the final approach course from which a normal descent from MDA to touchdown may be commenced, provided the required visual reference is established. The VDP is usually established by distance measuring equipment (DME) or a marker and normally provides for an approximately 3° descent path. Reaching the VDP prior to MDA should alert the pilot to the possibility of a missed approach.

4. Use the estimated ground speed and determine the time from final approach fix to MAP. This is indicated on the bottom left-hand corner of the approach plate in relation to various ground speeds. However, missed approach points are not always determined by time. They are sometimes established by DME, a marker, or a VOR radial. In any case, the time to the MAP is an important factor in establishing a target rate of descent. However, I recommend descending at *1,000' per minute* in all nonprecision approaches from final approach fix to MDA.

5. The pilot not flying should keep track of the time from FAF to MAP and advise when reaching MAP. I recommend that the pilot keeping the time monitor it in such a manner that it is possible, upon request, to inform the other pilot of the time remaining rather than elapsed time in the approach. The pilot requesting time, with 2 minutes 18 seconds estimated for the approach, is much better informed when told, "1 minute 33 seconds to go," than if told, "45 seconds elapsed time."

6. Excellent electronic navigation calculators are

available that take only seconds to compute true airspeed, wind vector solutions of ground speed, time to missed approach, and drift angle in the approach. I strongly recommend their use.

Recommendations for each type of nonprecision approach follow.

Localizer No-Glide Slope

The course information is displayed on the flight director. When ILS DME fixes are needed during the approach, they should be monitored and called out by the pilot not flying. The effect of the wind in the approach in estimating or computing ground speed is all that is necessary to determine the time to missed approach and descent rate, since reliable course information is provided by the ILS. It is a relatively simple and easy approach to execute.

VOR Approaches

VOR approaches have course-line information, may be flown to or from a VOR station with or without DME, and often involve step-down fixes determined either by DME or cross-radials from another VOR. Ground speed for the time to missed approach is not as important when DME from the VOR providing the inbound radial determines the MAP; in other cases it may be. The pilot duties and responsibilities vary somewhat, however, in relation to cross-fixes and how they are identified.

When step-down fixes are determined by radials from another VOR, they must be verified by reference to appropriate raw data information. The pilot flying the approach should use the flight director for course alignment. The pilot not flying should tune, establish, and verify cross-radials from raw data, calling out, "Descend," to whatever the next cross-radial altitude may be, and set up for the next fix.

With DME from the inbound radial determining the MAP, both pilots should be set up on the flight director; the pilot not flying should call out DME fixes and the next DME fix and its altitude in the same manner.

If cross-radials are not required except for the final approach fix and DME is not applicable, the pilot not flying should call out the arrival at the fix. Then that pilot should tune the VOR to the approach VOR and be in the same NAV mode as the pilot flying the approach, making descent callouts as required (usually at 1,000', 100' above MDA), keeping time in the approach, and being prepared to give the time remaining to the MAP upon request. In this case ground speed establishes the time to the MAP. I recommend that a time to the MAP be computed in *every* case.

To illustrate the manner in which VOR approaches using step-down fixes should be flown, look at Figure 19.1. I've seen this approach used many times, both in the simulator and in the aircraft for training and check rides. It appears to be a simple procedure, but it requires some preplanning.

The weather is 500' measured overcast, wind from 160° at 10 knots, visibility 1½ miles. You're flying a category C simulator with a V_{ref} of 130 knots and will use 135 knots indicated airspeed in the approach. With this wind, which remains constant in the simulator, the ground speed will be approximately 124 knots, and there will be a 2° drift correction. The DME is out of service; cross-radials must be used for step-down and time for the MAP.

The maneuver speed for the max landing flap configuration may be flown to the FAF, requiring a high rate of descent from HONDA to MEELY, descending from 3,000' to 2,100'; 1,000' per minute will accomplish it with only a few seconds to spare. I would advise going to landing flaps immediately upon leveling at 2,100', reducing to approach speed, and descending to 1,500' to cross HORSS, with a rate of descent in excess of 600' per minute but less than 1,000' per minute. From HORSS to the MAP is only 4 miles. You want to be at MDA at or before the visibility limit (1½ miles), and the descent should be made at 1,000' per minute.

Back Course ILS

The back course ILS approach, while it provides course information, is much more difficult to fly in course alignment. Look at any LOC (Back CRS) approach plate and pay particular attention to the narrowness of the localizer at its point of origin, the transmitter. It is no wider than the antenna at the transmitter, maybe 10–20' (I've never really made an effort to determine the actual width of an antenna but know that it is relatively narrow.)

If you measured the angle of divergence as depicted on an ILS course, you would find it to be very near 3° and adjusted so as to be slightly more than runway width at the threshold of the *front course* runway. This adjustment is not present in the back course, although it still is 3° on each side from the center line. If 1° equals 1 mile in 60, 1/60 of a degree or about 6,080', it follows that each degree of divergence will equal 101.3' per mile, and that times 3 = about 304'. Therefore, when first reaching a full-scale deflection at a final approach fix 5 miles out, the aircraft would be approximately 1,520' from the center line.

Course alignment is still relatively easy at that point, but I'll assure you that it becomes increasingly more difficult as the approach progresses and the localizer narrows if you attempt to fly the approach

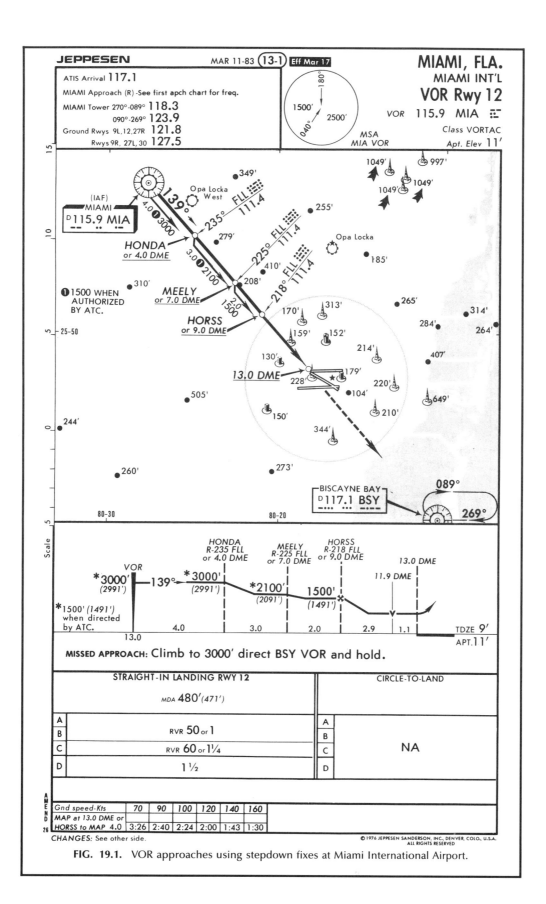

FIG. 19.1. VOR approaches using stepdown fixes at Miami International Airport.

with localizer reference only. I've observed a great number of approaches in the simulator where the wind remains constant that attempt to do so and miss the approach. Pilots have estimated ground speed and time to missed approach, but failed to compute a drift angle and heading to fly during descent to MAP; they drifted beyond full-scale deflection and overcorrected or were correcting in the wrong direction because of the characteristics of the back course ILS.

I recommend flying a constant heading, based on a computed or estimated drift angle, and using the ILS indicator as a *backup* to try to stay within less than full-scale deviation to MDA. If you can stay within full-scale deviation, you will be 304′ from runway alignment when visual reference is achieved 1 mile from the runway.

Automatic Direction Finder (ADF) or Nondirectional Beacon (NDB) Approach

The NDB or ADF approach seems to be the most difficult of all approach procedures for most pilots. The directional information isn't as accurately defined and the approach may be from a beacon at the FAF, a beacon positioned at some point on the field, or a beacon located at a position off the field. The latter two cases are generally never the case at major airports; but, in any case, maintaining the desired course involves the process of "tracking," flying a heading that will keep the ADF needles indicating the desired track.

You must track to the transmitter, not use it as a "homing" station. Homing is simply keeping the needle pointing to the station, right on the nose or heading. It will take you to the station; but without regard to the effect of wind, your heading will change and you cannot maintain a desired course or track. It is easy to "home" on a station, but flying a desired "track" is a different and much more difficult procedure. Tracking is apparently a lost art. An ADF approach is a tracking procedure, and tracking should be reviewed for better understanding of the manner in which it is best accomplished.

When tracking, the position of the aircraft in relation to the desired track or course is depicted visually by the radio magnetic deviation indicator (RMDI) needles. If there was absolutely no wind, it would be simple; make the RMDI indicate the desired track, with the heading the same. But winds do exist, and the problem becomes to establish a heading that keeps the aircraft on the desired track regardless of wind direction and velocity.

To see how it works tracking inbound, suppose the inbound course is 90° and, when turning inbound for the approach, you turn to 90° and that heading places the RMDI needles on 90° to the marker beacon. Holding this heading, a crosswind is clearly indicated

when the needle moves to the left or right. When the change becomes distinct (2–5°), make a turn in the direction the needle is moving for a correction in order to reintercept the desired track. If the needle moves left, turn left to intercept; if the needle moves right, turn right. The angle of interception should be toward the desired track and *must always* be greater than the number of degrees drifted. How far you turn to reintercept depends on distance from the station, true airspeed, wind, and how quickly you want to return to the course. Since you are usually close to the station in approach, doubling the deviation from the track is usually sufficient.

The needle of the radio compass will always point to the station on the RMDI, indicating the present track to the station; the same instrument will also indicate the aircraft heading, showing the degrees difference between the present track and the desired track. You should begin tracking when turning inbound from a procedure turn or when cleared for the approach from a vector. Try to nail down the heading to maintain the desired inbound track as quickly as possible, note your heading to see the drift angle, and try to be right on track when crossing the beacon.

Do not use the ADF needles immediately after passing the final approach fix; fly your computed or estimated heading for the track for one-third the time it will take in descent (usually 20–30 seconds), check the needles at that point, and correct the heading as necessary if track correction is needed. Remember, after passing the beacon you are tracking *outbound* from it. Do not turn *to* the tail of the needle; turn away from it. The needles are pointing toward the station, indicating the track back to it. The tail of the needle indicates the reciprocal, your track away from the station. Pull it in the direction you want it to go.

The approach can be flown using the needles for course alignment only, of course, but the estimation of wind and its effect is also important to compute ground speed, time to MAP, and a wind angle for course. Many pilots, however, estimate their ground speed and heading, fly to the beacon, and turn to their estimated heading and fly it to MDA; that works too. It is much better than turning toward the tail of the needles after passing the beacon.

There are several acceptable methods to estimate drift angle. You may fly the heading that maintained the inbound track to the beacon if you established it; you may compute an average wind from the altitude of the FAF to that reported if you have navigational equipment that will give you wind direction and velocity in flight; you may turn the surface wind right 23°, double its velocity, and average it with the reported wind and use that; or as many pilots do, you may simply use the reported wind and use indicated airspeed (IAS) for speed computation.

Using the NDB approach at Miami International Airport (Fig. 19.2), the inbound track is 90° and the

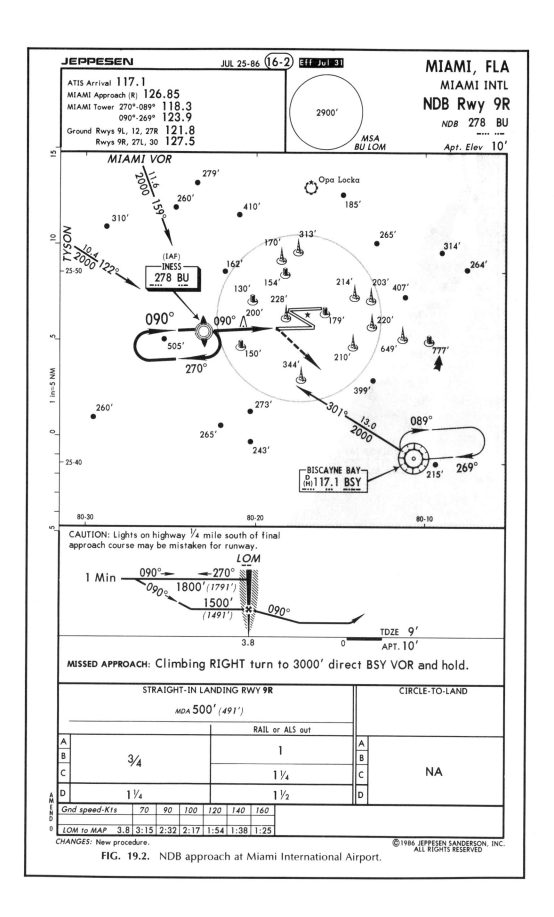

FIG. 19.2. NDB approach at Miami International Airport.

distance from the outer marker to missed approach is 3.8 miles. The weather reports the temperature is 86°F (30°C), the ceiling is estimated at 600', visibility is 1½ miles, and the wind is 130° at 12 knots.

Let's see how these different methods might affect the approach when flying an aircraft in category C with a 140 IAS in approach. With a standard adiabatic lapse rate, your average true airspeed would be approximately 145 knots.

Your first consideration is the time to missed approach, that is, ground speed. Equally important is drift angle for heading to maintain track. If you used the actual 145 true airspeed and the method of turning the wind right 23°, doubling its velocity, and averaging, you would use a wind from 153° at 18 knots in computation. Your ground speed would be 136 with a 6° drift correction to the right and a 96° heading to fly after the FAF. Time to missed approach would be, using a calculator as recommended, 1 minute 41 seconds.

If you used IAS in computation rather than true airspeed, turned the wind 23° to the right, and averaged the direction and velocity, ground speed would be 131, that is, a 7° right drift correction and time to the MAP of 1 minute 44 seconds. With the most common method, using IAS and reported wind, ground speed would be 131, and time to the MAP 1 minute 44 seconds with a 3° drift correction.

I've seen all of these methods used by other pilots, used them myself, and found them all reliable enough for desired performance as well as acceptable. As you can see, ground speed varies about 3 knots and estimated heading to maintain course varies 3–7°. Use whatever method you choose. It is, after all, a nonprecision approach and either of these methods will get you into a position to establish visual reference to correct course as necessary to land.

The minimum on this approach is ¾ mile, 4,560' from the threshold of the runway. If you were 5° off at this point, you would be approximately 380' from the runway center line, and that's close enough for government work. At 1 mile, you would be 507' from the center line. In either case, a course correction for landing is easily accomplished.

To descend to minimum altitude at 1,000' per minute would only take 1 minute in this approach, traveling approximately 2.3 miles at 136 knots, and would give you 1.5 miles and more time to establish visual reference before reaching the ¾-mile visibility limit. The idea of any nonprecision approach is to get down to landing minimums (with good control of approach track, speed, and attitude) and a rate of descent to establish visual reference and to land straight in with as little maneuvering as possible. I have just outlined the easiest way to accomplish this.

In review:

1. Precompute dead reckoning heading and drift for the course or track.

2. Estimate or compute ground speed and determine time to the missed approach.

3. Cross all crossing and final approach fixes at the correct altitude.

4. Descend so as to be at MDA either before or at the visibility limits, preferably before.

5. Use the precomputed heading for drift correction inside the FAF; make only small heading changes if they are necessary to correct the track.

Circling Approach

Circling approaches in a visual simulator, although they may be accomplished, are difficult, especially in a night environment; there is too little visual reference available. Some operators still do them, but as I have stated, most major carriers have abandoned the circling approach for many reasons, the primary ones being that large airports have many runways; most of them have ILS approaches to these runways, and even a nonprecision approach is rare; jet transports normally make all approaches in landing configuration; and the drag factor increases the difficulty of circling.

The circling-to-land procedure frequently may be required for private operators and supplemental and nonscheduled carriers. When it is required, it should be made in the maximum takeoff flap configuration.

The requirements for a circling approach follow:

1. You should be within visibility limits of a recognizable feature of the airport, be in visual contact with the airport and ground, be beneath the clouds and not below MDA, establish landing pattern and runway alignment, not descend below MDA until within 30° of the runway heading on final approach, and have runway alignment prior to 300'.

2. The turning radius of an airplane is related to the square root of the true airspeed and angle of bank. It is not necessary to compute it, but remember that at high-altitude airports the true airspeed will be considerably higher than indicated airspeed and require greater bank angles. The angle of bank used should be such as to keep within visibility limits of the airport but never in excess of 30°. As a rule of thumb, use 20° of bank for speeds 100 knots or below, 25° for speeds of 100–120 knots, and 30° for speeds in excess of 120 knots.

3. Maintain your configuration and speed and try to fly exactly at MDA. The ceiling, though it may be estimated 50–100' higher than MDA, may be lower in the circle; consider that above MDA you would theoretically be back in the clouds. Stay on the correct altitude and speed and keep the airport and desired runway in sight.

4. If circling to land on the same runway but in

the opposite direction of the approach, double your drift correction downwind so as not to drift in too close to the airport.

5. Extend your downwind leg 10–15 seconds, depending on speed, beyond the threshold of the landing runway.

6. Use a 30° bank turning base and make a constant turn at a constant bank to the final holding altitude and speed, and minimum landing or "approach" flaps may be extended in a tailwind condition. In a 0 wind condition, you would be within 45° of the runway heading 1 mile out, intercepting an approach angle of approximately 2.5° to landing. With a tailwind, extending approach flaps will have the same effect. Landing flaps should be extended at this 45° position. However, full landing flaps may be extended at 90° to the runway heading, and speed should be allowed to bleed off to V_{ref} + 5 if necessary to tighten the radius of turn due to wind.

7. The runway should be in sight of 45° of the runway heading or before; you may use visual reference for alignment, adjusting the bank as necessary (never steepening beyond a 30° bank) and adjusting the airspeed and rate of descent to roll onto the runway center line in the correct landing configuration and slot.

8. Do not descend below 400′ until within 30° of the runway heading; runway alignment must be established at 300′ minimum.

9. Be careful of high sink rates, and do not get on the back side of the power curve when going to full flaps.

10. Establishing a circling approach landing pattern for a particular runway is a matter of common sense and obstruction and traffic requirements. Normally the circle will be in a left-hand traffic pattern with the captain flying. Since the approach pattern is at a low altitude with a 30° bank, the copilot should be alert for too steep a bank, getting too high or too low, or getting too slow and on the back side of the power curve. The captain is flying about half visual and half instrument. At points in the pattern where the captain might be looking more at the airport and not at instrumentation, the copilot should monitor the instruments and not let the aircraft get into a dangerous situation.

20

Holding

■ Of all the airway traffic procedures, holding is perhaps the easiest. In a typical situation you are cleared to a fix—either a primary fix such as a VOR, outer compass locator, or homing beacon; a fix at the intersection of two VOR radials; or a distance measuring equipment (DME) fix on a radial—and you simply adjust your speed to arrive at the fix at the proper speed, flying a racetrack pattern of either right or left turns as indicated in your clearance, for the proper amount of time in the pattern.

Like everything else in a check ride, holding patterns are as much a test of judgment as a flying skill, and the problems of execution are multiple. The pilot must adjust the speed at the proper time to arrive in the fix, determine and execute the proper entry, time the pattern correctly, set up the radios properly, intercept and track the correct radial inbound to the fix, and at the same time continue to fly the plane on the correct speed and altitude. Many FAA examiners and company check pilots are proficient in clearing a pilot to a fix where these problems are difficult—not with malicious intent but to assess the pilot's ability to cope with a situation requiring more skill and sagacity than usual.

These problems are easily solved by a perceptive and well-trained pilot. All the information needed is contained in the clearance. From the clearance may be determined the outbound heading and inbound track, whether to hold with right or left turns, and the proper time and maximum speed in the pattern. The clearance is the key, and knowing how to use the information it contains is the solution for the problems we've mentioned so far.

These are the mechanical problems—technique and procedure—but there are other considerations of equal importance from a practical standpoint. In addition to your clearance, it is well to consider whether your hold is part of a departure or en route operation or a delay before arrival at your destination.

Beginning with these last items—arrival and departure holds (actually the first of your considerations)—let's take holding patterns completely apart for a better understanding of them.

Departure Holding

When you get a clearance to hold, consider first whether it's a departure or arrival hold. This may have considerable influence on your aircraft configuration, speed in the holding pattern, fuel consumption, etc. You may have to climb or descend in the pattern, accelerate to cruise from a departure or en route hold, decelerate in descent for approach from an arrival hold; both fuel consumption and aircraft performance will vary greatly with aircraft configuration.

The regulations (*Airways Traffic Procedures Manual*) state: "Turboprop aircraft may operate at *normal* climb IAS while climbing in a holding pattern, and turbojet aircraft may operate at *310 kts* IAS *or less* while climbing in a holding pattern." You can see that this may have its effect on your speed, depending on your altitude on arriving at the fix and the altitude to which you have been cleared. If you must continue to climb, 310 knots is maximum speed, but the 250-knot rule below 10,000' still applies.

As an easily remembered rule of thumb, I'd suggest climbing at normal speed in piston and turboprop aircraft and reducing to best rate of climb speed in turbojets. This speed in turbojets is the speed at which you have the best lift/drag ratio, resulting in less fuel burn-off as you climb while going nowhere, and is usually the same speed as that recommended for turbulence penetration.

Climbing in a holding pattern usually occurs in a departure or en route hold, when thought must be given to the fuel to take you to your destination. So fuel consumption is the primary factor to be considered when selecting a holding configuration in departure and en route holds, particularly in turboprop and turbojet aircraft.

When established in the holding pattern at the altitude specified in your clearance, hold at the proper maneuvering speed for your aircraft in the *clean* (zero flap) configuration (provided this speed is within the speed boundaries of the pattern altitude); your fuel burn-off will be less than it would be with the drag induced by flaps. When you are cleared for en route, you will be able to accelerate to a normal climb or en route speed much more quickly and thus will consume less fuel in the acceleration process.

Holding for Arrival

Arrival holds, loosely defined as a hold within the terminal area of arrival prior to approach, are different only in that they are normally done in a takeoff flap configuration and at the proper maneuver speed for the particular aircraft in that configuration. This is because, for all practical purposes, you are at your destination and in range of landing.

Fuel consumption, while still a factor to be considered, is not as critical now that you have arrived at your destination and intend to land within a reasonable length of time. But if the delay is lengthy or the weather doubtful, determine as early as possible the maximum time you can hold before fuel requirements will make it necessary to proceed to your alternate.

For turbojet operation, I would suggest holding clean (if maneuvering speed in that configuration would be within the speed requirements for that altitude and if ATC didn't request a slower speed that would require flaps) above 6,000' and beginning the flap extension schedule when cleared below that altitude. From that point on, particularly in high-density areas that are likely to hold you at a high altitude close to the field to allow departing traffic to pass underneath, it is reasonable to expect an approach clearance within a short time and to start setting up an approach configuration.

Maximum Speed in Holding Patterns

Regulations set forth maximum speeds in holding patterns predicated on type of aircraft and altitude:

1. Propeller-driven aircraft—175 knots indicated airspeed (IAS).

2. Civil turbojets have three altitude and speed structures. The *Airways Traffic Procedures Manual* refers to propeller-driven and turbojet aircraft only; therefore, I consider the speed restrictions to be the same as in FAR 91.70, which separates aircraft into categories of reciprocating engine aircraft and turbine-powered aircraft. Turboprop aircraft are also turbine powered and should be considered in this category.

    a. Minimum holding altitude to and including 6,000'—200 knots IAS.

    b. 6,000 to 14,000'—210 knots IAS.

    c. Above 14,000'—230 knots IAS.

Airspace protection for turbulent air holding is based on a maximum of 280 knots IAS or Mach 0.80, whichever is lower. ATC should be advised immediately if any increased airspeed is necessary due to turbulence or if you are unable to accomplish any part of the holding procedures. When higher speeds are no longer necessary, reduce to the proper speed for your aircraft, operate according to the above speed restrictions, and notify ATC. The use of turbulence speeds can create havoc with the flow of traffic in high-density areas, so use them discreetly and only when necessary.

Though it won't concern your operation unless you fly in the reserve or get into a stack with military aircraft, the military has a different set of speeds. Their propeller-driven and turboprop aircraft operate according to the above speed structures, but their turbojet maximum speed at all altitudes is 230 knots IAS except for fighters, which may operate at 265–310 knots depending on type.

SLOWING TO HOLDING SPEED

The *Airways Traffic Procedures Manual* states: "Cross holding fix initially at or below maximum holding speed. Effect speed reduction within 3 minutes prior to estimated initial time over the holding fix." All you have to know is where you are in relation to the fix, how fast you're going in ground speed to figure an estimate, and the length of time it will take to slow your aircraft from its present speed to the desired speed.

You might determine this 3-minute point by your position in relation to the fix in DME distance, by an estimate based on your last known position or checkpoint and ground speed (using cross radials from another VOR) or from a radar position from the ground controller. As a last resort on a check ride, though I wouldn't advise it unless absolutely necessary, you might ask the check pilot to act as ATC and give you your radar distance from the fix.

No matter how you determine the 3 minutes, the key is in two phrases—effect speed reduction *within 3 minutes prior to the fix* and *cross the fix initially* at or below maximum holding speed. If you cross the fix at or below the maximum speed at your desired speed, you've accomplished the speed reduction correctly. It's not necessary to fly the last 3 minutes prior to the fix at the maximum holding speed; that time is allowed to adjust your speed, taking into account possible errors in your estimate.

ATC Holding Clearances

The next problem of the holding pattern is the correct entry, and this is where most failures occur. Nearly all holding patterns are a simple procedure of flying up to a fix and making a 180° turn outbound. This isn't always true, of course, and the holding clearance for most instrument checks will not be so simple. You can expect a difficult entry problem, and this is where you had better really understand your holding clearance. A legal holding clearance, one containing all the information required to make it a valid clearance, tells you everything you need to know about your holding pattern and contains the key to the correct entry into the pattern upon your arrival at the fix.

ATC clearance requiring that an aircraft be held at a holding point includes the following information:

1. The direction to hold *from* the holding point. Herein lies the key to the entry. This will be stated as a direction to hold with relation to the holding fix and will be specified as one of the eight general points of the compass.

2. Holding fix. This will be whatever phraseology is necessary to describe the fix.

3. Specified radial, course, magnetic bearing, and airway number of jet route.

4. Outbound leg length if DME is to be used. This may also be expressed or specified in minutes if holding at a DME fix but not using a DME leg length outbound.

5. Left turns, if a nonstandard pattern is to be used.

6. Time to expect further clearance or time to expect approach clearance.

These are general holding instructions and may be accompanied by more detailed instructions in relation to DME pattern time instead of leg length or specifying right turns. Formerly, pilots were always expected to hold with right turns unless left turns were specified. However, at consistently used holding patterns, which are depicted on your area chart, you are now expected to hold in the depicted pattern unless otherwise instructed. If the pattern drawing shows left turns, use left turns in the pattern; if it shows right turns, use right turns *unless cleared to hold in a pattern other than that shown at the fix.*

If a consistently used holding pattern is not depicted on your chart, you are expected to hold with right turns in the pattern unless left turns are specified. Therefore, (1) hold in the pattern as shown on the area chart unless specifically cleared otherwise, or (2) hold with the right turns in the standard pattern at the fix where a pattern is not depicted unless left turns are specified.

Examples of holding are best described by the drawings in Figure 20.1.

Holding Pattern Entry Procedures

Just how to enter a holding pattern may not be as easy as it first appears. Holding pattern airspace protection is based on the airspeed limitations and entry procedures. The entry procedures contained in the *Airways Traffic Procedures Manual* are the only procedures for entry and holding recommended by the FAA.

For a graphic description of a holding pattern and related nomenclature, study Figure 20.2.

The *Airways Traffic Procedures Manual* states: *"Determine entry turn from aircraft heading upon arrival at the holding fix.* Plus or minus 5° in heading is considered to be within allowable good operating limits for determining entry." Thus the proper entry into a holding pattern is determined from one source (aircraft heading) and at one place (the fix). If an FAA inspector or an examiner asked you when you determine the entry method for entering a holding pattern, the only correct answer would be "Upon arrival at the fix." The *only* way of correctly determining the proper entry into a holding pattern is from the aircraft heading when first arriving at the fix. It is not always possible to refer to your chart and determine the entry method, but it can be done right at the fix almost instantly by referring to your heading. You may have to look at your chart for frequencies, radials, DME mileage, and general location of the fix (in many instances the proper entry may also be determined by visualizing your aircraft's position on the chart and en route to the fix), but the actual entry *must* be determined upon *arrival* at the fix and can best be done from your instrumentation.

But until you have your clearance, it isn't always easy. Figure 20.3 is a reproduction of an area chart; from these diagrams you are expected to figure out correct holding pattern entry procedures. Everything you need is right there. Figure 20.4 shows the standard pattern from the *Airman's Information Manual.* The explanation that accompanies it is as follows:

1. Parallel procedure—Parallel the holding course, turn left, and return to the holding fix or intercept the holding course.

2. Teardrop procedure—Proceed on an outbound track of 30° (or less) to the holding course; turn right to intercept the holding course.

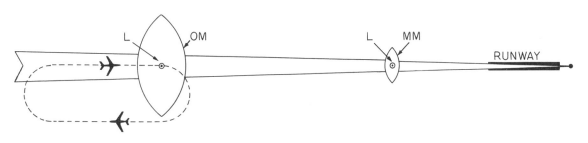

A. TYPICAL PROCEDURE ON AN ILS OUTER MARKER

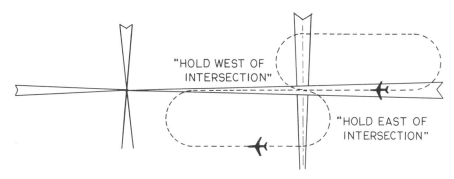

B. TYPICAL PROCEDURE AT INTERSECTIONS
OF RADIO RANGE COURSES

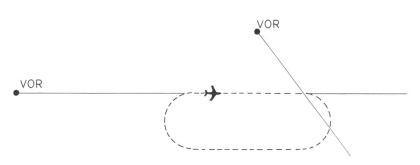

C. TYPICAL PROCEDURE AT INTERSECTION
OF VOR RADIALS

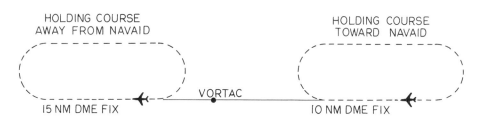

D. TYPICAL PROCEDURE AT DME FIX

FIG. 20.1. Examples of holding.

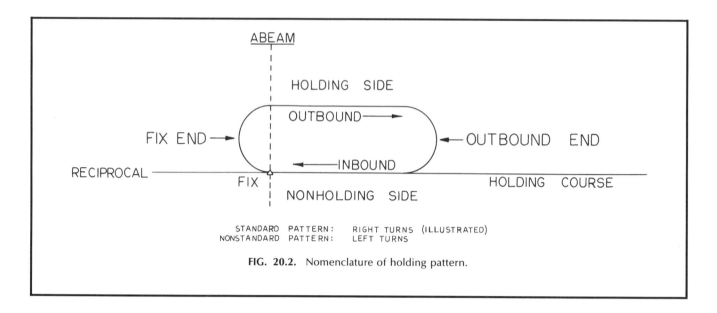

FIG. 20.2. Nomenclature of holding pattern.

3. Direct entry procedure—Turn right and fly the pattern.

That's the way the area charts and *Airman's Information Manual* show and describe holding pattern entry procedures. They show you a circle, with three quadrants from which the aircraft is approaching the fix, and a holding pattern. Note that the quadrants are different in size; one is 70°, another is 110°, and another is 180°. These three different quadrants from which an aircraft approaches a holding fix determine the holding entry, but how?

Look again at the diagrams for holding pattern entry as shown on area charts (Fig. 20.3). Figure 20.4 shows the same thing in a different form, but those in Figure 20.3 are cockpit charts and really show the in-

formation more clearly than the drawing from the manual. (See Fig. 20.2.) The manual, other than naming the three different procedures, doesn't really explain the diagram for the standard pattern. But if you read further, you will find the key in the statement the manual makes regarding nonstandard holding patterns. The manual states: "Fixed end and outbound end turns, in a nonstandard pattern, are made to the left. *Entry procedures are oriented in relation to the 70° line on the holding side just as in the standard pattern.*"

All the diagrams show the aircraft approaching the fix, at the center of the circle, from three different quadrants. But the *key* is the 70° quadrant in which the holding pattern is drawn. Note the right turn entry

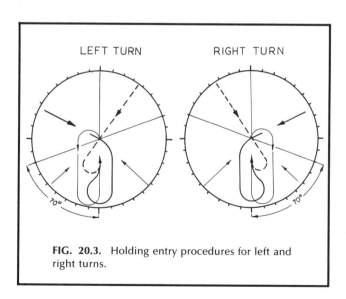

FIG. 20.3. Holding entry procedures for left and right turns.

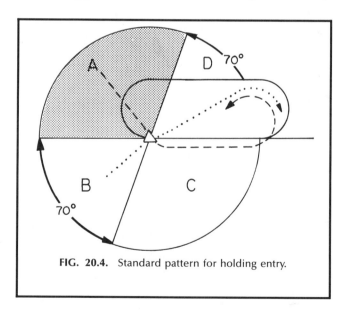

FIG. 20.4. Standard pattern for holding entry.

procedure shown in Figure 20.3. It shows that if the *outbound heading* in the holding pattern lies within a quadrant of no more than 70° to the *right* of the aircraft heading at the fix, then the entry procedure is a teardrop entry and would be entered with a right turn to a heading of no more than 30° to the outbound course on the holding side.

It also shows the aircraft approaching the fix from a 110° quadrant and the outbound heading from the fix lying in a 110° quadrant, relative to the aircraft heading, to the left and on the nonholding side. This pattern would then be entered with a left turn to the outbound heading and would be a parallel outbound procedure.

That leaves only one 180° quadrant or aircraft heading relative to the outbound heading in the pattern and would require a direct entry into the holding pattern.

So we've established three quadrants, all related to the outbound heading in the holding pattern and the aircraft heading upon arrival at the fix. Thus far, we've used only the right turn pattern in our discussion, but the same things apply to the left turn pattern except that the holding side and the 70° quadrant are to the left instead of to the right.

It takes a little imagination and thought, but the above information and the information you receive in your clearance may be transposed to your aircraft instrumentation and correct entry procedures determined immediately upon arrival at the fix.

Your clearance *must* contain a direction to hold *from* the fix. This will be a general direction. But the clearance must also contain a specified radial or airway, and this will give you an exact heading away from the fix. As an example: "Hold west of the fix on the 262 radial." The outbound heading to hold west would be 262° from the fix and the inbound track would be the reciprocal 82°.

Now it becomes simple. You have all the information necessary to execute a correct entry every time. Your clearance, which is the key, contains the outbound heading in the pattern; the direction of turns in the pattern (either right or left for determining the holding side and the 70° quadrant) and your compass show the aircraft heading upon arrival at the fix. Merely visualize the 70° quadrant on the holding side in relation to the heading on your compass and the 110° quadrant on the opposite side; the remaining 180° quadrant completes the circle of 360°. Whatever quadrant the outbound heading lies in determines the entry procedure.

You just transpose the picture to your radio magnetic indicator and see where the *outbound* heading is in relation to your heading; the quadrant in which it lies determines the entry.

You've probably mastered this simple method of determining entry procedures, but I've seen pilots be-

come confused, not really understanding how it's done, and enter patterns incorrectly so many times that I'd like to give some practical examples before we wind up this discussion.

EXAMPLE 1

Let's say you've just passed over the Miami VOR-TAC, southbound on V-157, en route to Key West. For some reason—traffic delays, perhaps a disabled aircraft on the runway at Key West, or anything that would cause the Center to give you a hold—you are given the following clearance: "Douglas 30 Whiskey, you are cleared to hold east of the Harvey intersection, on the Biscayne Bay 262 radial; maintain 7,000; expect further clearance at 1730 Zulu." This is a legal clearance. It contains all the items we've previously discussed; gives the direction to hold, describes the fix, specifies a radial, and indicates a time to expect further clearance.

To visualize the pattern, you might refer to a Miami area chart. Figure 20.5, however, contains the basic information you'll need. The holding pattern is drawn in solely for your information; the area chart does not contain one, since this intersection is not a frequently used holding fix.

Let's assume there is no wind, which would affect heading by drift, and then use Figure 20.6 as radio magnetic deviation indicator (RMDI) or heading information. Your heading will be the same as the airway radial 222°. No mention of a nonstandard or left turn pattern was made in the clearance, and no commonly used pattern is depicted at the fix on the chart, so it

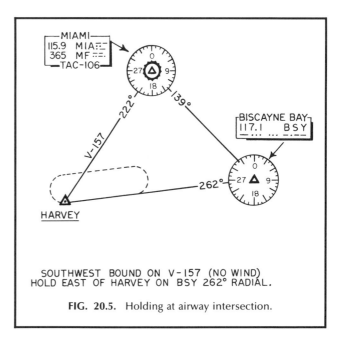

SOUTHWEST BOUND ON V-157 (NO WIND)
HOLD EAST OF HARVEY ON BSY 262° RADIAL.

FIG. 20.5. Holding at airway intersection.

will be a standard right turn pattern. The 70° quadrant for the teardrop entry is to the right; the 110° quadrant for the parallel outbound on the nonholding side entry is on the left; and the remaining 180° quadrant contains the outbound heading in the pattern and indicates a direct entry into the pattern. Therefore, to hold as cleared (east of Harvey) on the Biscayne 262 radial, you would fly a heading of 082 *away* from the fix and track the 262 radial of the Biscayne VOR inbound to the fix. Figure 20.6 shows where your outbound heading lies in relation to your aircraft heading upon arrival at Harvey. Upon arrival at the fix, you would make a *right* turn to a heading of 082° for a direct entry.

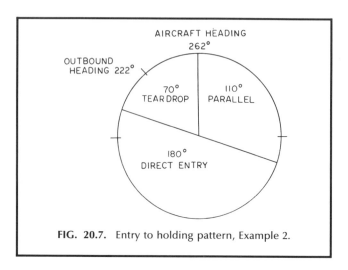

FIG. 20.7. Entry to holding pattern, Example 2.

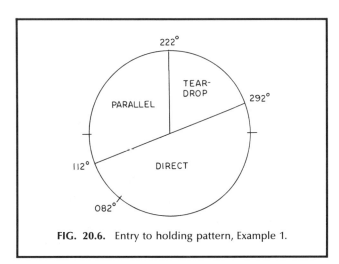

FIG. 20.6. Entry to holding pattern, Example 1.

EXAMPLE 2

This time you have come in from east of Miami, over the Biscayne Bay VOR, and you're proceeding out the Biscayne 262 radial to intercept Victor 157 southbound at Harvey intersection. You are cleared to hold southwest of Harvey, on V-157, with left turns. The hold is all we're interested in, so just assume the clearance is legal and contains all the required items. Assume also that you are again in a no-wind condition and your heading is the same as the airway radial—262°. Figure 20.7 is an illustration of how this would look on your RMDI as you reach the fix.

The outbound heading in the pattern, 222°, lies in the 70° quadrant on the holding side, so it's a teardrop entry. When you reach the fix, you'd turn left to a heading of 252°, fly the proper amount of time, then turn left back inbound to intercept the Miami 222 radial or 042 track inbound.

EXAMPLE 3

This time you're approaching to land at Miami from the north. You've passed the VOR and are cleared down the 222 radial to hold at Jersey intersection (the intersection of V-157 or the Miami 222 radial and the Miami ILS) west of Jersey on the Miami ILS with left turns. The ILS is 087° inbound, so to hold west of the intersection the outbound heading would be 267°. Since the clearance specifies left turns, the 70° quadrant is to the left (Fig. 20.8 shows how this would look on your RMDI at the fix).

The aircraft heading, compared to the outbound heading in the holding pattern, with the 70° quadrant always on the holding side from your heading, will indicate the proper entry method every time.

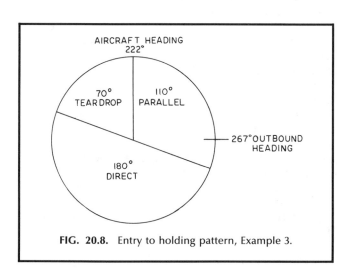

FIG. 20.8. Entry to holding pattern, Example 3.

Timing the Pattern

Concerning the rate of turns to use in a pattern, the *Airways Traffic Procedures Manual* contains the following statement: "Make *all* turns during entry and while holding at: (1) standard rate turns of 3° per second, or (2) a 30° bank angle, or (3) a 25° bank angle provided a flight director system is used, whichever requires the least bank angle. Compensate for *known* effect of wind, except when turning."

Concerning timing a pattern, the *Airways Traffic Procedures Manual* states: "The *inbound* leg *shall* be one (1) minute in length at or below 14,000 MSL and one and one-half (1½) minutes above 14,000 feet." This shows that the *inbound* leg is the most important and must be of the correct length in time.

The initial outbound leg should be flown for 1 or 1½ minutes, whichever is appropriate to altitude; the inbound leg back to the fix should be timed, and the timing for subsequent outbound legs should be adjusted as necessary to achieve the proper inbound time.

The question then arises, When does the initial outbound time start? The outbound time begins *over* or *abeam* the fix, whichever occurs later. If the abeam position cannot be determined, start timing when turn to the outbound heading is completed.

As a rule of thumb: Start the initial time over the fix for teardrop and parallel on nonholding side entries; start time either abeam the fix (if it may be determined) or on completion of the turn at wings-level on the outbound heading in direct entry. You'll never fly out of holding pattern reserved airspace, and you will also be more accurate and correct in timing in every case.

DME Holding

DME holding is subject to the same entry and holding procedures except that distances in nautical miles are normally used in lieu of time values. But time values may sometimes be substituted in the clearance for the end of the outbound leg. The outbound course of a DME holding pattern is called the outbound leg of the pattern, and the length of this leg will always be specified in the clearance. The end of the outbound leg is normally determined by the DME odometer reading.

As an example, when the inbound course is toward the NAVAID and the fix distance is 10 NM and the leg length is 5 NM, the end of the outbound leg will be reached when the DME reads 15 NM (Fig. 20.9a).

When the inbound course of the pattern is away from the NAVAID and the fix distance is 28 NM and the leg length is 8 NM, the end of the outbound leg will be reached when the DME reads 20 NM (Fig. 20.9b).

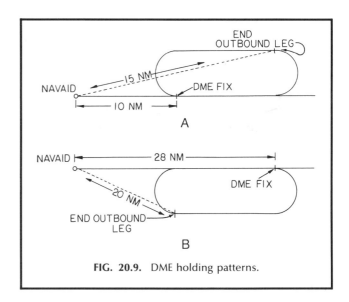

FIG. 20.9. DME holding patterns.

Holding with a Flight Director

This easy method of using a flight director system may not agree with the method the flight director manufacturer recommends, but it is easy to use and it works.

In most cases you'll be tracking a radial en route to your holding fix, and the flight director should be in the NAV/LOC mode so that your steering needle is giving you steering reference to track the radial. If your clearance is a cross-radial fix (at the intersection of two radials), have the pilot set the radio to the cross-radial and inform you when you are there.

Set your heading bug to the outbound heading in the pattern (this will not interfere with operation of the flight director in the NAV/LOC mode after capture of the radial) and compare this heading to your heading to determine the quadrant the outbound heading lies in and entry into pattern.

If your entry is to be either direct or parallel outbound, leave the bug set on the outbound heading. If it is to be a teardrop entry, set the bug on the heading you wish to turn within 30° of the outbound heading.

When you arrive at the fix, simply switch to the heading mode and turn in the proper direction to the heading selected. You may have to turn away from the steering needle or command bar reference if this heading is more than 180° away from the aircraft heading (as in the example for a direct entry), but just make a 25° banked turn and then follow the steering reference of the steering needle as it swings up into formation with your turn after you are out of 180° ambiguity.

When you first arrive at the fix, you're established in the holding pattern and should so report to ATC.

Start your time initially outbound at the fix for a teardrop or parallel outbound entry and either abeam the fix or wings-level outbound in a direct entry.

Have the pilot who is not flying retune the NAV radios if necessary and reset the correct inbound track on your flight director. The pilot should also be instructed to set the proper station and cross-radial on the radio and course indicator and inform you when you are back at the fix inbound.

Fly outbound for the proper time, start your time inbound, and switch the flight director back to NAV/LOC. You may again be turning away from the indication of the command bars or steering needle for a moment in the beginning of the turn, so just make a normal 25° banked turn.

As you swing back closer to the inbound track and heading, the command bars will show a capture of the inbound track, you are now using the flight director again for reference to adjust your turn to intercept the desired radial. However, *always* compare raw data information on your course indicator to the steering reference of the flight director and be *certain* you are definitely on the desired track prior to reaching the fix, even if it means flying against the steering reference a bit. The capture of the radial indicated by the flight director may be aiming you for a capture far beyond the fix, and you'd never actually arrive back at the fix, particularly in a strong wind, unless you also use the actual course indication.

From then on, all you have to do is switch back and forth from NAV/LOC to heading when changing direction of flight from inbound to outbound (adjusting the heading if necessary due to wind and drift) and from the heading to NAV/LOC turning back inbound.

Flying and its procedures are processes pilots must teach themselves. Use any method or technique that is easiest for you, as long as it works and doesn't violate any rules. If any of my suggestions make it easier for you, then I am satisfied that I have accomplished some useful purpose.

21

Radio Failure

■ When operating under instrument flight rules in controlled airspace, there are certain requirements that must be met regarding the malfunctions of navigational radio equipment or the loss of two-way radio communications. In nearly every instrument check or rating ride, the FAA inspector asks questions about these procedures. Most pilots have a general knowledge of these rules and know what they are to do in the event of radio failure, but a general knowledge may not be enough.

To refresh your memory, I'll give you a condensation of the FARs applicable to radio malfunctions—91.127 and 91.129.

Navigational Radio Failure

Air carrier rules require that pilots, if they have only the VOR, request a hold at a primary fix. It is far better to hold over the station or on a DME fix than to try to hold at the intersection of two radials where you must constantly switch back and forth from one station to another.

FAR 91.129 has to do with malfunction reports and is very simple. It requires an *immediate* report to ATC on the loss or malfunction of VOR, TACAN, or ADF; complete or partial loss of the ILS receiver or low-frequency navigation receiver; or loss of air-to-ground communication capability. In other words, if you lose *any* radio equipment, notify ATC immediately. The loss of air-to-ground communication capability referred to here is for the partial loss of your normal capability, such as the failure of one communication set.

In your report give aircraft identification or flight number, the equipment affected, and how this affects your capability to navigate and communicate; request

whatever special assistance or handling you may require.

Communication Failure

FAR 91.127 has to do with total loss of two-way radio communications, delineating the procedures pilots should follow in the unlikely event that they lose all communication capability but retain navigational radio. The regulations deal with failure of navigation and communication radios separately. No attempt at rule making is found for the total loss of all radio capability, but the suggested procedure to follow is given in the first statement in FAR 91.127: "If VFR, or later encounter VFR conditions, continue VFR and land as soon as practicable!"

If you are not in VFR conditions when the communication failure occurs, you have five things to consider: (1) route to fly, (2) altitude to fly, (3) when to climb, (4) when to descend, and (5) when to leave a holding fix if the failure occurs in a holding pattern.

Using commonsense language based on 91.127, let's break these down and see how it should be done. If you follow the procedures correctly after losing two-way communication, you may be assured that ATC will provide the necessary protection for your flight. They will probably monitor your progress along your route of flight on radar, and you may have confidence that you are flying in safe airspace. But to be assured of this protection, you must carry out the correct procedures. Though they may be watching your route on radar, they will have no way of knowing your altitude and when you begin to climb or descend. ATC will protect your altitude based on the assumption that you will begin to climb and descend at the proper time.

We'll look at the five different considerations separately.

1. Route:

   a. By the route assigned in the last ATC clearance received.

   b. If being radar vectored, by the most direct route from the point of communication failure to the fix, route, or airway specified in the vector clearance.

   c. If you have no assigned route, by the route advised in "May be expected in further clearance."

   d. If you have neither assigned route nor expected route, by the route you filed in your flight plan.

2. Altitude—At the *highest* of the following altitudes or flight levels:

   a. The altitude or flight level assigned in your last ATC clearance.

   b. The minimum altitude or flight level (FAR 91.81c) for IFR on the route or airway.

   c. The altitude or flight level that ATC has advised may be expected.

Note that no reference is made to your filed altitude.

3. Climb—If necessary to meet the altitude requirements above:

   a. Climb to altitude or flight level in accordance with your last clearance.

   b. Climb to the minimum altitude for IFR operation for time and place.

   c. Climb to the altitude or flight level contained in expect-further-clearance at the time and place contained in expect-further-clearance.

4. Descent—Begin descent from en route altitude or flight level upon *reaching* the fix from which an approach would begin, but not before:

   a. The expected approach clearance time (if you received one).

   b. If no expected approach time is received at your estimated time of arrival or at your filed time en route in the flight plan or as it may have been amended en route.

5. Leave holding fix—If holding, leave the fix at the time of expected approach clearance or expect-further-clearance time. If approach clearance or an expected time for approach clearance has been received, leave the holding fix in order to arrive over the fix to begin approach as close to the expected approach clearance time as possible.

FAR 91.81

Minimum flight level structure (FAR 91.81c) may be ignored if the altimeter is higher than standard and applies only when the altimeter is lower than 29.92. Flight levels begin at 18,000' when the barometric pressure is 29.92 or higher; 18,000', or flight level 180, is the lowest usable flight level. However, the lowest usable flight level must be higher when the current altimeter setting is lower than 29.92: flight level 190 for an altimeter setting of 29.42–28.92; 200 for 28.42–27.92; 210 for 27.42–26.92. The simplest way to use these flight levels is to add 1,000' of altitude to 18,000' for every inch of mercury the altimeter is less than 29.92. Fortunately, altimeter settings below 28.92 rarely exist, but you should be familiar with the method of determining minimum flight levels.

22

The Oral

Researchers in the FAA's medical department have found that pilots taking examinations and flight checks to obtain various licenses and ratings experience psychological pressures equivalent to those of military cadets or combat pilots preparing for a combat flight. As a result, their performance in these examinations or check rides is not equal to their actual ability; most pilots lose 20–30% efficiency by being observed and examined.

I never propose students for an oral or flight check until I'm sure they are ready for it. Even more important, I never send students up for a check until *they* know they are ready and are thoroughly confident of their ability to pass it. Pilots must not only be able to fly but must also have confidence in their ability to perform on every flight. This self-confidence can be developed by the instructor who spends as much time and effort in psychologically preparing the student as in teaching about the airplane.

FAA inspectors in an airline training operation have a tremendous responsibility. They are the watchdogs of public safety and must set high standards in examining pilots for ratings in air carrier operations. When a pilot becomes involved in an accident, the National Transportation Safety Board investigators turn first to the FAA inspector who granted the rating and next to the flight instructor for comments on the pilot's judgment and ability that might not have been indicated on the rating grades. It's a responsibility no one can take lightly, and no inspector likes to issue a license to a below-average pilot.

Inspectors are also an integral and important part of the airline training department to which they are assigned. They not only examine pilots for ratings but are also responsible for the entire training program being conducted along the lines of the FAA-approved training curriculum. They become familiar with the airline's supervisory personnel, check pilots, and flight instructors; the success of any training program becomes a common goal for which all are striving—that of turning out highly qualified pilots.

FAA inspectors—at least those in air carrier operations—are for the most part well-qualified and experienced pilots. Some of them are ex-airline people, but most of them are ex-military who were too old for airline employment upon retirement from the service. The training and retraining they receive in the FAA subjects them to a great deal of pressure, and they experience tension when they take a check themselves, so they are well aware of the human tendency to suffer from "checkitis." Most of them are masters at putting applicants for a rating at ease and adroitly bringing out their best. They are good psychologists who can quickly determine a person's knowledge, confidence, judgment, and flying ability.

Students who have studied for the oral and flight training as I recommended earlier will have little trouble if the instructors have also done their job. Students should have the operational knowledge when they go for the oral and be confident of that knowledge almost to the point of being cocky. They will know the operational and certification limitations perfectly, have all emergency and abnormal procedures down pat, be able to use all the performance charts in the aircraft manual, and have a good operational knowledge of every system in the airplane. In short, they will know all the answers and will recognize the questions no matter how they are phrased.

The inspector will probably want to use your manual for the oral and will thumb through it, going from section to section while asking you questions. Put yourself in the inspector's place for a moment. You know what the job is and that there is only a short time to assess your knowledge. There will be a few basic questions on each system; if you don't know the answers, it follows that your general knowledge is poor. As the inspector thumbs through the manual you have used in ground school, self-study, and flight training, a subconscious impression will register about you as a pilot. If important passages are underlined, every page is well thumbed, good notes appear in the margins of the pages and perhaps a more comprehensive set at the beginning of each section, and sample problems are drawn out on every performance chart, the inspector's subconscious impression will be a good one. If the manual is unmarked and perfectly clean, its condition may portray just the opposite impression. Give that a bit of thought; then make up your own mind as to the type of manual you want to take to your oral.

In the past, orals were more in a system construction format. Some people still believe that system knowledge is best determined from familiarity with system functions (over which you have no control) such as design features or pressures and temperatures where an automatic function occurs. Most inspectors realize, however, that you are not going to redesign a system or fix it in flight; they try to determine your knowledge of a system from your ability to operate it, troubleshoot it, or shut it down and get along without it. They know that a pilot who knows the aircraft limitations, emergency procedures, performance, and normal operating procedures will certainly be able to fly the aircraft within safe parameters of operation. Therefore, most orals today are operational.

The examiner will verbally portray an in-flight situation involving a normal, abnormal, or emergency system operation and then ask how you'd take care of the situation. Blown-up pictures and dummy panels of the cockpit are widely used in air carrier ground training. This is where the study of systems I've suggested comes in handy. You can answer any operational question by pointing to the various system controls; state what is happening within the system as you move or position the knob, switch, lever, etc.; point out the instrument upon which system operation is indicated; mention the instrument readings or annunciator panel lights that indicate malfunction or abnormal operation; and state the control positions you'd use and instrument and annunciator panel indications you'd expect in order to correct a malfunction or to operate the system in a manner other than normal. Your knowledge of what's happening within the system as you move a control, how its normal or abnormal operation is indicated, will prove to the inspector that you have sufficient understanding of the system to operate it and troubleshoot it.

If you don't know the answer or don't understand the question, don't guess at an answer. Even if you guess correctly, the examiner will soon realize that you are guessing and are weak in knowledge in that particular area and will ask further questions about the same thing, trying to plumb the actual depths of your knowledge. Be sure you understand the question; if you don't, ask for it to be restated. If you don't know the answer, just say so. That may startle the examiner, who then will probably have a tendency to instruct you. All examiners like to feel they have taught the applicant something, so give them the opportunity. Of course this applies only to minor things; an examiner can't instruct you in the things you absolutely must know.

Examiners may also use another trick. After you have given an answer they may say, "Are you sure? Is that right?" Your answer may actually be correct. The examiner is just trying to determine if you're parroting answers committed to memory or basing your answers on actual system knowledge and trying to see if you're unsure of yourself or confident and secure. Pilots often have to make split-second decisions and take immediate action based on those decisions. Those who are hesitant or unsure are more likely to make mistakes. When you know you are right, look the examiner in the eye and say, "Yes, I'm right!"

Summary

Have a well-used manual indicating that you have studied in preparation; exhibit the confidence expected of a command pilot; have good operational knowledge of your aircraft—its systems, limitations, performance, and emergency procedures.

If you are to be certified for Category II operation, you should also review TERPS (terminal instrument procedures) and regulations pertaining to such operation.

I've written this primarily slanted toward the captain student applying for a type rating in the aircraft, but it applies equally to copilots and first officers. The only difference is that the examination may be given by a company check pilot rather than an FAA inspector. A copilot, who may someday be a captain, is expected to exhibit confidence, judgment, and knowledge. Act like a captain!

23

The Flight Check

The flight check for both captains and copilots is actually in two parts—a preflight check and the flight check. The preflight check is normally a duty of the second officer or flight engineer (the copilot or first officer in two-pilot crews), but I'm old-fashioned enough to believe that every pilot that flies a plane should be able to visually inspect it on the ground to check its general condition and airworthiness before flying it. The FAA apparently agrees with me and has made the preflight check a required part of an initial checkout and instrument check.

Flight checks are usually given in the simulator by major airlines, but I'm sure that there may be times when they are conducted in the aircraft and the following would apply. The material is patterned on the DC-9, the walk-around or preflight, etc., but the principles and pattern involved are essentially the same for any aircraft.

The sequence of maneuvers is of a similar nature. It is merely an outline and obviously may be varied in sequence; it will probably require a one-engine approach in a three-engine aircraft and a Category III approach in an aircraft certified for Category III.

Preflight

Preflight may be waived for the six months' instrument check for a pilot already qualified in the aircraft. Even if waived, I'd suggest that you make a quick walk-around before the flight, checking important items. I've known of check pilots that have waived the preflight and then, if the pilot being checked didn't take the time to check the aircraft's general condition (tires, brakes, struts, etc.), would give a thorough and comprehensive preflight after the flight.

Many pilots are critical of the requirement that they make the check. They are of the opinion that maintenance has certified the airworthiness of the plane, the second officer has checked it, and they as pilots shouldn't have to be involved. This is ridiculous! The captain has to make the decision of accepting the plane. The second officer reports anything wrong to the captain, and captain should be capable of viewing the reported discrepancy and recognizing it. The captain may be at a nonmaintenance base and have to decide if it's all right to fly or be able to adequately describe findings to maintenance personnel on the phone so they will send the right part to fix the aircraft.

The preflight check varies from aircraft to aircraft, but the general procedure may be broken down into the following broad classifications: (1) cockpit check, (2) emergency equipment check, and (3) external walkaround.

As the applicant for the rating, you should, insofar as possible, take command of the situation and tell the examiner what you're doing, checking as you do it. Keep walking and talking, making a thorough check as you go, and try not to miss anything. This way the examiner will be getting a look at your ability to perform this function and will not have an opportunity to get in a question. There are some things the examiner is bound to ask about, but you want to keep questions to a minimum. Exhibit self-confidence and try to cover everything, but don't point out anything with which you are not familiar. It's better to miss some small item than to get trapped into answering questions concerning something you know very little about.

A good starting point is the aircraft log book. Check to see that all required maintenance checks have been accomplished and all items cleared by maintenance action.

Check the cockpit, particularly the items that may require maintenance, servicing, or replacement—fuel pumps, battery, electric hydraulic pumps, fuel, oil, hydraulic quantity, oxygen pressure, cockpit emergency equipment, checklists, and any manuals and charts that may be required in the cockpit.

Make sure that the proper emergency equipment is aboard in the cabin. Here's where your knowledge of the number and location of different items is useful. I've suggested that you separate your emergency equipment into types and learn number and location for each item. For example, you may be asked how many fire extinguishers you have. You would impress your examiner if you answered the correct number and types along with their location. Your answer for a particular airplane might be, "Five. Two CO_2 extinguishers, one in the cockpit and one in the right-hand coat compartment; and three water bottles, one in the left-hand coat compartment and two behind the last row of seats on the left side." You've answered all possible questions (except perhaps size of extinguishers) with one answer. This same type of answer would apply to any multiple units of emergency equipment.

An operational walk-around is to check the general condition of the aircraft for damage or leaks. Common sense dictates checking brake wear, strut inflation, tire condition, and fluid leakage, but a rating preflight goes a little deeper. You will be expected to identify certain features and items of the particular aircraft, explain their purpose, and tell the examiner what you're looking for as you check them.

I can't give you a specific walk-around to fit all aircraft (there's too much variation in design characteristics), but I can give you a general flow pattern that works well for most airplanes. The inspection will vary with location of engines and features, but the flow pattern and general methods of inspection are similar for all aircraft.

The flow pattern of the walk-around should start at the entrance door, proceed forward along the fuselage to the nosewheel well and nose of the aircraft, aft along the right side of the fuselage to the wheel well, out the right wing leading edge to the tip and back along the trailing edge to the fuselage, and along the aft section to the right side of the fuselage to the tail. Complete the preflight by moving forward along the left side from the tail and inspecting the wing and wheel well, then forward ahead of the wing back to the starting point at the entrance.

The external walk-around I outline for my DC-9 students is as follows:

1. Stairway—Check the handrails and general condition. Check the door, door well, and latching mechanism. Check the mirror for internal inspection of latches from the electronic compartment and the safety latch. The door cannot be retracted with the lock on.

2. Electronic compartment access door, beneath fuselage—closed and latched.

3. External power receptacle—Wheel well light on. This permits inspection of lights used to inspect gear locks if necessary in flight.

4. Radome fasteners—Down and latched. Explain that radar antenna and glide slope antenna are contained within the radome.

5. Pitot tubes, top of nose—Left, captain; center, air data computer; right, copilot. No obstructions; condition good.

NOSEWHEEL WELL

1. Nose gear strut—Inflation normal (about 4 inches); no leaks.

2. Nose gear tires—Proper chines, condition, and inflation (115 lb. max pressure, 105 minimum).

3. Nosewheel—Condition good; all bolts holding wheel together in place and secure.

4. Forward electronic compartment door, upper rear of nosewheel well—Closed and secure.

5. Steering mechanism—Good condition; no leaks.

6. Down locks, indicators, and gear pin—Point out down lock microswitch, visual viewing indicator, and gear pin not installed.

7. Forward nose gear doors—Closed and latched. Explain the inclinometer in the nosewheel and its use in determining the fuel quantity using dripless sticks.

8. Steering bypass control—Released and operating freely.

9. Nose gear ground shift mechanism linkage—Connected and secure.

10. Nose lights—Checked and unbroken.

RIGHT FORWARD FUSELAGE

1. Ram air temperature probe (free air temp)—Condition good; no obstructions.

2. Oxygen blowout disk—In place.

3. Galley door—Closed and secure.

4. Galley doorsill drain—Open.

5. Antennas—Condition good. Identify all antennas and give locations.

6. Stall warning vane (DC-9-31)—Condition good.

7. Right alternate static port—Clear; condition good; heated.

8. Right static port cluster—Identify static port. Clear; condition good; heated.

9. Forward cargo compartment door—Closed

and latched. If you open it, check floors and walls for tears or holes. It's a class D compartment, with no fire warning or protection, and holes destroy class D integrity.

10. Water service panel—Door closed. Open the door and explain servicing, heating, rubber drip shroud, quantity indication, and air connection to pressure tank. Reclose.

11. Wing leading edge floodlight, upper fuselage—Condition good.

12. Right ground floodlight (bugeye)—Condition good.

RIGHT WHEEL WELL

1. Gear door hydraulic bypass—Secured (extended if entering wheel well).

2. Center tank dripless stick and sump drains—Point out location. Check if desired.

3. Right wheel well door—Closed (after inspection if opened).

4. Hydraulic reservoir—Quantity checked.

5. Hydraulic clog filters—Checked.

6. Brake and system accumulators—Checked.

7. Spoiler bypass handle—Safetied on.

8. Auxiliary and alternate hydraulic pumps—Identify; check for leakage of hydraulic fluid.

RIGHT WING

1. Stall strip (DC-9-14)—Condition good. Explain function.

2. Leading edge slats (DC-9-31)—Condition good; check for hydraulic leakage.

3. Vortilon (DC-9)—Condition good. Explain function.

4. Stall fence (DC-9-14)—Condition good. Explain function.

5. Cross-feed vent, wing center section—Open.

6. Fueling panel doors—Closed and latched. Open and explain fueling, defueling, possible fuel transfer on the ground, and power source for operation. Reclose.

7. Fuel dripless sticks—Secure. Pull one and demonstrate use.

8. Main fuel vent—Open. Explain fuel venting.

9. Tip ram air vent—Open. Explain location of flux-gate compasses.

10. Stall warning transducer (DC-9-14), system 2—Condition good.

11. Tip lights (green and white)—Checked; fairings and glass good. Explain that the white light is the tail light and oscillator.

12. Landing light—Retracted; condition good; speed restriction, V_{MO}.

13. Anticollision light, atop fuselage—Condition good.

14. Static discharge wicks—Condition good (minimum two each wing tip, two each side horizontal stabilizer, and two on vertical tail assembly).

15. Aileron—Full throw clear; condition good. Explain how ailerons are tied together and actuated; explain viscous dampeners.

16. Aileron tabs—Condition good; no looseness. Explain functions.

17. Flaps—Condition good. Inspect hinge points for hydraulic leakage.

18. Spoiler panels—Down and faired.

19. Vortex generators (DC-9-14)—Condition good. Explain function and number required.

20. Bent-up trailing edge panels—Secured.

21. Cabin emergency exits—Flush and secure.

RIGHT MAIN LANDING GEAR ASSEMBLY

1. Wheels and tires—Condition good; inflation normal (125 lb. max, 115 min, DC-9-14; 145 max, 135 min, DC-9-31). Check speed rating of tire.

2. Hydraulic shimmy damper—Secure; level checked; no leaks.

3. Brakes—No leaks; hose condition good; wear indication checked.

4. Brake retainers—In place (two may be missing if not adjacent to each other).

5. Strut—Inflation correct (about 4 inches).

6. Overcenter markers, viewer light—Condition good; clean.

7. Downlatch—Microswitch and gear pin hole.

8. Hydraulic lines and electrical conduit of anti-skid—Attached; no leaks.

9. Fuel shroud drain—Checked. Explain function and maximum fuel allowed on check.

10. Belly collision light—Condition good.

ENGINE AND RIGHT REAR FUSELAGE

1. Cowl area—Unobstructed. Point out heated areas and Pt-2 probe.

2. Cowl heat overboard—Open.

3. Fuel heat intake and overboard—Open.

4. Leakage check—No drips.

5. CSD inspection door—Secure and latched. Open and check oil quantity.

6. Reverser safety latch indicator—Reversers stowed; indicator flush.

7. P & D valve—Open. Explain function.

8. Reverse accumulator pressure—Checked.

9. Reverser safety pin and door—Safety pin stowed.

10. Fuel shroud drain—Clear; no drips.

11. Oil service—Door closed. Explain.

12. Manual start valve access—Door closed and latched. Explain.

13. Oil vent—Clear.

14. Right rear toilet vent—Open.
15. APU vent—Open.
16. Tail bumper and strike indicator—Checked; parallel to fuselage.
17. APU doors—Point out.
18. Ground air conditioning connection—Closed.
19. Pylon ram air vent—Open.

UPPER REAR FUSELAGE AND TAIL ASSEMBLY

1. APU exhaust vent—Open.
2. APU exhaust—Open.
3. Air cycle ram air overboard—Open.
4. Ram air intake, vertical fin—Open; heated. Explain.
5. VOR antennas, vertical fin—Identify.
6. Rudder limiter probe—Heated. Identify.
7. Rudder and tab—Condition good. Explain control function in power mode and manual.
8. Horizontal stabilizer—Condition good; fairings in place. Explain stabilizer trim operation, elevator control function, and tab functions.
9. Static wicks—One on tail cone, one on upper vertical fin, two on each side of the horizontal stabilizer. All are required.

CONDUCT CHECK ON THE LEFT SIDE OF AIRCRAFT AS ON THE RIGHT SIDE PLUS THE FOLLOWING

1. Tail cone access door—Closed and secure.
2. Tail cone external release—Flush, stowed, and safetied.
3. Ground air connection—For ground start, identify and check that it is closed.
4. Tail compartment temp sensor and flapper—Identify and explain function.
5. Toilet service door—Closed. Explain vent under the door.
6. Cabin pressure outlets (outflow valves)—Clear and open; linked together. Explain function.
7. Radio rack venturi—Open. Explain function.
8. Forward toilet drain (DC-9-31)—Cover closed; no leaks.
9. Cabin pressure regulator safety valves—Closed. Explain function.
10. Main cabin doorsill drain—Open.
11. Airstair door safety—Unlocked.
12. External power receptacle—Door closed and secure; light switch off.

The flow pattern is similar for any airplane; the walk-around is actually based on common sense and aircraft knowledge. Just remember that the preflight is an important part of your rate ride. Keep walking and talking, use a planned flow pattern method, and you'll be surprised at how easy it is.

The Rating Ride

The hard part is over. You've completed the oral and the preflight (a combination operational walk-around and extension of the oral); all that remains is the flight portion of the rating.

A company check pilot will actually conduct the test, acting as your copilot, although actually the pilot in command. The format of the flight check has been agreed upon between the FAA and the company and will contain nothing that hasn't been covered in your flight training. The FAA will evaluate your performance and grade the flight for license purposes, the check pilot will also grade you for your company file, but there will be no surprises. When you are ready for your check, you should be thoroughly confident of your ability to fly the aircraft and perform anything required to better than minimum standards on every flight, so don't be concerned about your grading at this point.

Take command of the aircraft from the beginning, and use the check pilot like a brand new copilot on a first trip. The check pilot can't do anything to help you other than what you command. So keep this in mind and don't rely on reminders to correct any oversights. Half of your grade will be on command ability and judgment, so act like a captain. An experienced flight instructor, check pilot, or FAA examiner can very nearly tell what type of pilot you are by the time you have started the engines and taxied out for takeoff. A good early impression is important. Try to project yourself as able, confident, sure of your actions, and thoroughly in command of the airplane at all times.

Don't rush through anything, but don't be too slow. Be thorough, unhurried, and deliberate in your actions. Do whatever is appropriate, such as setting up the cockpit before starting the engines, and then call for the proper checklist. Checklists are not work lists but a check to assure that the proper procedures have been followed. Be sure you call for the appropriate checklist at the right time, and make your responses to the checklist challenges correctly.

Taxi smoothly in the center of the taxiway and use power and brakes judiciously for a safe speed as if you had flight attendants walking up and down the aisles checking seat belts and you didn't want to throw them off balance or cause injury. Make turns smoothly and at a speed that considers the people in the cabin behind you. The same applies on long straight taxiways; use a speed that would permit a safe emergency stop in a reasonable distance.

Be sure you understand correctly any clearance the inspector or check pilot may give you. You can't fly a clearance correctly unless you understand it.

Don't try to read a chart and fly at the same time.

Your flying will suffer, or you'll read the chart incorrectly, or both. If you want to take a look at a chart, have your copilot fly for a minute and give instructions to follow the clearance.

If you are not given the weather before each takeoff and approach and landing, ask what the weather is. The weather you expect to have will establish your possible need for a takeoff alternate, runway markings, etc., and will determine the type of approach.

The sequence of maneuvers will vary due to traffic, actual weather, the particular aircraft, etc., but don't be too concerned with the sequence. You've practiced everything in the check ride and more in your flight training. You can do anything the aircraft is capable of that may be required at any given time.

I'll outline a more or less standard sequence of maneuvers for a rating ride. It is patterned for the DC-9 but is very much the same check ride you may expect in any air carrier aircraft.

1. Rejected takeoff—If done in the aircraft, it must be accomplished at a speed *not* in excess of 50% of V_1.

2. Engine failure on takeoff (usually critical engine at V_1).

3. ILS with simulated engine-out to minimums (may be 100′ for Category II) and missed approach.

4. VFR zero flap approach and go-around.

5. Planned ILS or nonprecision approach (with a circling approach if your company still does them) with a touch-and-go landing.

6. A Category II (and a Category III if approved) to a full-stop landing.

7. Normal takeoff with hood up below 100′.

8. Area departure (merely flying a clearance as if departing the area).

9. Holding pattern.

10. Steep turns.

11. Emergency descent.

12. Approach to stalls.

13. Normal and abnormal operation of systems.

14. Emergency procedures.

15. Normal approach, either instrument or visual, to rejected landing.

16. Landing with simulated (in aircraft) 50% power loss in two- or four-engine aircraft and a single-engine approach in a three-engine aircraft.

The above, with the simulator and flight training in an approved airline training program, is all that is normally required. If you're rating in an aircraft from a nonapproved airline program or an approved manufacturer's or distributor's training school, I would suggest that you review FAR 121-F1. The requirements set forth in this regulation have been covered in the above check and approved training program, but the inspector will probably require in-flight demonstration of such things as instrument approaches using all the facilities and radio aids you have. These may be done in training (in a simulator or in flight) if they are observed by an approved check pilot or an approved instructor in the controlled and approved training program.

No matter what type of program you have been involved in, all the maneuvers are thoroughly covered in Chapter 6 of this book.

24

Weight and Balance

This book is not written as a manual for the inexperienced pilot to use in obtaining the initial Airline Transport Rating but rather as a guide to better understanding and practical application of the professional pilot's knowledge and skills. For this reason I will not attempt to write a complete manual on weight and balance in this chapter. This could be an entire book in itself, and I refer any that need a refresher for the ATR oral or written test to any of several good ATR study guides and manuals available on the market.

It must be assumed that the professional pilot already has an understanding of the principles, terminology, basic measurements and computations, weighing procedures, and loading and weight distribution. Therefore, I will confine this chapter to weight and loading procedures for large aircraft and illustrate a simple method for determining weight and balance for a jet transport (DC-9-31) in an air carrier fleet operation.

It is important that the aircraft be loaded properly, that the distribution of its weight be such that the center of gravity will fall within the envelope of CG limits set forth and described in the limitation section of the aircraft; otherwise, critical and adverse flight characteristics may result. An error in weight and weight distribution may result in the aircraft's being unable to rotate and take off, which could be very serious.

Even when loaded properly, weight is important. In previous chapters we discussed how stall speeds and takeoff performance are affected by weight. The total gross weight must not be in excess of the maximum for takeoff as determined by runway length, temperature, pressure altitude, runway conditions, and wind. There is no question of the importance of weight and balance, and it is always the pilot's responsibility to be sure that the aircraft is loaded properly and not in excess of maximum weight.

Large aircraft may be defined as aircraft with a gross weight of 12,500 lb. or more. In general use these include airline aircraft (scheduled and nonscheduled, cargo operation as well as passenger service), business and executive aircraft, and some cases of privately owned aircraft.

The methods of determining empty weight, useful load, and the effects on moment and arm with the addition of fuel, oil, and hydraulic fluid and the addition of passengers and cargo in various fuselage stations are essentially the same for all aircraft. The one exception for large aircraft is the method of handling the aircraft and the equipment required to weigh it. This presents a problem due to size and weight of the aircraft, although the basic procedure itself remains unchanged. The weighing procedures are usually done by the manufacturer before delivery of the aircraft; the airline operator at major maintenance stations has the equipment required, but the private operator who finds it necessary to reweigh the aircraft may have an expensive problem.

The necessity for reweighing a large aircraft after it has been placed into flight service depends on its use. Privately owned or executive aircraft owned by corporations are not required by regulation to be weighed periodically to ascertain any weight change during any specific period of operation. Minor changes in weight and balance may be calculated (such as changes in radio equipment), but aircraft should be reweighed after extensive alterations (such as complete redesign and installation of a new cabin interior) or after any modification that might create some doubt as to the accuracy of a computed weight and balance.

Airline aircraft are subject to more stringent and comprehensive rules and regulations regarding weight control. The professional pilot should be familiar with these regulations for a better understanding of the cargo and weight manifest and the weight and balance procedures used by the airline. All the rules and requirements are outlined in excellent detail in the FAA's *Advisory Circular AC 121-5—Air Carrier and Commercial Operations Aircraft Weight and Balance Control.* It covers the material so well that I am going to reproduce it in its entirety, with any remarks or comments of my own in brackets.

A. PURPOSE. This Advisory Circular provides a method and procedures for weight and balance control.
B. REFERENCES. This document is appropriate for the guidance of that segment of the public which operates or plans to operate aircraft in accordance with Federal Aviation Regulations, Part 121. [This would also now include club operations under Part 123.]
C. GENERAL. The operator may submit for inclusion in the operation specifications any method and procedure by which it can be shown that the aircraft is properly loaded and will not exceed authorized weight and balance limitations during operation. By whatever method used, the operator should account for all probable loading conditions which may be experienced in service and show that the loading schedule will provide satisfactory loading. Loading schedules may be applied to individual aircraft or to a complete fleet. [The airlines use a fleet method.] When an operator utilizes several types or models of aircraft, the loading schedule, which may be index type, tabular type, or a mechanical computer, should be identified with the type or model of aircraft for which it is designed.
D. LOADING PROVISIONS. All seats, compartments, and other loading stations will be properly marked, and the identification used should correspond with the instructions established for computing the weight and balance of the aircraft. When the loading schedule provides blocking off seats or compartments in order to remain within the center of gravity limits, effective means will be provided to assure that such seats or compartments are not occupied during operations specified. Instructions should be prepared for crew members, cargo handlers, and other personnel concerned, giving complete information necessary regarding distribution of passengers, cargo, fuel, and other items. Information relative to maximum capacities and other pertinent limitations affecting the weight or balance of the aircraft are to be included in these instructions. When it is possible by adverse distribution of passengers to exceed the approved CG limits of the aircraft, special instructions should be issued to the appropriate crewmembers so that the load distribution can be maintained within the approved limitation.
E. TERMS, DESCRIPTIONS, AND GENERAL STANDARDS.
1. Empty weight. The empty weight of an aircraft is the maximum gross weight less the following:
   a. All fuel and oil, except system fuel and oil. System fuel and oil is that amount required to fill both systems and the tanks, where applicable, up to the outlets to the engines. [Tanks are not usually applicable.] When oil is used for propeller feathering, such oil is included as system oil.
   b. Crew and baggage.

c. Drainable antidetonant injector and deicing fluids.
d. Passengers and cargo [revenue and nonrevenue].
e. Removable passenger service equipment, food, magazines, etc., including drainable washing and drinking water.
f. Emergency equipment [overwater, tropical, frigid].
g. Other equipment, variable for flights.
h. Flight spares [spare parts flyaway kits of spark plugs, wheels, tires, brake cylinders, etc.].
2. Operating weight. The basic weight established by the operator for a particular model of aircraft should include the following standard items of the operator in addition to the empty weight of the aircraft unless otherwise specified by the operator. [Most airlines use this last clause to add the weight of fuel, additional crewmembers in excess of the normal crew complement, and any special passenger service items such as champagne and liquor kits, etc., to determine a basic operational weight.]:
   a. Normal oil quantity.
   b. Antidetonant injector and deicing fluids.
   c. Crew and baggage.
   d. Passenger service equipment including washing and drinking water, magazines, etc.
   e. All other items of equipment considered standard by the operator.
   f. Emergency equipment, if required for all flights.
3. Aircraft, zero fuel weight. The zero fuel weight of an aircraft is the maximum weight authorized for such aircraft without fuel. The weight of fuel carried in the fuselage, or equivalent locations, will be deducted from such maximum. When zero fuel weight limitations or equivalent restrictions are specified, proper provision for loading will be made by the operator so that such structural limitations are not exceeded.
F. AIRCRAFT WEIGHTS. Aircraft weight and balance control systems normal contain provisions for determining weight in accordance with the following procedures:
1. Individual aircraft weight and changes. The loading schedule may utilize the individual weight of the aircraft in computing pertinent gross weight and balance. The individual weight and balance of each aircraft should be reestablished at the specified reweighing periods. It is normally reestablished whenever the accumulated changes to the operating weight exceed plus or minus one-half of 1 percent of the maximum landing weight or the cumulative change in CG position exceeds one-half of 1 percent of the mean aerodynamic chord (MAC).
2. Fleet weights, establishment and changes. For a fleet or group of aircraft of the same model and configuration, an average operating fleet weight may be utilized if the operating weights and CG position are within the limits established herein. The fleet weight will be calculated on the following basis:
   a. An operator's empty fleet weight is usually determined by weighing aircraft according to the following table: for fleet of 1 to 3, weigh all aircraft; for fleet of 4 to 9, weigh 3 aircraft plus at least 50 percent of the number over 3; for fleet of over 9, weigh 6 aircraft plus at least 10 percent of the number over 9.
   b. In choosing the aircraft to be weighed, the aircraft

having the highest time since last weighing should be selected. When the average empty weight and CG position have been determined for aircraft weighed and the basic operating fleet weight [winter and summer, if applicable] established, necessary data should be computed for aircraft not weighed but which are considered eligible under such fleet weight. If the basic operating weight of any aircraft or the calculated basic operating weight of the remaining aircraft in the fleet varies by an amount more than plus or minus one-half of 1 percent of the maximum landing weight from the established basic operating fleet weight, or the CG position varies more than plus or minus one-half of 1 percent of the MAC from the fleet weight CG, that aircraft should be omitted from the group and operated on its actual or calculated operating weight and CG position. If it falls within the limits of another fleet or group, it may then become part of that operating fleet weight. In cases where the aircraft is within the operating fleet weight tolerance, but the CG position varies in excess of the tolerance allowed, the aircraft may still be utilized under the applicable operating fleet weight but with an individual CG position.

  c. Reestablishment of the operator's empty fleet weight or the operating fleet weight and corresponding CG positions may be accomplished between weighing periods by calculation based on the current empty weight of the aircraft previously weighed for fleet weight purposes. Weighing for reestablishment of all fleet weights is normally conducted on a 2-year basis unless shorter periods are desired by the operator.

3. Establishing initial weight before use in air carrier service. Prior to being placed into service, each aircraft should be weighed and the empty weight and center of gravity location established. New production transport category aircraft delivered to operators normally are weighed at the factory and are eligible to be placed into operation without reweighing if the weight and balance records have been adjusted for alterations or modifications to the aircraft. Aircraft transferred from one operator to another need not be weighed prior to utilization by the latter unless more than 24 calendar months have elapsed since last weighing.

4. Periodic weighing, aircraft using individual weights. Aircraft operated under a loading schedule utilizing individual aircraft weights in computing the gross weight are normally weighed at intervals of 24 calendar months. An operator may, however, extend this weighing period for a particular model of aircraft, when pertinent records and routine weighing during the preceding 24 months of operation show that weight and balance records maintained are sufficiently accurate to indicate aircraft weights and CG positions are within the established limitations. Such applications should be limited to increases in increments of 12 months and should be substantiated in each instance with at least two aircraft weighings. In-

creases should not be granted which would permit any aircraft to exceed 48 calendar months since last weighing.

5. Periodic weighing, aircraft using "fleet weights." Aircraft operating under fleet weights should be weighed in accordance with procedures outlined for the establishment of fleet weights. Since each fleet weight is normally reestablished every 2 years and a specified number of aircraft weighed at such periods, no additional weighing is considered necessary. A rotation program should, however, be incorporated so all aircraft in the fleet will be weighed periodically.

6. Weighing procedure. Normal precautions, consistent with good practices in the weighing procedures, such as checking for completeness of the aircraft and equipment, determining that fluids are properly accounted for, and that weighing is accomplished in an enclosed building preventing the effect of wind, should prevail. Any acceptable scales may be used for weighing providing they are properly calibrated, zeroed, and used in accordance with the manufacturer's instructions. Each scale should have been calibrated, either by the manufacturer or by a Civil Department of Weights and Measures, within 1 year prior to weighing any aircraft for this purpose unless the operator has evidence which warrants a longer period between calibrations.

G. PASSENGER WEIGHTS. The operator may elect to use the average passenger weight to compute passenger loads over any route, except in those cases where nonstandard weight passenger groups are carried. Both methods may be used interchangeably provided only one method is used for any flight from beginning to terminating point of the particular trip or flight involved, except as indicated in subparagraph (2). Provisions should be incorporated in the load manifest to clearly indicate to personnel concerned whether actual or average passenger weights are to be used in computing the passenger load.

1. Average passenger weight. An average weight of 160 pounds [summer] may be used for adult passengers during the calendar period of May 1 through October 31.

      An average weight of 165 pounds [winter] may be used for each adult passenger during the calendar period from November 1 through April 30. [Most airlines add 5 pounds to these minimum average weights and use 165 pounds for summer weight and 170 pounds in the winter period.]

      An average weight of 80 pounds may be used for children between the ages of 2 and 12. Children above 12 years of age are classified as adults for the purpose of weight and balance computations. Children less than 2 years old are considered "babes in arms."

      The above passenger weight includes minor items normally carried by a passenger.

2. Nonstandard weight group of passengers. The average passenger weight method will not be used in the case of flights carrying large groups of passengers whose average weight obviously does not conform with the normal standard weight. Actual weights will be used when a passenger load consists to a large extent of athletic squads or other groups which are

smaller or larger than the U.S. average. Where such a group forms only a part of the total passenger load, the actual weights may be used for such groups and average weights used for the balance of the passenger load. In such instances, a notation should be made on the load manifest, indicating number of persons in the special group and identifying the group, i.e., football squad, Blank Nationals, etc.

Actual passenger weight may be determined by scale weighing of each passenger prior to boarding the aircraft, and such weight is to include minor articles carried on board by the passenger. If such articles are not weighed, the estimated weight is to be accounted for. The actual passenger weight may also be determined by asking each passenger his weight and adding thereto a predetermined constant to provide for hand-carried articles and also to cover possible seasonal effect upon passenger weight due to variance in clothing weight. This constant may be approved for an operator on the basis of a detailed study conducted by the operator over the particular routes involved and during the extreme seasons when applicable.

H. CREW WEIGHT. For crewmembers, the following approved average weights may be utilized:
1. Male cabin attendants 150 pounds; female cabin attendants 130 pounds.
2. All other crewmembers 170 pounds. [Most airlines use a 200-pound weight for cockpit crewmembers in addition to the normal crew, such as jumpseat riders, to also include their baggage.]

I. PASSENGER AND CREW BAGGAGE. Procedures should be provided so that all baggage, including that carried on board by the passengers, is properly accounted for. If desired by the operator, a standard crew baggage weight may be used. [Most operators use a standard weight of 30 pounds.] The following average passenger baggage weights may be used in lieu of actual weights:
1. For domestic operations.
   a. For each piece of checked baggage, an average of not less than 23.5 pounds.
   b. For each passenger boarding the aircraft, an average of not less than 5 pounds shall be added for hand baggage whether or not such baggage is carried by the passenger. [This is the 5 pounds added to average passenger weight I mentioned earlier. It is shown on the load manifest simply as "passenger weight."]
2. For trans-Atlantic flights between U.S. and Europe or trans-Pacific flights via Manila, Tokyo, Hong Kong, or Australia.
   a. For each piece of checked baggage, an average of not less than 26.5 pounds.
   b. For each passenger boarding the aircraft not less than 5 pounds shall be added for hand baggage whether or not such baggage is carried by the passenger.
3. For trans-Pacific flights originating and terminating at U.S., Honolulu, or Alaska.
   a. For each piece of checked baggage, an average of not less than 23.5 pounds.
   b. For each passenger boarding the aircraft not less

than 5 pounds shall be added for hand baggage whether or not such baggage is carried by the passenger [the same weight averages as for domestic operation].

Average passenger baggage weights shall not be used in computing the weight and balance of charter flights and other special services involving the carriage of special groups. [The 5 pounds for passenger hand baggage is used, but the baggage must be actually weighed.]

J. CENTER OF GRAVITY TRAVEL DURING FLIGHT. The operator will show that the procedures fully account for the extreme variations in center of gravity travel during flight caused by all or any combinations of the following variables:
1. The movement of a number of passengers and cabin attendants equal to the placarded capacity of the lounges and lavatories from their normal position in the aircraft cabin to such lounge or lavatory. If the capacity of such compartment is one, the movement of either one passenger or one cabin attendant, whichever most adversely effects the CG condition, will be considered. When the capacity of the lavatory or lounge is two or more, the movement of that number of passengers or cabin attendants from positions evenly distributed throughout the aircraft may be used. Where seats are blocked off, the movement of passengers and/or cabin attendants evenly distributed throughout, only the actual loaded section of the aircraft will be used. The extreme movements of the cabin attendants carrying out their assigned duties will be considered. The various conditions will be combined in such a manner that the most adverse effect on the CG will be obtained and so accounted for in the development of the loading schedule to assure the aircraft being loaded within approved limits at all times during flight. [These are design problems and most aircraft are so designed that any number of passengers, seated in any combination or configuration within any of the compartments in the cabin, or any movement of the passengers or flight attendants in flight, will not cause the CG to travel to such an extent as to cause it to exceed the CG limits if the cargo in the baggage compartments is loaded correctly according to cargo and baggage weight bin limits for varying groups of passengers.]
2. Landing gear retraction. Possible change in CG position due to landing gear retraction will be investigated and results accounted for. [This is another design factor and is shown in limitations if applicable for certain weights or passenger or cargo weights in varying locations.]
3. Fuel. The effect of the CG travel of the aircraft during flight due to fuel used down to the required reserve fuel or to an acceptable minimum reserve fuel established by the operator will be accounted for.

K. RECORDS. The weight and balance system should include methods by which the operator will maintain a complete, current, and continuous record of the weight and center of gravity of each aircraft. Such records should reflect all alterations and changes affecting either the weight or balance of the aircraft, and will include a complete and current equipment list. When fleet weights are used, pertinent computations should also be available in individual aircraft files.

L. WEIGHT OF FLUIDS. The weight of all fluids used in aircraft may be established on the basis of actual weight, a standard volume conversion, or a volume conversion utilizing appropriate temperature correction factors to accurately determine the weight by computation of the quantity of fluid on board. [The standard volume conversion is used.]

M. CONTENT OF OPERATIONS SPECIFICATIONS PROCEDURES FOR AIRCRAFT WEIGHT AND BALANCE CONTROL. The Operations Specifications, as submitted by an air carrier, will contain the procedures used to maintain control of weight and balance of all aircraft operated under the terms of the operating certificate which assures that the aircraft, under all operating conditions, is loaded within the gross weight and center of gravity limitations. This description should include a reference to the procedures used for determining weight of passengers/crew, weight of baggage, periodic aircraft weighing, type of loading devices, and identification of the aircraft concerned.

There is a great deal of good information in that circular, but I find that most pilots are concerned about the average weights applied to passengers and baggage. Everyone agrees that the actual aircraft weights are reasonably accurate and that a reasonable margin has been allowed in determining average fleet weights to account for dirt and dust that will accumulate in the aircraft and add to its weight; but there is some concern over the practice of estimating and averaging passenger, baggage, and cargo weights. However, these averages have been determined by studies over a long period of time and found to be fairly accurate with a reasonable tolerance of error, and these estimates and averages affect the payload only. Though the airlines are allowed now to estimate cargo weight, the percentage of error here is fairly small in that the weight of most cargo is actually known. So let's see how estimating and averaging passenger and baggage weight may affect the actual weight of the aircraft. Table 24.1 shows the weights applicable to aircraft in a typical airline fleet and the percentage of gross weight in payload.

This isn't intended to be a chart of aircraft efficiency, although the percentage of its total weight that may be carried in payload is some measure of an aircraft's work capability; but it shows the percentage of the maximum gross weight that may be put aboard in payload. It has been said that a reasonable tolerance of error in estimating weights would be 10%, a very liberal figure; such an error in the DC-9-31 would affect the total gross weight by about 2.5%. A 10% error in estimating the payload of a Boeing 727QC, however, would affect the total gross weight by 1.42%. The actual error is much closer to 5% than 10%, and the actual aircraft performance is not critically affected. However, in the higher-performance aircraft, those whose payload may exceed 20% of total weight, there is a definite need for careful loading and weight control.

The airlines comply with the provisions of *Advisory Circular AC-121-5* by the use of a cargo and weight manifest form. This is a form with which the pilot should be familiar and should use to check the aircraft for proper loading and weight to determine takeoff speeds. It is a multipurpose form used to record the total weight and bin location of cargo for each destination, to verify that the aircraft is loaded in proper balance, to compute allowable and actual payload, to compute actual gross weight and maximum allowable takeoff weight, and to calculate advance load planning. It also has other uses, but we're concerned only with the weight and balance functions. A typical cargo and weight manifest is shown in Figures 24.1 and 24.2.

Let's simulate a DC-9-31 flight from Miami to New Orleans and follow the preparation of the cargo and weight manifest form from the beginning until the time it is handed to the copilot just prior to departure. The preparation of the cargo and weight manifest begins in the operations office before the crew arrives at the field. The forms agent uses the *Gross Weight/Performance and Planning Manual* for the aircraft to determine maximum allowable takeoff and second-segment limiting climb weights for the takeoff power settings and flap setting that may be used. These are entered in the boxes in the upper center of the cargo and weight manifest. The agent takes the latest local weather and enters the temperature, wind, and runway that will provide the greatest wind and tempera-

TABLE 24.1. Weights for Representative Aircraft

Type Aircraft	Seating Configuration	Landing Weight (lb.)	Max Takeoff Weight (lb.)	Payload (lb.)	Payload Percentage of Max Gross Weight
DC-9-31	18/70	98,100	105,000	27,060	26
DS-9-14	16/50	81,700	90,700	21,930	24
Boeing 727	23/75	135,000	152,000	22,300	14.3
Boeing 727	23/75	137,500	160,000	26,300	16.2
Boeing 727QC	23/75	142,500	169,000	24,370	14.2
DC-8-21	26/111	199,500	275,500	36,980	13.6
DC-8-61	18/185	240,000	325,000	64,530	20.2
DC-8-63	28/167	275,000	355,000	64,650	18.8

CARGO AND WEIGHT MANIFEST

TRANSMIT TO (SYMBOLS)	FLIGHT/ORIG. DT.	MON./YR.	CARGO AND WEIGHT MANIFEST 24-00-0010 50-66 PRINTED IN U.S.A.	STATION	PLANE NR.	GROUP
//MSY CC//	//8Jun 16//	3/	①	MIA	948	

APPLICABLE TO DC-9-14

RWY	TEMP	WIND	CLIMB LIMIT TOGW
POWER	FLAP	TOGW	
D-5	20°		
	10°		
D-1	20°		
	10°		

⑥

APPLICABLE TO DC-9-31

RWY.	TEMP.	WIND		CLIMB LIMIT TOGW
27R	50°	320/5		
POWER	FLAP	TOGW		
D-1	15°	104400	98800	
	5°	108500	106400	
D-7	15°	105000	99900	
	5°	110 000	106900	

ESTIMATED LOAD ③

	THRU	
O N	EXPRESS	50
	FREIGHT	800
	COMAT	450
	MAIL	200
	BAGGAGE	2270
TOTAL CARGO		3770
PSGRS. 8/46	PSGR. WGT.	9180
TTL. EXTRAS		
MAX. T/O (ITEM 13)		105000
PAYLOAD		12950
ALLOWABLE OPR. WGT.		
MIN. FUEL		
MAX. FUEL		
SUGGESTED MAX. FUEL		
FUEL USED TO NEXT STA. PLUS MIN. OUT NEXT STA.		

CARGO LOCATION

TOTAL CABIN WGT. C		B-3	B-4									

CARRY-ON	THRU												
	ON												
	TOTAL												
REAR BIN	THRU												
	ON												
	TOTAL	2400	1700	700									

CAPTAIN GIVEN

✓	WEATHER FCST.	
✓	WEATHER	
✓	WINDS FCST.	
✓	WINDS	
		FAM

⑦

TO→ LBS.	MSY	DAL									
THRU Weight											

O N	T H I S	S T A T I O N										
EXPRESS		4	4									
FREIGHT	753	753										
COMAT	415	15	400									
MAIL	24	11	13									
F/C MAIL	173	70	103									
CHKD. BAGS	2350	1880	470									
CARRY-ON BAGS												
TOTAL	3219	2773	986									

CALCULATION OF ALLOW. PAYLOAD OPER. WEIGHT ④

MAX. P/L AIRPLANE	27060
MINUS EXTRAS	200
ALLOW. P/L AIRPLANE	26860
JUMP SEAT	200
ADDTNL. F/A(S)	
LIQUOR	
CHAMPAGNE	
OTHER	
TOTAL EXTRAS	200
BASIC OPR. WGT.	84592
OPR. WT. (USE FOR ITEM 6)	84792

⑧

		CABIN WGT.	BIN WEIGHT	TOTAL REAR WGT.	ALLOWABLE REAR		WGTS.	GALS./LBS. FUEL
					MAXIMUM	PSGR. GROUP	MINIMUM	
2	BALL B		CABIN WGT.	BIN WEIGHT				
3	TOTAL CARGO WEIGHT	3719	F 8 Y 44	3219	2400	3900	F 7-12 Y-41-60	1080	24652
4	PSGR. WEIGHT	8840	52	NR. PSGRS.				
5	TOTAL PAYLOAD	12559	OPERATIVE ADI	AUTO. FTHR.				
6	OPR. WGT.	84792						
7	ACT. T/O GROSS WGT.	97351	SPEEDPAK ATTACHED					

8		WIND	RUNWAY	FLAP	WIND ADJ. T/O WGT.	13	MAXIMUM T/O GROSS WEIGHT THIS FLIGHT (LEAST OF 10, 11 OR 12)	105000	⑤
9	TEMPERATURE	STAND.	ACT.	CRIT.		14	OPERATIONAL WEIGHT	84792	
				(LBS.) X		15	ALLOWABLE PAYLOAD (13 MINUS 14)	20208	
10	WIND & TEMP. ADJUSTED MAX. T/O GROSS WEIGHT ②				105000	16	ALLOWABLE PAYLOAD OF AIRPLANE	26860	
11	MAXIMUM ALLOWABLE T/O GROSS WGT. FOR ARPT.				105000	17	MAXIMUM ALLOWABLE PAYLOAD THIS FLIGHT (LESSER OF 15 OR 16)	20208	
12	MAXIMUM T/O WEIGHT FOR LANDING NEXT STOP				111500				

REMARKS

STATION SYMBOL	J	INITIALS SURNAME	W. PARKER FAA
	S	CLASS DEST.	C-1
		AGENT SIGNATURE	

FIG. 24.1. Cargo and weight manifest.

MAY THRU OCT. USE SUM. WGTS. **PASSENGER WEIGHT** NOV. THRU APR. USE WIN. WGTS.

SUM.	NO.	WIN.	SUM.	NO.	WIN.	SUM.	NO.	WIN.	SUM.	NO.	WIN.
165	1	170	6765	41	6970	13365	81	13770	19965	121	20570
330	2	340	6930	42	7140	13530	82	13940	20130	122	20740
495	3	510	7095	43	7310	13695	83	14110	20295	123	20910
660	4	680	7260	44	7480	13860	84	14280	20460	124	21080
825	5	850	7425	45	7650	14025	85	14450	20625	125	21250
990	6	1020	7590	46	7820	14190	86	14620	20790	126	21420
1155	7	1190	7755	47	7990	14355	87	14790	20955	127	21590
1320	8	1360	7920	48	8160	14520	88	14960	21120	128	21760
1485	9	1530	8085	49	8330	14685	89	15130	21285	129	21930
1650	10	1700	8250	50	8500	14850	90	15300	21450	130	22100
1815	11	1870	8415	51	8670	15015	91	15470	21615	131	22270
1980	12	2040	8580	52	8840	15180	92	15640	21780	132	22440
2145	13	2210	8745	53	9010	15345	93	15810	21945	133	22610
2310	14	2380	8910	54	9180	15510	94	15980	22110	134	22780
2475	15	2550	9075	55	9350	15675	95	16150	22275	135	22950
2640	16	2720	9240	56	9520	15840	96	16320	22440	136	23120
2805	17	2890	9405	57	9690	16005	97	16490	22605	137	23290
2970	18	3060	9570	58	9860	16170	98	16660	22770	138	23460
3135	19	3230	9735	59	10030	16335	99	16830	22935	139	23630
3300	20	3400	9900	60	10200	16500	100	17000	23100	140	23800
3465	21	3570	10065	61	10370	16665	101	17170	23265	141	23970
3630	22	3740	10230	62	10540	16830	102	17340	23430	142	24140
3795	23	3910	10395	63	10710	16995	103	17510	23595	143	24310
3960	24	4080	10560	64	10880	17160	104	17680	23760	144	24480
4125	25	4250	10725	65	11050	17325	105	17850	23925	145	24650
4290	26	4420	10890	66	11220	17490	106	18020	24090	146	24820
4455	27	4590	11055	67	11390	17655	107	18190	24255	147	24990
4620	28	4760	11220	68	11560	17820	108	18360	24420	148	25160
4785	29	4930	11385	69	11730	17985	109	18530	24585	149	25330
4950	30	5100	11550	70	11900	18150	110	18700	24750	150	25500
5115	31	5270	11715	71	12070	18315	111	18870	24915	151	25670
5280	32	5440	11880	72	12240	18480	112	19040	25080	152	25840
5445	33	5610	12045	73	12410	18645	113	19210	25245	153	26010
5610	34	5780	12210	74	12580	18810	114	19380	25410	154	26180
5775	35	5950	12375	75	12750	18975	115	19550	25575	155	26350
5940	36	6120	12540	76	12920	19140	116	19720	25740	156	26520
6105	37	6290	12705	77	13090	19305	117	19890	25905	157	26690
6270	38	6460	12870	78	13260	19470	118	20060	26070	158	26860
6435	39	6630	13035	79	13430	19635	119	20230	26235	159	27030
6600	40	6800	13200	80	13600	19800	120	20400	26400	160	27200

REMOVALS

	TOTAL WGT.	DEST. ON THIS FLT. —WGT.	DEST. ON THIS FLT. —WGT.	DEST. ON THIS FLT. —WGT.	DEST. ON THIS FLT. —WGT.	DEST. ON THIS FLT. —WGT.	DEST. ON THIS FLT. —WGT.	DEST. ON THIS FLT. —WGT.
EXPRESS								
FREIGHT								
COMAT								
AIR MAIL								
F/C MAIL								

FIG. 24.2. Reverse side of cargo and weight manifest.

ture-adjusted takeoff weight. This is marked with a circled figure 1 in Figure 24.1.

The actual takeoff capability of the aircraft on this runway and in these conditions of wind and temperature may exceed the certificate limits. But this is the information the pilot will use to determine the flap setting and power for takeoff, considering mostly climb limit or second-segment takeoff gross weight.

The agent then drops down to the bottom center of the form (marked 2 [circled]) to items 8 through 12. Items 8 and 9 are not applicable to this aircraft and are left blank. Items 10 and 11 (wind and temperature adjusted max takeoff weight and maximum allowable weight for the airport) have been determined from the *Gross Weight/Performance and Planning Manual* as being maximum, and are entered as 105,000 in these two boxes.

For item 12 (maximum takeoff weight for landing at the next stop) the agent must refer to (1) the weight chart for the aircraft and (2) the clearance or dispatch release message from the dispatcher for the planned fuel burn-off.

The maximum landing weight is 98,100 lbs. for the aircraft; the agent again checks the *Gross Weight/Performance and Planning Manual* for the maximum landing weight at New Orleans and finds it also to be max landing weight.

The agent now checks the dispatch release, which reads as follows: CLR 526 TO DAL VIA MSY. IAA BTR FOR MSY. GSW FOR DAL. FUEL 24,600 MIA. 18,600 MSY. 1130 Z TURLINGTON B/O 13,400/26 8,400/28. This translates: Clear flight 526 to Dallas via New Orleans, instrument flight authorized, Baton Rouge as an alternate for New Orleans and Ft. Worth alternate for Dallas. An estimated fuel burn-off of 13,400 lb. is computed for flight at 26,000′ to New Orleans and 8,400 lb. at 28,000′ from New Orleans to Dallas. The agent now adds the dispatcher's fuel figure, 13,400 lb., to the maximum landing weight, 98,100 lb., and enters the result in item 12—111,500. The sole purpose of this figure is to assure that the aircraft will be at max landing weight or below upon arrival.

Next the agent checks the number of passengers booked and the freight and cargo planned for the flight, and prepares the estimated load portion of the cargo and weight manifest in the upper right hand corner (marked 3 [circled]). The estimate is that the flight will depart with 50 lb. express, 800 lb. freight, 450 lb. company material, 200 lb. mail, and 2,270 lb. checked baggage for a total of 3,770 lb.

There are 8 first-class passengers booked and 46 coach passengers for a total of 54. The reverse side of the cargo and weight manifest (Fig. 24.2) gives the passenger weight breakdown of 9,180 lb. at winter weights. This gives a total estimated payload of 12,950 lb.

At this point an FAA inspector requests the jump

seat to Dallas; the agent, ready to calculate the allowable payload and operational weight, comes down the right side of the cargo and weight manifest to the area marked 4 (circled) and enters 200 lb. in the appropriate box and also shows 200 lb. as the total extra weight.

The agent refers to the weight chart (Fig. 24.3) and enters the max payload of the aircraft, 27,330 lb., minus the 200 lb. of the jump seat rider as extra weight, for an allowable payload of 26,860 lb. The basic operating weight of the aircraft (empty weight plus fuel) from the weight chart is 84,592 for the basic operating weight plus the 200 lb. for the jump seat and enters 84,792 as the operating weight.

In the area marked 5 (circled) the maximum takeoff gross weight of 105,000 lb. (the lesser of the weights shown in area 2 [circled]) and the operational weight of 84,792 are entered. Subtracting the operational weight from the maximum takeoff weight, the agent gets the allowable payload of 20,208 lb.

From area 4 (circled) is the calculated allowable payload, 26,860 lb. is obtained and entered in area 5 (circled). Comparing the two payload figures, the lesser of the two is taken for a maximum allowable payload for this flight, 20,208 lb.

Going again to the weight chart in Figure 24.3, the agent preplans loading for CG balance for the passenger grouping and cargo and baggage weight. Finding the total bin load at the position marked 1 (circled), it is noted that an estimated 3,770 lb. falls in the box between 3,500 and 3,999 lb. The agent then goes to the forward grouping of passengers, 8 being between 7 and 12, which is marked 2 (circled), and 46 in the rear between 41 and 60. Where these two columns intersect, it can be seen that the maximum rear weight would be 3,900 lb. and the minimum 1,080. This chart has been prepared by engineers; if the rear bins are loaded correctly, with all other weights in cargo and baggage loaded in the forward bins, the aircraft will be within CG limits for takeoff and during flight to landing. Therefore, the agent enters on the cargo weight manifest (Fig. 24.1) in area 6 (circled) 2,400 lb. to be loaded in the rear bins with 1,700 lb. in bin 3 and 700 lb. in bin 4. This provides ramp service with weight information to load the cargo and baggage.

The operations agent is now finished, and the estimated loading of the aircraft has been ascertained to be such that the weight and CG limits will be met and still provide some flexibility for additional passengers or cargo.

The cargo and weight manifest now leaves operations and is sent to the agent at the gate who supervises the loading of the aircraft and informs ramp service how the aircraft is to be loaded (weight distribution required as to maximum and minimum rear weight loading): 2,400 lb. is to be loaded in the rear

DC-9-31

12 - 3 - 70

Airstair removed:
a. Increase Payload to 27570.
b. Reduce B.O.W. 240.
c. Place first 420 Lbs. in Bin 1 as balast and use this same chart until revision is issued.

(D11)

EXTRAS

J/S RIDER	200 LBS.
4th F/A	180 LBS.

FUEL (POUNDS)	BASIC OPERATIONAL WEIGHT (POUNDS)
0	59,670
8,000	67,670
8,500	68,170
9,000	68,670
9,500	69,170
10,000	69,670
10,500	70,170
11,000	70,670
11,500	71,170
12,000	71,670
12,500	72,170
13,000	72,670
13,500	73,170
14,000	73,670
14,500	74,170
15,000	74,670
15,500	75,170
16,000	75,670
16,500	76,170
17,000	76,670
17,500	77,170
18,000	77,670
18,500	78,170
19,000	78,670
19,500	79,170
20,000	79,670
20,500	80,170
21,000	80,670
21,500	81,170
22,000	81,670
22,500	82,170
23,000	82,670
23,500	83,170
24,000	83,670
24,500	84,170
24,652	84,322

MAXIMUM PAYLOAD
CREW OF 5
(2 COCKPIT; 3 F/A)

27,330 POUNDS

REAR COMPARTMENT WEIGHTS

TOTAL BIN LOAD LBS.	NBR. OF PSGS.	0 ◄ FORWARD ► 6 AFT				7 ◄ FORWARD ► 12 AFT				13 ◄ FORWARD ► 18 AFT			
		0-20	21-40	41-60	61-70	0-20	21-40	41-60	61-70	0-20	21-40	41-60	61-70
0	MAX	0	0	0	0	0	0	0	0	0	0	0	0
	MIN	0	0	0	0	0	0	0	0	0	0	0	0
1	MAX	499	499	499	499	499	499	499	499	499	499	499	499
499	MIN	0	0	0	0	0	0	0	0	*0	0	0	0
500	MAX	999	999	999	999	999	999	999	999	999	999	999	999
999	MIN	0	0	0	0	0	~0	0	0	240	0	0	0
1000	MAX	1499	1499	1499	1499	1499	1499	1499	1499	1499	1499	1499	1499
1499	MIN	0	0	0	0	0	0	0	0	530	*0	0	0
1500	MAX	1999	1999	1999	1999	1999	1999	1999	1999	1999	1999	1999	1999
1999	MIN	0	0	0	0	*180	0	0	0	810	480	220	0
2000	MAX	2499	2499	2470	2499	2499	2499	2499	2499	2499	2499	2499	2499
2499	MIN	0	0	0	0	460	*120	0	0	1100	760	510	0
2500	MAX	2999	2999	2490	2999	2999	2999	2999	2830	2999	2999	2999	2999
2999	MIN	*40	0	0	0	750	410	*160	0	1380	1050	790	*50
3000	MAX	3010	3010	2810	3040	3499	3499	3499	2890	3499	3499	3499	3499
3499	MIN	330	*0	0	0	1030	700	440	0	1670	1330	1080	340
3500	MAX	3320	3320	3140	3200	3999	3999	3999	3060	3999	3999	3999	3650
3999	MIN	610	280	*30	0	1320	980	730	*0	1950	1620	1360	620
4000	MAX	3650	3650	3460	3380	4499	4499	4499	3240	4499	4499	4499	3770
4499	MIN	900	560	300	0	1600	1260	1010	270	2240	1900	1650	910
4500	MAX	3970	3980	3790	3550	4995	4995	4990	3410	4995	4995	4970	3950
4999	MIN	1180	850	590	0	1890	1550	1300	560	2520	2190	1930	1190
5000	MAX	4300	4300	4110	3730	4995	4995	4995	3590	4995	4995	4980	4120
5499	MIN	1470	1130	870	*140	2170	1830	1580	840	2810	2470	2220	1480
5500	MAX	4620	4480	4290	3900	4995	4995	4995	3760	4995	4995	4995	4300
5999	MIN	1750	1420	1160	420	2460	2120	1870	1130	3090	2760	2500	1760
6000	MAX	4950	4650	4470	4080	4995	4995	4995	3940	4995	4995	4995	4470
6499	MIN	2040	1700	1440	710	2740	2400	2150	1410	3380	3040	2790	2050
6500	MAX	4995	4830	4640	4250	4995	4995	4995	4110	4995	4995	4995	4650
6999	MIN	2320	1990	1730	990	3030	2690	2440	1700	3660	3330	3070	2330
7000	MAX	4995	4995	4810	4430	4995	4995	4995	4290	4995	4995	4995	4820
7499	MIN	2610	2270	2010	1280	3310	2970	2720	1980	3950	3610	3360	2620
7500	MAX	4995	4995	4990	4610	4995	4995	4995	4460	4995	4995	4995	4995
7999	MIN	2890	2560	2300	1560	3600	3260	3010	2270	4230	3900	3640	2900
8000	MAX	4995	4995	4995	4780	4995	4995	4995	4640	4995	4995	4995	4995
8499	MIN	3180	2840	2580	1850	3880	3540	3290	2550	4520	4180	3930	3120
8500	MAX	4995	4995	4995	4950	4995	4995	4995	4810	4995	4995	4995	4995
8999	MIN	3460	3130	2870	2130	4170	3830	3580	2840	4800	4470	4210	3740
9000	MAX	4995	4995	4995	4995	4995	4995	4995	4990		4995	4995	4995
9499	MIN	3750	3410	3150	2420	4450	4110	3860	3260		4750	4500	4030
9500	MAX	4995	4995	4995	4995	4995	4995	4995	4995	NOT		4995	4995
9999	MIN	4030	3700	3440	2700	4740	4400	4150	3680			4780	4310
10000	MAX	4995	4995	4995	4995		4995	4995	4995				4995
10999	MIN	4600	4270	4010	3550		4970	4720	4250	AVAILABLE			4880
11000	MAX		4995	4995	4995	NOT			4995				
11999	MIN	NOT	4840	4580	4390				4820				
12000	MAX				4995	AVAILABLE							
13425	MIN	AVAILABLE			4995								

BIN WEIGHTS LIMITS

BIN NUMBER	TOTAL CAP. LBS.	MAX. FLOOR LOADING LBS./SQ. FT.
FWD BELLY NO. 1 & 2	8,430	150
AFT BELLY NO. 3 & 4	4,995	150

IF J/S RIDER IS ABOARD:

(a) ALL MINIMUM REAR WEIGHTS SHOWN IN COLUMN AT ASTERISK (*) AND BELOW MUST BE INCREASED 200 POUNDS.

(b) MAXIMUM REAR WEIGHTS MAY BE INCREASED 230 POUNDS.

FIG. 24.3. Weight chart.

bins with 1,700 in bin 3 and 700 in bin 4. Later the agent receives the actual breakdown of total bin weight, as shown in area 7 (circled), and enters the total bin weight of 3,719 lb. along with its breakdown according to destination. This information will be sent to New Orleans so that they can prepare an estimated load considering through weight to Dallas plus whatever they may load.

After all the passengers have boarded, the gate agent prepares area 8 (circled) in the manifest by showing the total cargo weight: 8 passengers in front and 44 in the rear, the total bin weight with the total rear weight, and the maximum and minimum rear weights along with the passenger grouping that was used to determine these weights. The flight crew should check this carefully to be sure that the computations are correct and the weight charts were used correctly.

The agent now enters the correct passenger weight, taken from the reverse side of the cargo and weight manifest (Fig. 24.2). This, combined with the total cargo weight—8,840 plus 3,719—gives the total payload of 12,559. The total payload added to the operational weight—12,559 plus 84,792—results in the actual ramp weight of the aircraft—97,351 lb.

It is now reasonable to assume that the aircraft is not overweight and that the CG will fall within limits. However, it is still necessary to actually determine the exact CG, verify the weight, and use these to compute a stabilizer trim setting for takeoff. This may be done in a few seconds using a gross weight and CG chart and a weight and CG nomograph for the aircraft (Figs. 24.4 and 24.5).

The gross weight and CG chart is prepared by engineers by taking the known CG at the empty weight of the aircraft and constructing a graph for the effects of fuel, total cargo, rear cargo weights, and passenger groupings for the front and rear sections of the aircraft. Starting at the left of Figure 24.4, you see that the range of percentage of mean aerodynamic chord (MAC) is very wide—between 6 and 34.7%. It is known that at an empty weight of 60,000 lb. (rounding off the 59,940 actual) the CG is at 20.5% MAC. The effect of fuel upon both weight and CG is also known, so it is easy to construct the curve shown in the black line in the center of Figure 24.4. This establishes the ratio of distance between the weight lines (vertical lines on

the chart) and the expansion of the MAC lines according to weight.

After preparing this chart, a nomograph (Fig. 24.5) may be prepared to fit the scale of the gross weight and CG chart to draw in the effects of passenger and cargo weight.

To use the combination, the total bin weight must be interpolated (1 [circled] on Fig. 24.5 to use the numbers of flight 526 from the cargo and weight manifest), and this figure is placed contiguous to the total fuel aboard. Care must be taken that the vertical line of the nomograph is parallel to the gross weight lines of the chart, and a line is drawn upward from the total cargo weight to the apex of the vee and down again to the interpolated position of the total rear weight. This rear weight was 2,400 lb. and is marked with a 2 (circled) on the nomograph.

We then move the nomograph to place the number of passengers in the forward or first-class section adjacent to the end of the line drawn for the cargo weight and draw a line up to the vee and down to the number of passengers in the rear cabin (8 and 44, marked 3 [circled] and 4 [circled]).

Now we remove the nomograph and survey our line. We find that the line ends at a point just over 97,000 lb. (our weight sheet showed 97,351), that the CG is just under 19% MAC, and that the stabilizer should be set about 6.2 nose-up units for takeoff.

We can also look at the fuel line and subtract our planned burn-off from full tanks and find that the CG would be the same for about 11,000 lb. of fuel as it is for full tanks; therefore our CG will be exactly the same for landing as it is for takeoff. During flight, while burning from the center tank, it will move aft about 2.2% and then begin to move forward as fuel is taken from the main tanks.

I would advise you to check your weight and CG very carefully. Remember that the agents loading the aircraft know only how to use the charts and figures; they may make an error that would put you out of limits or in a CG situation that would cause poor performance by being slightly tail or nose heavy. If you have an accident and the aircraft is found to be loaded incorrectly, the pilot is at fault. This is another important reason why all airline aircraft should be equipped with both CG and gross weight instrument readout capability.

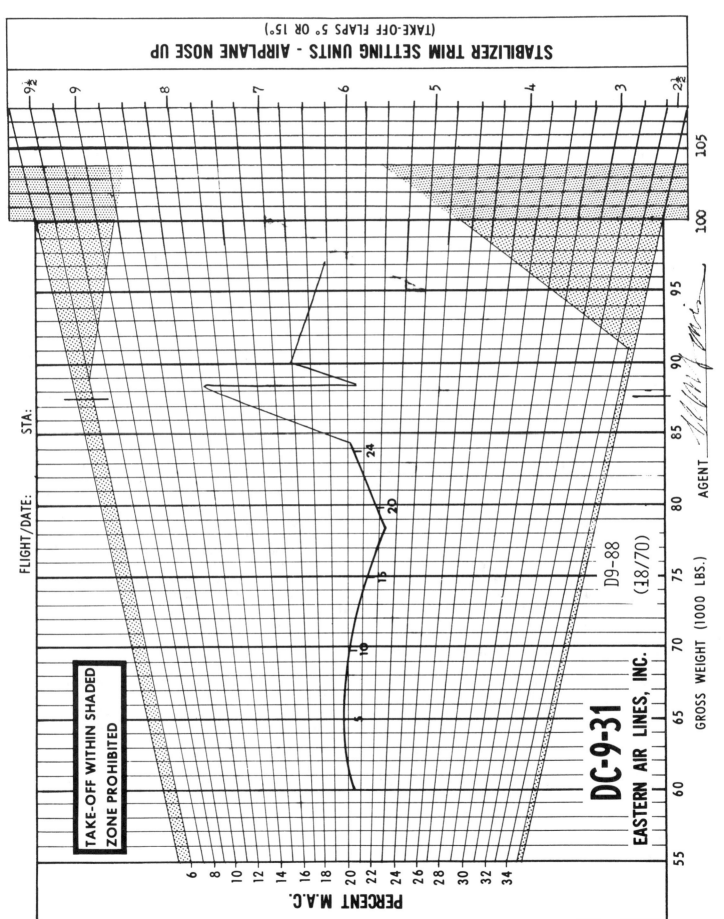

FIG. 24.4. DC-9-31 gross weight and center of gravity chart.

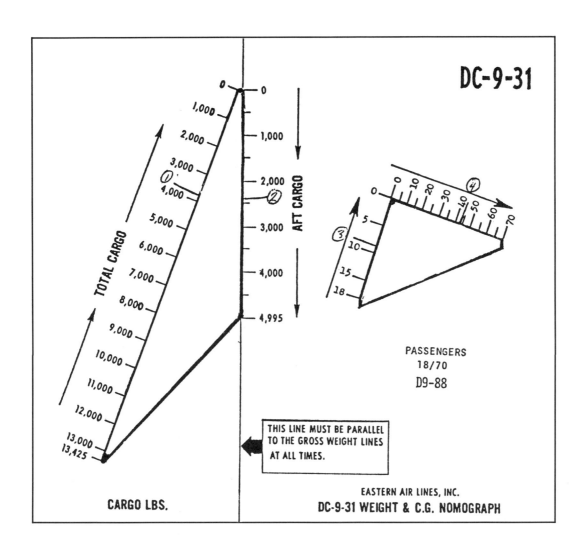

FIG. 24.5. DC-9-31 weight and center of gravity
nomograph.

25

Flight Planning

The ability to actually plan a flight with a high degree of accuracy is a good indication of a pilot's professional status, but with computerization this is becoming a lost art. Many airlines use the Center-stored flight plan system. Flight plans are prefiled for every route and aircraft type and include altitude, airways routing, and a standard true airspeed to the nearest rounded-off number ending in 0 or 5. Then, for every route over 430 nautical miles and many shorter route segments, the flight is given a computer flight plan, based on the routing of the Center-stored flight plan, with alternate routes provided for weather and winds and with suggested altitudes considering gross weight, winds, temperature, and operating costs.

It is unusual for a pilot to actually plan a flight—to select the proper altitude and route considering winds and weather, to file a correct true airspeed and an accurate off-to-on time, and to know with a reasonable degree of certainty what the fuel burn-off will be. A really competent pilot, familiar with the airplane, can perform the planning functions with a high degree of accuracy offhand without ever opening a book.

The computer flight plans—programmed for winds, temperature, altitude, gross weight, and aircraft performance—are highly accurate. With these advanced aids the airline pilot has an advantage over other pilots flying high-performance aircraft. However, flight planning by rule of thumb is easily done with 98% accuracy. I will give you these rules of thumb and then prove how accurate they are by comparing them to a computer flight plan followed by the actual flight performance, using both the rule of thumb and computer figures. To begin, let's isolate the factors we need:

1. Altitude—Without considering weather, there is an easily computed optimum altitude in relation to route distance.

2. True airspeed—This can be accurately determined for either a Mach cruise or constant indicated airspeed cruise based on temperature and altitude and is important for ground speed determination.

3. Time en route—This may be figured with a high degree of accuracy based on true airspeed and average wind component for route distance.

4. Fuel burn-off—This may be accurately computed in pounds per minute for ground operation, off-to-on time, and any anticipated holding delays.

Altitude

This is the easiest of all. Taking weather and cloud tops into consideration, simply use an altitude that would be the route distance times 100 for short-range flights. A flight of 60 nautical miles, for example, would be 60 × 100 or 6,000'. (These rules of thumb are mostly for jet operation but also hold true for prop and turboprop operation up to limits of operating altitudes, particularly up to the best or more efficient altitudes.) This distance times 100 holds true up to 250–300 miles; then higher altitudes must be considered for weather, winds, and fuel consumption. You couldn't reasonably (at least not yet) plan a 1,000-mile flight at 100,000'. But the rule of thumb will hold true for selecting optimum altitudes for flights of a distance that would compare to the maximum altitude of your aircraft. Flights of more than 250–300 miles will normally require altitudes at least above 25,000'.

True Airspeed

This is easy too. For props, you should know the indicated airspeed your aircraft would have at an average operating weight at the cruise power planned (normally max cruise in airline operation) at sea level standard. You merely use that speed as a base airspeed, divide your altitude by 1,000 and double that figure, and add that to the base speed.

For example, a DC-7 with an average load on a standard day indicates 250 knots. The true airspeed for max cruise at 10,000′ would therefore be 270 knots. The only variable that affects this speed is weight; temperature doesn't affect it a bit. The whole thing is strictly a function of gross weight, power, and altitude.

The formula for jet operation is somewhat different. There are actually three phases of airspeed considerations in a jet: (1) Mach cruise at altitude, (2) constant indicated airspeed cruise below the altitudes where Mach number is no longer the control factor, and (3) constant 250 indicated below 10,000′. Knowing as exactly as possible what your actual true airspeed will be is very important in flight planning. If you don't know, you can't determine an estimated average route ground speed; therefore, your time en route and fuel burn-off will be wild guesses at best.

MACH CRUISE

Most jet operation is at a Mach cruise number, defined as the ratio of the speed of the plane to the speed of sound at its altitude. If you plan to cruise at Mach 0.78 at flight level 300 (30,000′), the true airspeed would be 78% of the speed of sound at that altitude.

But don't get the impression that the speed of sound and therefore your cruise speed is controlled by altitude. The speed of sound depends on the temperature of the air and nothing else. The speed of sound at the North Pole, at say −66°F at sea level, is exactly the same as it would be at an altitude of 35,000′ with a standard temperature. The speed of sound varies with altitude only as temperature varies with altitude.

As a result of this relationship, the speed-of-sound curve follows the temperature curve, decreasing from 661.7 knots at sea level (for a standard day at 59°F) to 573.8 knots at the top of the troposphere. The speed of sound remains nearly constant, as does the temperature from 35,000′ to 105,000′, then rises to 729 knots, reverses, and falls to 603 knots at the top of the stratosphere. The curve then begins its final climb through the ionosphere.

We aren't concerned with much above 105,000′ yet, so we'll keep our figures down in the troposphere. Table 25.1 shows variation of temperature and speed of sound with altitude in the standard atmosphere.

The table ends at 40,000′ because all the factors, both standard temperature and therefore the speed of sound, remain the same (−56.5° and 573.8 knots) above that altitude, and this is the altitude structure in which we will be flying for a long time; 40,000′ will

TABLE 25.1. Temperature and Speed of Sound in Standard Atmosphere

Altitude (feet)	Temperature °F	Temperature °C	Speed of Sound (knots)
Sea level	59.0	15.0	661.7
5,000	41.2	5.1	650.3
10,000	23.3	−4.8	638.6
15,000	5.5	−14.7	626.7
20,000	−12.3	−24.6	614.6
25,000	−30.2	−34.5	602.2
30,000	−48.0	−44.4	589.6
35,000	−65.8	−54.3	576.6
40,000	−69.7	−56.5	573.8

be the upper limit I'll consider, and it will give you a good picture of operation up to 50,000–60,000′ if you can go that high. I mentioned earlier that the speed of sound and temperature remain nearly constant above 35,000′, but you will notice that the table shows differences of 2.2°C in temperature and 2.8 knots in the speed of sound between 35,000 and 40,000′. This is due to the slight temperature variation associated with the tropopause and is significant. This temperature variation will be discussed later; right now its significance is in the relationship of 2.2° to 2.8 knots change in the speed of sound, or about 1.27 knots speed change per degree of temperature change.

The temperature change in Celsius is roughly 10° for every 5,000′ of altitude change, or approximately 2° per 1,000′. Standard temperature may thus be determined for an altitude (within ½°) by dividing altitude by 1,000, multiplying by 2, and applying the +15° of sea level standard temperature algebraically. Thus, the standard temperature at 25,000′ would be 2 × 25, or −50°; then apply +15° to the −50° for approximately −35°. This rule of thumb may be used up to 35,000′.

The speed of sound from 20,000 to 25,000′ decreases 12.4 knots, or 1.2 knots per degree of temperature change; from 25,000 to 30,000′, 12.6 knots, and from 30,000 to 35,000′, 13.0 knots. This is very near the average of 1.27 knots per 1,000′ of altitude change we previously noted in the relationship of temperature and the speed of sound change from 35,000 to 40,000, or 2.2° as it relates to the 2.8-knot speed change.

We now have the tools for rule-of-thumb computations. Celsius temperature changes approximately 2° per thousand feet, and the speed of sound varies from 1.2 to 1.3, or 1.25 knots per degree of change.

To use this rule of thumb, we must first construct another table with some rounded-off numbers. Beginning at 20,000′, we will round off the standard temperatures and the speed of sound for every 5,000′ level, then extrapolate from Table 25.1 the temperature and speed of sound for every flight level between 20,000 and 35,000′, and use a rounded-off standard for flight levels above 35,000'. This table may also be used for reference for true airspeed for Mach cruise versus temperature, so I'll include some popular Mach numbers.

Indicated airspeed cruise is normally V$_{MO}$ − 10 to prevent nuisance overspeed warnings. You merely go into your aircraft performance charts one time to determine where your indicated cruise speed will cross your Mach cruise on a standard day and use Table 25.2 for the true airspeed for the Mach number at the nearest flight level. For example, the DC-9 cruises at 340 knots indicated, or Mach 0.78, which cross at 23,700' on a standard day. Thus 24,000', or flight level 240, becomes the nearest flight level, and Mach 0.78 at that altitude standard is 472 knots. That standard speed now becomes a *base speed*, and you merely subtract 2 knots per thousand feet (as revealed in Table 25.2) for every thousand feet you plan to fly above 240.

Now comes the important part. Either look on your computer flight plan at your wind and temperature variations from the standard averaged for your route or actually determine from the upper forecasts the actual average temperature for the route; then apply a temperature correction to the true airspeed. For example, we wish to cruise at DC-9 at flight level 310, which is 7,000' above our base of 240 at 472 knots for Mach 0.78 on a standard day; the average temperature from standard is forecast to be +12. We subtract 14 (2 × 7) from 472 for 458, and then add 12 knots for a true airspeed of 470.

To show you how accurate this is, look again at Table 25.2. The standard temperature at 310 is −47°C, but the temperature will average 12° *above* standard. Therefore, the actual outside air temperature will be −35°C; the speed of sound for −35°C is 603 knots; 78% of that is 470 knots. The actual temperature determines the speed of sound; you can file a true airspeed with a little knowledge of your aircraft performance and the charts I've shown you simply by memorizing base speeds for the lowest altitude of your

Mach cruise, taking off 2 knots per thousand above that altitude, and adding or subtracting temperature variations from standard.

The temperature variations from standard are very accurate in forecast, due to the number of flights reporting them all over the country with great frequency; so this method of computing and filing correct true airspeeds works very accurately—within 3 knots—every time.

BELOW MACH ALTITUDES

For computing true airspeed below your Mach cruise altitudes, the altitudes where you will be limited by indicated airspeed, subtract 5 knots (because now the relationship is to air density, temperature, and constant indicated airspeed rather than a speed of sound) for every thousand feet below your Mach cruise base and apply the following temperature variations from standard: 1° = 1 knot to flight level 180; from 18,000 to 10,000', 2° = 1 knot as the air is rapidly becoming denser. Below 10,000', for 250 indicated cruise, figuring it on the computer is the easiest way.

Time en Route

Time en route may be estimated very accurately by taking your true airspeed at your planned cruising altitude and applying the wind aid or retard factor forecast for your altitude for a ground speed at altitude. (If the wind aid/retard forecast is not available, you may have to "eyeball" the winds and figure your own wind component.) Then, instead of working a climb and descent problem, merely figure your time en route over the entire route distance at your average ground speed figure and add half your altitude divided by 1,000 in minutes for an off-to-on figure. This will take into consideration the factors of noise abatement, 250-knot climb to 10,000', normal climb cruise and

TABLE 25.2. True Airspeed

Flight Level	Temperature (°C)	Mach							
		1.0	0.70	0.72	0.74	0.78	0.80	0.82	0.84
200	−25	615.0	431	443	455	480	492	504	517
210	−27	612.6	429	441	453	478	490	502	515
220	−29	510.2	427	439	452	476	488	500	513
230	−31	607.8	425	438	450	474	486	498	511
240	−33	605.4	424	436	448	472	484	496	509
250	−35	603.0	422	434	446	470	482	494	507
260	−37	600.4	420	432	444	468	480	492	504
270	−39	597.8	418	430	443	466	478	490	503
280	−41	595.2	417	429	440	464	476	488	500
290	−43	592.6	415	427	439	462	474	486	498
310	−47	587.4	411	423	435	458	470	482	493
330	−51	582.2	408	419	431	454	466	477	489
350	−55	577.0	404	415	427	450	462	473	485
400	−57	575.0	403	414	426	449	460	472	483

descent back to 10,000', 250-knot descent below 10,000', and normal maneuvering for landing.

Let's construct a problem and then compare these rule of thumb computations to computer flight plans and actual flight performance. Let's plan a flight from Washington National (DCA) to St. Louis Lambert Field (STL): 643 nautical miles, at flight level 260, temperature forecast + 10° from standard, a wind retard of −24 knots, and planning to cruise at Mach 0.78.

Using our rule of thumb for true airspeed for Mach 0.78 in the DC-9, where V_{NO} of 340 knots normal cruise airspeed indicated crosses Mach 0.78 on a standard day for a true airspeed of 472, we subtract 2 knots per 1,000' we plan to cruise above 240 (4 knots) for a new base of 468 knots, to which we add 10 knots for the temperature variation above standard. The true airspeed for Mach 0.78 will therefore be 478 knots. Or we could use Table 25.2: 10° above standard for 26,000' would be 10° higher than the standard at 26,000', which is −37°, or −27°, and Mach 0.78 at −27° is 478 knots.

For the 643 nautical miles from DCA to STL (airport to airport) we subtract the 24-knot headwind for a ground speed of 454 knots. Running this through the computer for the entire route distance at a ground speed of 454 gives 1 hour and 25 minutes, to which we add one-half our altitude divided by 1,000 in minutes (½ of 26 = 13) for a total off-to-on time of 1 hour and 38 minutes.

It works with a high degree of accuracy. For cruise below 180 you figure true airspeed, use the wind component for a ground speed for route distance, but add two-thirds your altitude divided by 1,000 because of the limited climb, cruise, and descent above 10,000'.

Fuel Burn-Off

You should determine a fuel burn figure, in both pounds per hour and pounds per minute, for normal operation of your aircraft. Fuel burn in all jet aircraft is closely related to indicated airspeed; the fuel burn per hour at 300 knots indicated at 35,000' will be exactly the same as for 300 knots indicated at 10,000'. You need an average figure (in pounds per minute) for fuel requirements for off-to-on en route time, another figure for holding or en route delays (also in pounds per minute), and another for ground operation in taxi and ground delay. Of course, you can go into your charts and figure en route fuel requirements and time; but if you want to change your flight plan en route, you may not have the charts handy in flight or the time to read them. Rule of thumb knowledge is convenient at such a time and also helps to verify chart figures.

The DC-9, for example, is an easy machine for fuel planning. It uses about 100 lb. of fuel per minute for normal cruise at Mach 0.78, including climb and descent, 50 lb. per minute on the ground, and 80 lb. per minute holding in flight. This makes figuring fuel requirements without a chart a simple matter—so simple that I strongly recommend that you devise similar figures for the aircraft you may be flying.

From the previous en route time example, 1 hour 38 minutes or 98 minutes off-to-on, suppose you anticipated a 15-minute takeoff delay and a 15-minute en route holding delay and wished to arrive at your destination with 9,000 lb. of fuel on board. This would be minimum fuel to a 100-mile alternate after a missed approach, 10,000' cruise, and 5,300 lb. on board arriving over the alternate with zero wind. You always figure fuel requirements backward: fuel at alternate, fuel to alternate, fuel at destination, fuel en route, holding or contingency fuel, and taxi fuel. Then, to get your minimum fuel requirements (using the figures in the example), you merely add them up:

9,000	lb.	arrival at destination including alternate fuel
9,800	lb.	en route burn-off (98 minutes at 100 lb. per minute)
750	lb.	taxi fuel (15 minutes at 50 lb. per minute)
1,200	lb.	contingency or holding fuel (15 minutes at 80 lb. per minute)
20,750	lb.	minimum for the flight, which rounds out to 21,000 lb.

Now let's see how well these rule of thumb figures hold up by comparing the pilot's roughed-out figures with a computer flight plan for the flight and then the actual figures of the flight performance.

Flight Plan by the Pilot

I once flew flight 511 from DCA to STL in a DC-9-14. I chose to file Direct CSN, J-30 CRW, J-8 to STL, a distance of 643 miles, at flight level 310. The wind was forecast to average a 27-knot headwind and the temperature to average 9° above standard.

Applying the rule of thumb for true airspeed, I subtracted 14 from 472 (7,000' above base altitude of 240 at 2 knots per thousand) for a new base of 458, to which I added 9 knots (1° = 1 knot) for a filed true airspeed of 467 average for the route.

From this 467, I subtracted the average headwind of 27 knots for a ground speed of 440 knots average at cruise. Thus, 643 nautical miles at 440 knots is 1 hour 28 minutes plus one-half my altitude in minutes (15 minutes, since STL is 500' above sea level and DCA is sea level) for a total en route time of 1 hour 43 minutes (103 minutes).

This gave me an en route fuel requirement of

10,300 lb. at 100 lb. a minute, and the weather would be good on arrival; I took 1,200 lb. of taxi and contingency fuel with an estimated ground time of 13 minutes, and wanted to arrive at STL with 9,000 lb. on board. This was my method of computation to request a minimum fuel of 21,000 lb. (an extra 500 lb. for contingency) with a minimum fuel specified in my dispatch release of 20,000 lb. by the dispatcher.

Then I took a look at my computer flight plan, which I'll reproduce for you (Table 25.3) The first portion states trip number and date, route segment, type of aircraft, payload, arrival fuel, landing gross weight, distance, scheduled block time, and estimated ground time. The second portion indicates the route of flight.

board at the beginning of the takeoff roll and departure.

The computer flight plan called for a burn-off of 10,100 lb., which would give me a landing weight of 73,570 lb., or 2,570 lb. above flight plan weight. The computer flight plan (in a portion I haven't yet reproduced but will in another example) estimated that I'd burn 61 lb. per 1,000 lb. above computed landing weight. That would make my computer burn-off about 150 lb. more than en route burn-off, or 10,100 lb. plus 150 for a total fuel burn-off of 10,250 lb. My own figures, computed by rule of thumb, indicated a fuel burn-off of 10,300 lb. The computer and I were still very close in our figures.

TABLE 25.3. Computer Flight Plan

FLT 511/22 DCA-STL DC-9-14 Pyld 10,000 Arvl 9,000 LGW 71,100 Dist 643 SKED 1:57 EGT me 13 CSN J-30 J-8

To	Dist	MC	FL	OAT	Wind		Dft	TAS	G/S	DTGo	Tme	T/Tme	B/O	T/BO
TOC	188	267	31	−08	28	030	L01	378	351	453	32	32	048	048
CRW	037	266	31	−37	27	034	00	466	433	416	05	37	005	053
SDF	181	267	31	−37	28	038	00	466	439	235	25	1:02	022	075
BOD	121	280	31	−36	29	023	L01	466	444	114	16	1:18	015	089
STL	114	281	16	−05	30	018	L02	296	279	000	25	1:43	012	101

The main body of the computer flight plan shows: distance to checkpoints; MC, magnetic course; FL, flight level; OAT, outside air temperature; wind; Dft, drift; TAS, true airspeed; G/S, ground speed; DTGo, distance to go; Tme, segment time; T/Tme, total time; B/O, burn-off; T/BO, total burn-off.

I hadn't looked at the computer flight plan when I figured and filed my own figures. I took 310 as an altitude based on my own analysis of the weather, route, wind, and fuel considering payload and departure weight, which were estimated. For comparison, the computer figured 1 hour 43 minutes en route, and I figured 1 hour 43 minutes; the computer figured a fuel burn-off of 10,100 lb., and I figured 10,300 lb. and then added 1,200 lb. for taxi delay and contingency. Now let's look at the actual flight and compare my figures to the computer.

The flight left the ramp right on schedule, 9:05 A.M. EDT, with scheduled arrival in St. Louis of 10:02 CDT, or 1 hour and 57 minutes scheduled block-to-block time. I had a 21,000 lb. fuel load and 11,370 lb. payload (1,370 lb. above estimated payload) for a total ramp weight of 84,460 lb.—2,260 lb. more than the estimated gross weight indicated by the computer flight plan (landing weight plus fuel burn-off).

Takeoff time was 9:21 EDT, 1321 GMT (all times will be Greenwich mean time from here on); according to my fuel flow counters I had burned very nearly 800 lb. on the ground, leaving a total of 20,200 lb. on

Actual Flight

Now, let's look at the log of the actual flight (Table 25.4) and compare it to both the rule of thumb and the computer. (The first figures under each heading are from the computer flight plan, and the second figures are actual performance.)

After takeoff, the computer flight plan showed that I'd reach top of climb (TOC) at flight level 310 in 32 minutes, covering 188 miles at a speed at 351 knots with an average true airspeed in climb of 378 with an average 28-knot headwind.

The computer doesn't take into consideration such things as noise abatement, departure vectors, etc., and is based on a standard climb profile from 1,500′ above field level. The 250-knot climb below 10,000′ makes this impossible, so I reached top of climb in 20 minutes, or 12 minutes ahead of the computer flight plan. I requested direct to CRW when handed off to the Center by departure control to make

TABLE 25.4. Log of Actual Flight

To	Dist		Time		TAS		G/S	
TOC	188	118	32	20			351	354
CRW	037	107	05	15	466	471	433	428
SDF	181	181	25	23	466	473	439	470
BOD	121	125	16	16	466	473	444	472
STL	114	110	25	23			279	286

up for the 250-knot climb and reached top of climb at 1341 instead of the 1353 computer estimate. But, I'd covered only 118 miles instead of 188, and was 107 DME miles from CRW rather than 37. However, my average ground speed in climb had been 354 and very close to the computer's 351. I'd filed 469 for an average true airspeed, and the computer called for 466; actual was 471 on the static air temperature and true airspeed instrument with the temperature $-34°$ instead of $-37°$.

The computer flight plan called for being over CRW 37 minutes after takeoff, or 1358, but I was actually over at 1356. I was now 2 minutes ahead of the computer, but my ground speed from top of climb to CRW was 428, or 5 knots slower than the computer ground speed. I was encountering a headwind component of 43 knots rather than 33.

On the next segment, CRW to SDF, the computer said it would take 25 minutes at 439 ground speed. However, the temperature began to rise, going up 2° to $-32°$, or 15° above standard. My true airspeed went to 473, my drift began to be left instead of the zero indicated in the computer flight plan, indicating both a light chop and wind shift to a more northerly wind. The 181 miles of the leg was covered at a ground speed very near true airspeed. But this wasn't too surprising. It was apparent from the drift angle to hold the airway radial that the wind was shifting to the north, and the temperature rise coupled with the light chop coincided with the cirrus patterns with a base at my flight level. I was now 4 minutes ahead of the computer flight plan and revised my estimate to 1458, 6 minutes ahead of filed time, for landing at STL, assuming I'd pick up about another 2 minutes.

I flew the 16 minutes recommended by the computer to the beginning of descent and covered 125 miles rather than the 121, which confirmed my revised estimate. I began the descent, gave St. Louis a call, and estimated the ramp on schedule—at 1502. I'd been over SDF at 1419, began descent at 1435, and estimated landing at 1458 and parking at the ramp at 1502.

I landed at 1458 and parked at 1502, right on schedule. The actual flight time was 1 hour 37 minutes, 6 minutes faster than both computer and rule of thumb. This was due, no doubt, to the temperature variation from standard being more than forecast and the wind being just a bit different from the forecast.

The actual fuel burn from taxi-out to touchdown was 10,500 lb. The computer had called for a fuel burn of 10,000 lb. en route, but we'd actually burned 9,700 lb. off-to-on, or exactly 100 lb. per minute, plus 800 lb. in taxi before takeoff.

The point I want to make is the close comparison of computer flight plan time and rule of thumb figures. In this case it was exactly the same, although actual flight performance was a little faster. The rule of thumb fuel burn planning proved exact and every bit as accurate as the electronic brain.

I've checked actual performance against my own computations and those of the computer on every flight for years. It has proved to me that I can consistently and accurately plan true airspeed within 3 knots, off-to-on time within 3 minutes, and fuel burn-off within 300 lb. Sometimes the computer and my figures are exactly the same, sometimes not; but actual flight proves both my figures and those of the computer to be about equal in accuracy.

26

Icing

Icing conditions exist when two things are present: visible moisture and temperature within a range that would be capable of changing this moisture to ice. These conditions may be found during any flight, during all seasons of the year, and in any climate.

I'm not going into great detail on ice from the meteorological viewpoint. The libraries are full of books describing meteorological phenomena, including ice and icing conditions, but I've never found any really good information relating to actual flight operations in these conditions. Therefore, I'm going to stick to a cockpit viewpoint. You need the knowledge to pass written exams, but you also must know how to operate in icing conditions.

In the early days of instrument flight, when ice was first encountered as a hazard to flight, several precepts evolved regarding flight into known or possible icing conditions. They are still valid today, though in application they apply mostly to aircraft without deicing equipment or with boots only. They are briefly:

1. *Do not* plan or continue your flight into a region or altitude of known icing conditions.

2. *Do not* fly through rain showers or wet snow, if it's possible to avoid it, at a flight level where the temperature is near freezing.

3. *Do not* fly parallel to a front under icing conditions.

4. *Do not* fly into clouds at a low altitude above the crests or ridges of mountains. No less than 4,000', preferably at least 5,000', above the ridges should be your lowest altitude on instruments over mountains for several reasons; the possibility of ice is one.

5. *Do not* fly into cumulus clouds at low temperatures if they may be avoided. You might pick up a good load of heavy glaze ice.

These are basic reminders that apply to operation at slower speeds and are rather difficult to put into practice in today's crowded airways with the attendant problems of traffic control, but they are still good rules to remember. Even a jet is a slow aircraft in the takeoff and landing configurations.

Over the years, flying in all kinds of icing conditions in different types of aircraft (taking off and landing when icing conditions were reported as "light" and then getting clobbered, encountering some really heavy icing situations shortly after takeoff or on landing approach), will cause the experienced pilot to make a rather extensive study to determine the temperature zones where freezing precipitation will occur.

I will pass my version along—for whatever use it may be to you. It has been most useful to me in anticipating conditions I might encounter (based on surface observations, temperatures, and types of precipitation) in planning flight departures and approaches. I have also found it useful in flight where the type of precipitation may be identified and the exact temperature where it is encountered is known. A knowledge of the temperature zones where freezing precipitation occurs and an understanding of the characteristics involved in formation of different types of freezing precipitation provide the pilot with some rules of thumb for either avoiding or getting out of icing zones. Bear in mind that it is rule of thumb information, not necessarily exact in all cases; however, the incidence of conditions varying from those described is very rare.

This information and these figures were obtained from years of actual observation and experience, and the meteorological experts generally agree that they are valid and correct. But they won't commit themselves on sleet, saying that not enough is actually known about the formation of sleet to make any categorical statements. Sleet is an icing phenomenon that may be encountered in flight, and every pilot should know something about it.

Sleet

Sleet is the popular name for transparent, round, small ice particles that are actually frozen raindrops. Sleet, then, is ice pellets. These pellets are produced by raindrops formed in clouds in an upper warm-air layer, becoming frozen while falling through a lower air layer having a temperature well below freezing.

Observations based on flight experience indicate that rain must fall through about 4,000' of this lower air layer well below freezing to freeze and solidify. Therefore, if sleet is present, *there will be a layer of warm air about 4,000' above it* where it will be warm enough for the precipitation to be rain. This warm-air layer has been found to be another 4,000' in depth, so you may expect temperatures *above* freezing from 4,000 to 8,000' above the level where sleet is present.

If you climb through sleet, you must expect to encounter precipitation in the form of freezing rain in the upper portion of below-freezing air—anywhere from 2,000 to 4,000' above the level of the sleet. You would experience moderate to heavy icing and should either stay out of these levels or accelerate climb or descent through them.

Freezing Rain Mixed with Sleet

When this condition occurs, it is within the transitional zone of rain to sleet just described. As a rule of thumb, consider that the below-freezing temperatures in this zone will extend 2,500–3,000' *above* the level where both sleet and freezing rain exist. Again, expect to encounter moderate to heavy icing in climbing or descending through this layer of precipitation. Accelerate climb and descent and avoid level flight in this type of precipitation.

You could also expect to be in the warmer air above after 3,000' of climb, and this layer of warmer air will be at least 3,000' in depth. You want an altitude of at least 3,000', preferably 4,000', above the level where freezing rain mixed with sleet exists in the warmer air above it. For low-level flight, either stay in this warm air layer or climb into the colder air above it.

Freezing Rain

For this type of precipitation to exist in the form of water droplets, it must start as rain from a warmer air layer in the clouds above and fall into air below freez-

ing. It has been found, from both ground and flight observations, that freezing rain as a form of precipitation is commonly found in a temperature zone of about 28°F, or about −2.2°C. It hasn't had time to freeze into sleet and is a supercooled water droplet that will freeze on contact.

Heavy icing will be encountered in flight through freezing rain, even in accelerated climb and descent, and constitutes the great hazard associated with this type of precipitation. When freezing rain exists at the surface level, flight operations are normally suspended. However, if you fly an all-weather operation, it is something you will certainly encounter in flight.

You wouldn't take off from or make an approach to an airport reporting this condition, but you may encounter this condition in any phase of your flight even though not reported as an airport weather condition. If the weather is such that precipitation in the form of rain would be produced and this precipitation is encountered in the temperature zone of freezing rain (such as in scattered showers), the possibility of encountering icing conditions is ever present.

If freezing rain exists, the layer of below-freezing air where it is present is probably no more than 2,000' thick. Above that, just as for sleet and freezing rain mixed with sleet, there is an above-freezing zone of air approximately 4,000' in depth. That is to say that at 2,000–6,000' above freezing rain there is a warm layer of air containing precipitation in the form of rain. Sometimes the above-freezing layer is more than 4,000' in depth but rarely less.

The type of cloud that produces either sleet or freezing rain probably has snow in the below-freezing air above the layer of warmer air. The rule of not staying at the level of the freezing precipitation, climbing into the warm air layer above it, still applies. Or climb well into the much colder air above and expect icing conditions in snow when climbing above the warm air layer forming the freezing rain or sleet.

If freezing rain exists at the surface, for low-level flight you want an altitude of 3,000–6,000'. In general, you must assume that a second freezing level exists at about 6,000–7,000' above the surface, and you want to be in the warm air below it but above the freezing rain or else at an altitude well above the second freezing level.

Freezing Drizzle

This is much like freezing rain, commonly found with a temperature of about 28°F (−2.2°C), but the cloud formation producing it is different from that producing heavier precipitation.

When freezing drizzle is encountered, either in flight or at the surface, the layer of below-freezing air exists only about 1,000' above the level of the freezing drizzle. There will be a layer of warm air, just as for sleet and freezing rain, above this level that will be at least 3,000' and probably 4,000' in depth.

As an example, a departure in a light freezing drizzle would require an initial low-level altitude of 2,000-3,000' and then an unrestricted climb to well above the freezing zone above.

SUMMARY

1. Sleet—Plan flight, or climb, 4,000-8,000' (preferable 5,000-7,000') above the level where sleet is present. Expect heavy ice between 2,000 and 4,000' of climb in freezing rain.

2. Freezing rain mixed with sleet—Plan flight, or climb, 3,000-6,000' (preferably 4,000-5,000') above the level where freezing rain mixed with sleet is present.

3. Freezing rain—Plan flight, or climb, 2,000-6,000' (preferably 2,000-5,000') above the level where freezing rain exists.

4. Freezing drizzle—Plan flight, or climb, 1,000-4,000' (preferably 2,000-3,000') above the level of freezing drizzle.

In all of the above conditions you may climb well into the colder air above, being aware that a second freezing level will exist at higher altitudes, but expect icing conditions in snow during the climb above the warm air layer.

Snow

Snow is a solid form of water that grows while floating, rising, or falling in the free air of the atmosphere. Temperature is the primary condition controlling the formation of snow, along with the moisture present. But it is more complex that that. It takes a nucleus and water present at the proper temperature to form snow. The polluted air of today provides plenty of nuclei, so we need consider only the temperature and water.

Snow, which is actually ice crystals, forms at temperatures in a range of -40 to $0°C$. We see snow or ice crystals in cirrus clouds, for example, when the air contains the necessary moisture and the temperature is about $-30°C$. In clouds, visible moisture (snow) will form in the temperature range of -40 to $0°C$. The actual formation depends on the various chemical forms of the nuclei, forming snow and ice crystals at different temperatures for different nuclei, but we don't need to know too much about that. As pilots we are more concerned about the temperature than anything else; snow is dry ice crystals below $-15°C$ and a potential icing hazard above that temperature.

Most of the heavy to moderate rain forms in clouds that have great vertical development. This rain begins as snow and then melts, going through all the transitional phases of snow, sleet, freezing rain, and finally just rain. But we are most concerned with snow when it affords the greatest hazard to flight—when it is wet snow or heavy slush.

Wet snow is encountered in flight, as well as being observed on the surface, in one of three situations: (1) in a temperature slightly above freezing, (2) mixed with rain, or (3) in a temperature below freezing. Each presents a slightly different situation, but we know that snow is formed in temperatures of freezing and below; to become wet snow it must fall through a layer of warm air.

WET SNOW IN ABOVE-FREEZING TEMPERATURE

This usually occurs at a temperature of about $34°F$ $(1.1°C)$ on the surface as well as in flight.

At the level of wet snow at this temperature, you may expect the layer of above-freezing air to be shallow—probably no more than 1,000-1,500' thick. The trick here is to get above this temperature level. Climb at least 2,000' above this temperature and wet snow condition; below-freezing temperatures and dryer snow will be encountered. You could either cruise below the cloud bases or, if still in wet snow, climb well up into the cold air at higher levels. In climbing above the wet snow, you could expect to pick up light rime ice in the clouds.

RAIN MIXED WITH WET SNOW

This is a situation exactly like wet snow in temperatures above freezing with one exception: the warm layer of air in which it is present will be approximately 3,000' in depth.

WET SNOW WITH BELOW-FREEZING TEMPERATURE

This is an entirely different situation. The snow must form in very cold temperature, fall through a layer of warm air to melt slightly and become wet snow, and then fall into another layer of below-freezing air.

The layer of below-freezing air in which the wet snow is encountered would be very shallow—less than 1,000' if this condition exists at the surface. If encountered in flight, it could have slightly more depth, but wet snow will turn again to normal snow after falling approximately 1,000' in below-freezing temperature. Therefore, in wet snow at temperatures below freezing, you may expect a layer of warmer air about 1,000' above it.

The depth of this stratum of above-freezing air will depend on the temperature we find upon entering it. That is to say, this layer of warmer air above wet snow in below-freezing temperature will vary in depth according to its temperature. If the air in this warm layer was found to be 1°C upon entering it, it might be several thousand feet thick; if it is found to be 2–3°C, it will probably be only about 2,000' thick. Above this warm layer of air, whatever its depth, the temperature will drop again to well below freezing.

It's virtually impossible to cover all the situations and varying circumstances related to icing conditions, and the foregoing should not be regarded as absolutely correct in all conditions. However, the principles I've described will *always* be applicable—proven reliable by flight experience except in the most unusual situations—and should be considered as rules of thumb relative to the icing conditions I've listed. I've found them helpful for flight planning when these conditions exist at the surface, particularly when no reliable temperature-aloft reports are available. They are good, workable principles that may be used in flight to climb out of icing conditions where the type of precipitation is recognizable and the temperature known.

For those of you still flying conventional propeller-driven aircraft with reciprocating engines that do not have the speed and ability to operate at the altitudes of jets and turbine-powered aircraft (and therefore are exposed to icing more often and for longer periods of time), I would remind you that carburetor preheat, like engine anti-ice for the turbines, should be applied *before* entering conditions conducive to ice if possible.

Also, when flying in icing conditions, don't allow your indicated airspeed to drop more than 5–10 knots before applying more power. It is also good practice to keep your attitude as flat as possible and climb with higher airspeed than usual so that the lower surfaces of the aircraft will not be iced by flight at a high angle of attack.

There's also another principle that may be applied when the temperature of the outside air is at or close to freezing—frictional heat. At 5,000' with a true airspeed of 180 knots and a temperature of 0°C, the friction of the air passing over the wings and aircraft surfaces raises the temperature of the aircraft skin 4°C. Under the same conditions at a true airspeed of 230 knots, the skin temperature is increased about 6°C over the outside air temperature. At a true airspeed of 140 knots, the frictional heat rise is 2.5°C. By utilizing this principle, it is sometimes possible to "manufacture" heat rise that will take ice off or keep it from forming. Generally speaking, ice forms on aircraft surfaces in visible moisture or precipitation with a true outside air temperature between −4 and +1°C (26–34°F), and the surface of the aircraft must be at a temperature of freezing or below for it to stick.

The best rule of all is, Stay out of Icing Conditions. This, of course, is not always possible, so you must know how to operate your deicing and anti-icing equipment and then get out of the icing condition as quickly as you can.

27

Use of Anti-Ice and Deice

Ice and icing conditions do not present the same problems or hazards to today's modern aircraft that they do to aircraft with less capability. The Lockheed Electra (L-188), for example, has an anti-ice and deice capability allowing it to operate in virtually any icing condition. The turbojet aircraft operate at high speeds and are able to climb and descend through heavy icing levels with such great speed that icing problems are rare. But not all pilots operate these types of aircraft, and proper techniques in the operation of anti-ice and deice equipment is essential to safe operation. Even the jet pilots would do well to understand their equipment and use it intelligently. Improper use of engine heat in a jet engine is one of the causes of engine failure and shortens the span of time between overhauls.

Reciprocating Engines

Engines will not run without three things—fuel, air, and ignition. The fuel and air are furnished through a carburetion system; in conditions conducive to carburetor ice, such as heavy wet snow, the engines can ice up quickly. The proper use of preventive measures in such conditions is mandatory.

Carburetor ice may take many forms. It may block the throat of the carburetor, it may close off the fuel jets in the carburetor, or it may block off the air intake. This last is called screen ice; some engines have an alternate source of air, under-cowl air, so are less conducive to this type of ice.

The secret to using carburetor heat is to apply it *before* entering icing conditions. Go to a full rich mixture and apply at least 20°C preheat. The best way is to pull the heat full on, back it off to the desired amount, and then relean the engines. Then get to another altitude out of the ice! In a heavy precipitation the engines can still ice up.

Some engines are provided with alcohol as a deice measure. This washes down the carburetor and removes the ice but also cools the carburetor. It's kind of a last resort and not recommended for use in conjunction with heat. It upsets the complex functions of the fuel control, radically changes the temperature sensing, and should always be used with a rich mixture. If you ever have to use it, be sure to get out of the icing level at once, because the carburetor will probably ice up even more after its use.

Another system is the alternate fuel source (or prime all engines). If the engine quits after heat has been applied, fuel will be supplied from an alternate source and provide enough power to get more heat to the engines. Get the heat full on and change altitude after using the alternate fuel source.

One of the most aggravating and disconcerting things that can happen is to have the throttles stick so you can't move them. This usually happens to the unwary pilot who makes a long descent at reduced power. Heat robs an engine of power, but when you don't need power, heat is very cheap. If you're descending into a suspected condition of engine ice (particularly long low-powered descents), keep enough power on to make heat and apply it. Then move the throttles from time to time to keep them free.

The things to remember are: Apply the heat before entering icing conditions, and then get out of those conditions as quickly as possible!

Engine ice is a little different in turbine engines, more easily handled, but can be extremely serious if anti-ice procedures are not properly applied.

Jet Engines

The first consideration is fuel heat. You might find a fuel temperature gauge in your cockpit and possibly a fuel-filter-pressure-low light on your annunciator panel. There will also be fuel heat control switches located either in your anti-ice and deice control panel or near the ignition switch. The first two tell you when fuel heat is necessary, and the last gives you the capability of heating the fuel. This is done with a timer, usually for 1 minute, that bleeds a slight amount of air off your compressor and sends it through an air heat exchanger before the fuel goes into the fuel control. There are a filter, a pressure differential switch that turns on the fuel-filter-pressure-low light when the filter begins to clog, and a filter bypass to allow fuel to bypass the filter and enter the fuel control when the filter is clogged.

Kerosene is just a bit colder than its surrounding temperature and also has the characteristic of being homogenous with water, allowing water to mix with it rather than separate. When it is cold enough, ice crystals and slush form in the fuel and clog the filter. The filter bypass then allows the fuel, with its ice crystals and slush, to enter the fuel control—which brings about more problems. But when it begins to bypass, the pressure differential switch turns on the fuel-filter-pressure-low light and indicates the necessity for fuel heat.

If the ice and slush reach the fuel control in sufficient quantity, they may eventually cause the loss of the engine. In making power changes you might discover an iced-up fuel control, with the engine not responding to throttle or power control movement. The use of fuel heat is required in such a case; in an extreme case, close the throttle and snap the fuel control lever from off to on to shock the fuel control.

Here are some rules of thumb for the use of fuel heat in icing conditions:

1. Prior to takeoff, if the fuel temperature is less than 10°C, cycle the fuel heat but *do not have it on during takeoff.*

2. When beginning descent from high altitude where the wing is very cold and may sweat, thus dropping moisture into the kerosene in the tanks, cycle fuel heat.

3. Prior to approach for landing after operating at high altitude, cycle fuel heat but *do not have it on for landing.*

4. When fuel temperature is 10°C or lower in cruise, cycle fuel heat about once every 20 minutes.

5. Use fuel heat any time the fuel-filter-pressure-low light comes on.

The most serious ice for a jet engine would be that blocking the air intake, indicated by a rise in engine pressure ratio (EPR). In other words, the engine would be taking in less air and compressing it more to do the work. EPR is merely a readout of the difference in pressure between the air entering the engine and the air leaving it at the rear of the turbine. Blocking the airflow intake will cause the EPR as well as the fuel/air ratio to rise and will affect fuel flow and exhaust temperature accordingly (and possibly RPM). The best preventive measure is to use engine heat *prior* to entering icing conditions.

The entire cowling area is heated around the intake or cowl, in the inlet guide vanes, in the pressure intake probe, and usually in the intake "bullet" by bleeding off hot air from the compressor. The cowling, being the largest area and requiring the largest volume of air, is supplied hot air from one of the aft stages of compression; the intake probe and bullet and the inlet guide vanes are supplied from a more forward stage of compression. The hot air from the cowling is usually dumped overboard, but the air from the intake probe and inlet guide vanes is reingested into the engine.

The use of engine heat robs the engine output of about 6–10% of its power due to the reingestion of the hot air, so it is not recommended that it be turned on at takeoff power. It may be used for takeoff, however, by turning it on prior to applying takeoff power. Some engines then require a correction to takeoff power settings with engine heat on. You'll just have to know your particular engines and the corrections to apply for power settings with engine heat.

The best way to use engine heat—in fact the only way—is in such a manner as to prevent the formation of ice. It is an *anti*-ice system. Turn it on before entering icing conditions and *do not* use it as a deicer except in an emergency. If you let the ice accumulate and then turn on the engine heat, you would remove the ice. You would also throw large chunks of ice right into the engine, and the ingestion of foreign objects into a jet engine can cause great damage if not total destruction.

In aircraft without a static air temperature gauge corrected for ram rise, the temperature range where engine heat is required is hard to figure. The total air gauge is not really adequate for this function. Many times a pilot with a total air temperature gauge will not use engine heat when it is required except in the most obvious conditions and will pick up a slight amount of ice, which is invariably ingested into the engine. This materially shortens the life of the engine.

You need a good temperature gauge for flight in icing conditions, one that gives a static air temperature corrected for ram rise or one that gives a total air temperature and requires continual interpolation for actual outside air temperature, considering ram rise.

There are some good rules of thumb for the use of engine heat. Engine heat should be turned on before takeoff and on the ground when the temperature is 6°C or less and any kind of visible moisture is present.

Some engines will even ice at idle with no visible moisture present if the temperature is around 6°C and the temperature and dew point are within 5°. This is due to the pressure change and expansion of the air going into the engine and the air dropping to freezing near the inlet guide vanes.

In flight turn on engine heat whenever icing conditions are anticipated or when *outside ambient* temperature is between +6 and −15°C and visible moisture is present. Even the fluffiest cloud is moisture and a potential icing hazard in the temperature range conducive to ice.

Wing and airframe ice is not much of a problem in jet operation except at slow speed such as approach and landing, noise abatement climb procedures, and holding. Most jet flights are conducted at speeds producing a considerable friction and ram rise and at altitudes well above most icing levels where the moisture present is in the form of ice crystals already frozen. Nevertheless, a jet can still pick up a good load of ice while holding, flying at reduced speeds, and in climb and descent. So knowing how to use the airfoil anti-ice and deice systems is important.

Most jets have "thermal" anti-icing equipment or "heated" wings. This is the most effective system and is also used on such propeller-driven aircraft as the Convair 240, 340, and 440; the Martin 202 and 404; the Douglas DC-6 and 7; and the Lockheed Electra. Some of the smaller business jets (DC-3, Lockheed Constellation) may still be equipped with boots. The first rule in all cases, regardless of the type of equipment, is *do not stay at the icing level.* Get out of it as quickly as possible, but use your equipment properly while in the icing conditions.

If you are equipped with boots, let the ice build up to ¼–½″ and then turn on the boots to crack it off. As soon as the ice is gone, turn the boots off and let it build up again and repeat the process. Leaving the boots on continually creates the possibility of ice building up *around* the boots, causing them to operate ineffectively beneath the ice and giving you a serious problem in additional weight, drag, and change in airfoil.

Turbine-powered aircraft, which produce a great deal of heat in their engines, are designed to be able to bleed off a portion of the heated air from the compressor sections of the engine and use it to heat the wing. This is a very efficient and effective system but will cause a power loss of 3–5% of total thrust, perhaps even more in some engines. This results in a drop in engine pressure ratio, and you'll have to be familiar with your engines and their performance to know the power loss to expect in EPR and deduct it from your normal power settings.

Most of these heated wing arrangements are anti-ice systems, designed to be turned on before entering known or suspected icing conditions to prevent ice from forming. The precipitation striking the heated surface is vaporized and will not form ice. If the tail surfaces are not heated simultaneously (and usually they are not), you will have to use the tail heat periodically, during which period the wings will not be heated because the hot air from the engines is now rerouted from the wings to the tail surfaces. This is the weakness of such a system. Sustained flight in icing conditions will ice the tail. When you turn the tail heat on, you'll be using heated air as a deice system. Sustained flight in icing conditions will ice the tail. When you turn the tail heat on, you'll be using heated air as a deice system to remove the ice from the tail surfaces. Normally this is done with an automatic timer that actuates a valve rerouting the heated air from the wings to the tail for 2 minutes, which is sufficient. But the wings can cool considerably in 2 minutes in icing conditions, particularly at slow speeds such as in holding patterns. While the leading edge is still hot, moisture will not stick; as it cools off and no longer vaporizes the moisture, the freezing precipitation melts slightly and slides back to form a spoiler on the wing. In really heavy conditions, ice will just be forming on the wing when the wing heat comes back on with the automatic timer. As the wing begins to heat again, there is a possibility of ice melting and running back to form a spoiler. This is especially true at reduced power, which also produces less heat. Never stay in heavy icing conditions, and limit the use of tail deice while in them unless a buffet is noticeable.

Do not use spoilers for a speed brake in icing conditions! Avoid the extension of spoilers on top of the wing if at all possible. When spoilers are used as a speed brake in descent in icing conditions, the ice runback builds up beneath the spoilers and does not allow them to fully retract. This seriously increases stall speeds and makes approach and landing more hazardous.

When making an approach in icing conditions with the heated wing, try to stay above the icing condition as long as possible and have the wing well heated prior to descent and during all maneuvering until just prior to the outer marker. At the same time you extend the gear, push the tail heat button, deicing the tail for control function in approach and landing, and expect heat to be back on the wings prior to missed approach altitude.

There is another heated-wing system that is best used as a deicer rather than a anti-icer. The Lockheed L-188 is equipped with such a system. You use it just like boots. Let the ice build up to ¼–½″ and then turn it on. The ice will be gone in less than a minute— about the time it takes you to select the various wing sections on a temperature gauge; then turn it off and let the ice build up again. This is an excellent system unless you leave it on too long. Such a small portion of the leading edge is heated that runback over the wing is almost certain unless the system is turned off as soon as the ice cracks off.

The temperature range for wing ice is the same as for engine heat (about +6 to −10°C with visible moisture present), but the difference is that this is total air temperature, which takes into account ram rise.

Ice is not really a major hazard to flight except in some extreme conditions and at ground levels. Just stay out of it as much as possible, use common sense and your anti-ice and deice equipment judiciously, and you can cope with virtually every situation with ease.

28

Winter Operation

Winter, even to the lucky pilots based in sunny climes, means not only cold weather but all the accompanying problems of flight into the sections of the country afflicted with snow, ice, and freezing rain. Some of the common problems encountered during winter operation will be covered in this chapter; none of this information should be considered as replacing your own manuals but as supplemental information.

Planning

It is apparent that good flight planning is of the essence for any flight in all weather conditions, but this is particularly true for winter. In addition to studying the weather at your departure station, along the route of intended flight, and at your destination, study the latest reports on field and runway conditions at your points of intended landing and at alternates. This is easy in an airline operation; the dispatcher is kept informed of field conditions on the airline system by station personnel at the various fields. The airline pilot has only to confer with the dispatcher and the weather forecaster and to review all current field advisory and operational messages pertinent to the flight. Any station in the system having abnormal conditions during winter (or any time) will file a field report for the benefit of pilots, dispatchers, and any others directly engaged in the movement of aircraft.

Nonairline pilots do not have this advantage. They have to take whatever information they can get, apply their own analysis of probable conditions, and use the landing rules of thumb (Chapter 15) as a yardstick in assessing runway requirements for landing and stopping safely. Pilots should try to determine the follow-

ing information regarding destination and alternate fields:

1. Navaids—Find out if all are operating or which are out of service.
2. Runways—Condition: which ones closed; which open; how plowed; width and length; drifts and windrows; snowbanks at ends and sides; surface conditions such as smooth ice, rough ice, dry snow, wet snow, slush, water, layers, bond, depth and area covered, and braking action reported by aircraft.
3. Taxiways—Conditions as above.
4. Ramps—Same.
5. Lights—Many airports have landing minimums based on availability of various lighting aids. You may need to know the status of all navigation and landing aids including high-intensity runway lights, runway end identification lights, center-line lights, touchdown zone lights, approach lights, and condenser discharge lights (flasher).

Fuel Requirements

An important part of flight planning is fuel requirements. Compute the fuel load you'll need for the flight, including any anticipated delays and fuel to an alternate, and be sure you have enough fuel to conduct your flight with a safe margin. This could be considered a minimum for fuel. Many pilots, after computing minimum fuel for the flight, find they have considerable fuel capacity left and have a tendency to fuel to the maximum allowable for landing at their destination. Sometimes it is cheaper to freight low-cost fuel to a station where fuel costs are higher. However, there is one point I'd like to emphasize regarding freighting of additional fuel in winter. Carrying a considerable amount of fuel above your minimum requirements increases your flare speed or V_{ref} at your destination by increasing your weight. If the runways are long and dry, this presents no problems. But if the field reports indicate the landing runway to be ice covered, if the runway leaves little room to spare in runway requirements, or if the braking action is reported to be anything less than good, a heavier fuel load could be disadvantageous.

Ground Conditions

If your aircraft isn't hangared, a good set of canvas wing covers will save time and trouble in snow and ice removal. However, if snow and ice have accumulated on the airplane, either remove it yourself or watch to be sure it is done properly. Loose and dry snow should be removed from wings and tail surfaces with squeegees or soft-bristled brooms; adhering ice or snow may be removed by the application of hot air, glycol, or water, depending on the existing conditions and the facilities locally available. The main hazard of using water or glycol to remove snow and ice is that the resultant mixture of slush tends to enter openings between fixed and movable surfaces and lodge on critical components, reducing operational clearance. This might cause damage, mechanical failure, or control failure, so use caution.

When using glycol, be very careful that the glycol mixture is not splashed on cabin windows or windshield panels. It leaves an oily film that reduces visibility, but it can also be expensive in that it may cause crazing.

Pitot and static ports should be carefully checked and any ice removed. However, if a deicing solution is used, extreme caution is required to keep the fluids from entering the ports.

Be very careful in using glycol on landing gear. Glycol is detrimental to the life of seals and should never be used on exposed portions of strut pistons and assemblies.

If your aircraft is such that someone must make fuel and oil checks on top of the wing, make every effort to accomplish this check *prior* to glycoling because of the danger of slipping on the freshly glycoled surface. A freshly glycoled wing is even more slippery than ice, and you can hurt yourself badly by slipping off a wing. Professional flight engineers always carry a short screwdriver when on top of the wing under such conditions; if they slip, they can drive the screwdriver into the wing and hang on. It's cheaper to patch a little hole in the wing than to repair a broken leg or back.

After fueling, if the temperature is freezing or below and snow is falling or lying on the wings or in the area of the fuel tanks, *check for ice formation*. If the temperature of the fuel is above freezing, it can cause the snow to melt and then refreeze into a film of ice on the wing around the tanks. The reverse of this condition is also a possibility. Even though the temperature may be slightly above freezing, the temperature of the fuel may be below freezing. If the wings are wet or if rain is falling, below-freezing fuel can cause the temperature of the wing to drop to a point where the precipitation can freeze into a film of ice on the wing. Therefore, it's good practice to always check for ice formation around the fuel tanks after fueling.

Preflight

The airplane preflight inspection varies with different aircraft. In winter operation it should include a careful visual check of wings, control surfaces, and hinge points to determine that deposits of frost, slush, ice, and snow are cleared from these areas before each departure. Particular attention should be given to cove and balance bay areas, junction areas between flight control surfaces and their trim and flight tabs, upper wing surfaces between inboard nacelles and the fuselage, and the wing to fuselage fillet area. These areas should be be thoroughly inspected whenever an airplane has been subject to snow or ice accumulation, whether or not a deicing procedure has been used.

Ice accumulation on the wings in the area between the engine nacelles and the fuselage is very critical in some aircraft. Airplanes like the Martin 404; Convair 440; Douglas DC-4, DC-6, and DC-7; and Lockheed Constellation (probably most multiengine airplanes) may experience a severe buffeting of the tail with ice in this area. Be aware of the possibility of ice melting off the heated cabin, running down into this area of the wing, and refreezing. In addition to the top of the wing, the area between the fuselage and nacelles, particularly the wing fillet, must be absolutely clear of ice before takeoff.

The underside and bottom of wings and surfaces should be checked as well as the top. In checking the under surfaces, be certain that wings, tail surfaces, control surfaces, wing flaps, fillets, spoilers, nosewheel steering cables, etc., are free of frost, snow, or ice deposits.

The Boeing type of aircraft has balance panel bay areas for the control surfaces. Accumulations of ice, snow (wet or dry), and slush must be removed from these balance panel bay areas for all three controls—ailerons, elevator, and rudder. This may be accomplished by the application of glycol or the careful application of heated air. When an accretion is found in the gap between the fixed and movable surfaces, dry snow may be removed by glycol, heated air, or nitrogen directed into the air vent between the fixed and movable surfaces.

Some reciprocating engines have a tendency to load up and foul the plugs while idling on the ground; this may be prevented by leaning them out for ground operation, and in some conditions carburetor heat may also be applied. If carburetor heat has been used on the ground or the engines have been leaned out prior to takeoff, take the heat off and go to rich mixture for a sufficient time before takeoff to allow the automatic mixture control to stabilize at the lower carburetor air temperature in order to prevent lean mixtures during takeoff.

Some aircraft (L-188) have electrically heated propellers. The electrical power applied to the propeller ice protection systems and the heat induced in the propeller heating elements is high enough that circuit failures and separation of the heating elements are not infrequent. For this reason, these systems should *never* be used except when *flying* in weather actually conducive to ice formation. This, of course, does not prevent you from testing these circuits in the ground test procedure, which applies a reduced current to prove the circuit for only a few seconds.

In the event that electrical propeller ice protection systems are inoperative and icing conditions exist on your flight, the aircraft may be cleared for the flight by the application of Icex or some similar product to the prop blades. This is a normal procedure for props electrically deiced on turboprops, but there is no reason that the same procedure can't be used on reciprocating engines with inoperative alcohol deicing systems. Be careful, particularly with turbine engines, that not too much is applied. The surplus could be ingested into the turbine engine and gum up the compressor blades, resulting in a slight power loss and (perhaps more important from a standpoint of operating cost) special cleaning processes to eliminate the gum from the compressor.

Once application of Icex is effective for a total cumulative exposure of 6 hours of precipitation or 50 hours elapsed flight time, whichever occurs first. Too frequent applications are just as detrimental to the engines as an overapplication. When an Icex application is made, the time and date of application should be noted in the aircraft maintenance log. Then flight crew entries should be made in the log of total hours exposure to precipitation during any flight. By noting the time and date of last application and elapsed time or in-flight exposure to precipitation since the application, flight crews will be able to determine when another application is needed.

After any refueling when the temperature is at or near freezing or below, be sure to reinspect the wings and fuel tank area.

Aircraft that have an uplatch engaging lug on the gear should have a careful inspection of the gear struts. On the Lockheed L-188 Electra, for example, ice or slush frozen to the nose gear strut may prevent full nose gear retraction when it contacts the wheel well bumper blocks; as a result, the uplatch may not fully engage. If your study of field conditions prior to takeoff indicates that you will have slush-covered runway, you may reduce ice formation and adhesion during the takeoff roll by applying a liberal coating of thick grease (plain old axle grease at least ¼ inch thick) to the uplatch engaging lugs.

Due to the location of the down lock limit switches, landing gear safety switch, or gear light microswitches, they are exposed to ice formation during operation in slush. They are susceptible to freezing and should be checked and any ice removed prior to takeoff.

Simply stated, when there is a possibility of ice or slush deposits on the landing gear, landing gear latches, and indicating light switches, they should be checked before takeoff to assure proper operation of the gear and its indicating system.

Be sure that all pitot and static heat is operable and all pitot and static system openings are free and clear of any obstructions and covers.

And last, check that plugs have been removed from the intake ducts of jet engines.

Engine Starting

When starting a reciprocating engine that has been cold soaked, be aware that excessive priming may wash oil from the cylinder walls, causing piston scuffing and scarring of the cylinder walls.

When starting jet engines, especially those having more than a single stage of compression or rotors, be particularly vigilant of N1 (first stage of compression) RPM and fuel flow. Lack of N1 rotation could indicate that the lower blade tips of the compressor are locked to the compressor case with ice. *Do not attempt start without an indication of N1 rotation!* A sudden and abnormal fuel flow when the fuel control is actuated is a tip-off of possible frozen entrained moisture in the fuel control. If abnormal fuel flow behavior is observed, abort the start *immediately,* because you will not be able to avoid an overtemperature condition at "light off."

When the season of cold weather approaches, review cold weather starting and warm-up procedures as set forth in the airplane flight manual to help minimize cold weather problems.

There are many different jet engines, with different operating procedures and characteristics, but there are some things regarding fuel control and pneumatic starters that may be common.

In the Pratt and Whitney JT8D engines (commonly used in Douglas DC-9s and Boeing 727s) the starter valves have a tendency to freeze up when moisture is present and temperatures are well below freezing. This is an electrically controlled, pneumatically actuated valve typical of such valves for air-started engines. You probably will never encounter this difficulty with a starter generator system. Many times when this occurs, this valve may be made to function by holding the solenoid energized for an extended period; the heat from the coil melts the ice and the operation of the valve becomes evident by a duct pressure drop and engine rotation. If the valve does not function after holding the start switch energized for a warm-up period of 5 minutes, heat in the form of hot air should be applied by maintenance.

Of course you do not want to exceed starter duty limits, but there is no time limit on the valve and switch—only the starter. The starter isn't turning until the valve opens, and the starter duty limits begin from the moment duct pressure drops and rotation is observed.

Poor or No Throttle Response after Start

To most pilots the fuel control is a miraculous black box that nearly always functions perfectly in response to the fuel control lever and power lever. It senses a number of signals from engine operation and, from these signals and throttle position, controls the proper amount of fuel to the engine. In extremely cold weather there will be internal clearance changes within the fuel control itself as well as in some of the mechanical linkage between it and other portions of the engine. Interference may occur between the speed cam follower lever and the compressor inlet temperature power piston long rod; or ice may form in the burner pressure system. Either occurrence would result in no acceleration of the engine in response to forward movement of the power lever or throttle.

If you should be confronted with a no-response-to-power-lever situation with a cold soaked engine, merely apply fuel heat. Such an application, within the time limits of your particular engine, is usually effective in curing the problem.

This can also happen in flight—usually after long flight-idle descents or due to ice forming in the fuel, bypassing the fuel filter, and entering the fuel control. Fuel heat is again the answer. If this doesn't work, close the throttle and snap the fuel control lever from off to on rapidly to shock the fuel control and then again apply fuel heat.

Ground Operation

Taxiing away from the ramp may be a problem. Pressure decreases the freezing point of water, and due to the pressure a heavily loaded tire exerts on ice directly beneath it, the ice will melt. While the aircraft is parked, water thus formed flows out from under the tire and refreezes. Very heavy airplanes parked on an ice-covered ramp can literally freeze themselves to the pavement.

If you have trouble getting away from the ramp and suspect that your wheels may have frozen themselves to the ramp, try rocking the nosewheel back and forth gently, breaking it lose, while applying *moderate* taxi power. If this doesn't work, you'll have to get the ground crew to pour a bit of glycol on the ice around the wheels to free them. Whatever you do, be patient and do not exceed recommended taxi power. The use of excessive power in leaving the ramp is extremely hazardous, particularly in jet aircraft.

Use caution when taxiing, and remember that nosewheels and brakes are not very effective on slippery surfaces except at slow speeds. If the ramp markings are covered by ice or snow, proceed with extreme caution and have signal crew on the ground when working close to parked aircraft and assorted obstructions such as baggage carts in the ramp area.

Taxiways, even in the best conditions, are all too often marginal in width for the proper maneuvering of large aircraft. Where turns are required, many airports have small radius fillets—too square on the corners and not enough pavement—at the taxiway intersections, which result in the main gear cutting the corner while negotiating the turn and maybe dropping off into the mud. On such taxiways, follow the centerline markings closely and always remember how far back of the cockpit the main gear is traveling in the turns.

A pilot is always responsible for the aircraft at all times, and doesn't always get the expected cooperation and services from airport operators to make the operation safe. These operators clear the runways and taxiways so that aircraft can operate in and out, but areas on either side of taxiways are not always kept clear of obstacles that could interfere with the passage of wing tips, propellers, or engine pods when the outboard wheels of the aircraft are at or near either edge of the taxiway. Snowbanks are the principal offenders and account for sizable damage losses each winter. Obstacles along a taxiway that force a pilot to leave the center line of narrow taxiways sooner or later cause trouble. Always be alert to the possibility of hitting such obstacles; stay in the center of the taxiway as nearly as possible, and report obstacles so that they may be removed as quickly as possible.

On wet, slippery, and icy taxiways *taxi slowly!* Jet aircraft with tail-mounted engines occasionally have not turned on icy taxiways as well as aircraft with wing-mounted engines that may assist the turn. With any aircraft, especially aircraft with tail-mounted engines, taxi slowly and with caution.

Before Takeoff

Be sure you thoroughly understand your airway's traffic clearance before takeoff, and when weather conditions warrant (departure airport in instrument conditions) have your radios set up to facilitate a quick return and instrument approach to the airport in case an emergency situation develops shortly after departure.

Normally, pitot and windshield heat are on for every departure in jet aircraft, but be doubly certain to have your pitot heaters on and the windshield sufficiently heated to keep it clear if conditions should warrant.

If jet engine and nacelle anti-icing (engine heat) should be needed during or shortly after takeoff, it is important that the systems be turned on *before* applying takeoff power or just after the first power reduction. Turning them on while engines are operating at takeoff power may result in serious disturbance of engine operation by upsetting the inlet temperature sensing of the fuel control. Airfoil ice-removing systems should not be operated on the ground. Boots change airfoil characteristics, and thermal anti-icing will overheat the wing without airflow to cool it and will weaken the metal in the structure. However, wing heat and boots may be operated shortly after lift-off if required.

If you are flying a reciprocating engine, cycle the props from high RPM to low and back several times during run-up if the airplane has been standing idle very long in cold weather. This replaces the cold congealed oil in the prop dome with warm oil from the engines, which will be of adequate fluidity to eliminate sluggish propeller operation.

When taking off in slush or water, *be sure* that the maximum slush and water depth limitations for your aircraft are not present and will not be exceeded. The airlines have "snow" committees, composed of check pilots and various ground personnel, who rotate the duty on a 24-hour basis between the various airlines operating out of the airport. This committee inspects and measures the water and slush. They have the authority to suspend airline operations at the airport but not to close or suspend *all* operations; that's a responsibility of the airport manager. Nonairline pilots may be able to get this information from an airline operations office and should be wary of operating at an aiport where airlines have suspended operation. If the station is not a crew base of some airline, the committee will be made up of ground personnel, but they do a good job and may be relied on.

The takeoff limitation chart and instructions for slush, water, wet snow, and dry snow should be reviewed and the performance manual and charts should *always* be checked to determine what weight reduction will be required for the depth of the slush, water, and snow and its effect on performance.

Proper runway alignment, on the center of the runway and headed straight down the middle, should be accomplished before beginning takeoff. If possible, check nosewheel alignment and correct any misalignment as a result of plowing through the slush onto the runway. During takeoff, power application should be made smoothly and evenly to prevent any yaw tendency from unequal thrust.

If is helpful if you know the hydroplaning speed of both the nosewheel and main wheels and realize that control beyond a certain point (above nosewheel hydroplane) will be from aerodynamic forces only. Runway conditions have a marked effect on accelerate/stop performance (reflected in the aforementioned charts and discussed in Chapter 9), so ease off the forward pressure on the nosewheel just a bit, letting it float up naturally on the slush rather than holding it firmly down. This will greatly reduce the bow wave and not throw as big a rooster tail of water and slush, which may enter aft-mounted engines. You might also (depending on the design of your aircraft) be alert to the possibility of the static sources being covered by deflected slush.

Runway clutter such as standing water, slush, and snow can cause airframe damage as well as produce a detrimental effect on takeoff performance. There have been takeoff accidents, usually from an aborted takeoff attempt, where it was found that acceleration was normal at the lower speeds and then fell off considerably at speeds very near V_1 and between V_1 and the desired lift-off speed.

After takeoff from a slush-covered runway, the landing gear should be raised in the normal sequence and not lowered again until the plane is ready to make a landing approach. Instances have occurred wherein landing gear or brake operating difficulties have been encountered following takeoff from a slush-covered runway. In every instance, the pilots involved reported that they had either delayed raising the gear for an extended period (with the thought that airflow over and around the gear would blow off the slush that had accumulated on it) or, while they were at cruising altitude in subfreezing temperatures, had decelerated and lowered the landing gear for the same purpose. It doesn't work! Operating the gear in this manner makes the condition worse rather than alleviating it.

If you suspect that ice or snow has accumulated anywhere on the airplane prior to or during the takeoff, flap retraction should be initiated at higher than normal airspeeds and in small increments, thus reducing the tendency of the aircraft to buffet in some cases or settle as the flaps are raised and providing adequate time for acceleration.

En Route

Stay abreast of the situation and informed of en route, destination, and alternate weather. When the weather is questionable, review your plans with other crew members and keep them clued in on what you're thinking. If advisable, or when the destination goes below landing limits or is likely to prior to your arrival, have your copilot prepare a flight plan to the alternate. Know *exactly* your *maximum* holding time upon arrival at your destination before it will be necessary to proceed to your alternate. "Be prepared." No situation is so bad that it can't get worse, rapidly, and the pilot who thinks and plans ahead will never get caught short.

You need a "legal" alternate, with all the normal requirements for safety, which you can reach if all else fails. However, a "passenger" alternate—another airport of intended destination as near the original destination as possible—is reasonable and considerate.

The conditions required for ice to form on the surface of an aircraft in flight are: visible moisture, *true* outside temperature of between $-4°$ and $+1°C$ (26–34°F), and the surface of the aircraft at a temperature of freezing or below. However, when the aircraft is slowed and the ram rise decreases to a value of 5°C or less, the possibility of wing ice exists when the other requirements are met. Therefore, wing anti-icing should be turned on *before* slowing to flap extension speeds when you are flying through visible moisture and the true outside temperature is within the probable icing range. Watch engines and wings closely for indications of ice in precipitation, and use your ice protection systems in their prescribed manner.

Engine anti-ice is one of the first items on the taxi and before-takeoff checklists for good reason. At a temperature of 6°C and below with visible moisture present, ice may occur. But it is wise to use engine anti-ice while taxiing if the temperature is below 8°C (46.4°F) and visible moisture (fog, rain, snow, wet snow) is present and to use it for takeoff if the temperature is below 10°C. This is based on the following reasoning.

When taxiing in visible moisture and in temperatures of 8°C or below, the inlet guide vanes can be expected to pick up ice. This is also true at any low airspeed and high engine RPM. A jet engine requires vast amounts of air, gulping it in easily from ram air at high speeds, but actually sucking it in at low speeds when it is not adequately supplied by ram air. The suction of air into the inlet duct *increases* with decreasing airspeed, becoming greatest at 0 airspeed or in a static condition. (A static condition may be considered anything less than 60 knots.) The suction causes a drop of the static pressure of the air passing through the compressor inlet, especially in the vicinity of the inlet guide vanes. This decreasing pressure at the inlet is accompanied by a resultant drop in temperature of as much as 6°C (10°F).

This means that if visible moisture is present when an aircraft is operating on the ground (such as taxiing) or at low airspeeds and high RPM (such as during takeoff), ice may form in the compressor inlet even though the ambient temperature is several degrees above freezing. This would become apparent to the unwary pilot about the time the bleed valves close in power application by a compressor stall. Have the heat on at the proper times and spool up the engines; close the bleed valves and let the engines stabilize before releasing the brakes to begin the takeoff roll.

After a flight through icing conditions, the use of anti-ice during inbound taxi is a good practice for three good reasons: (1) It may prevent ice from forming in the inlet, which could lock the compressor blades at the next start; (2) if icing conditions exist during the next flight of the aircraft, it is a required minimum equipment item for dispatch and should be checked; and (3) if found inoperative, it provides time to repair the malfunction with minimum delay.

Icing conditions forming ice at the engine inlet will most likely be noted by a rise in engine pressure ratio and in exhaust temperature, usually accompanied by a loss of thrust and resultant loss of airspeed and possibly by compressor stall.

Inlet ice may form whenever the conditions described above are encountered, either in flight or during ground operation. Therefore, icing conditions should be anticipated and, whenever possible, the anti-icing system should be turned on in advance of flying into icing conditions in order to warm up the engine inlet *before* ice actually commences to form. When in doubt, use engine anti-ice. It is designed to *prevent* ice from forming and not for ice removal.

Always turn the ignition on before turning anti-ice on or off in order to prevent possible flameout due to disrupted airflow.

During a rapid descent through icing conditions, when the aircraft is operating with anti-icing on and at a high airspeed with low RPM on the engines, icing conditions may become severe. It is almost impossible to avoid low-power flight-idle descents, but an intermittent use of high RPM is advisable and will provide enough heat to keep the inlet warm even in severe icing conditions.

Instrument Approaches

The majority of the nation's airports are still served by nonprecision instrument approach facilities. Occasionally, during the winter and other periods of bad weather, this is brought to our attention by crashes on hilltops in procedure turns. The following suggestions are offered to help maintain safety standards.

Make certain before entering an approach control area that each crew member is familiar with the instrument approach procedure to be used and that all understand what they are to do and when and how to do it.

Be sure you know exactly where you are before starting a letdown. If in doubt, don't! You want to land on the airport, not on a hill. Double-check altimeter settings and strictly observe minimum altitudes. Tune both receivers to the ILS (or any other approach facility) and check one against the other. Check the glide slope by altimeters and marker beacons. Have both automatic direction finders tuned to the marker homer and check the localizer deviation by the ADF indication. Above all, don't crowd or go below your minimums unless visual contact is made and landing assured.

Landing under Adverse Conditions

Due to the reduced effectiveness of braking and steering, caution should be used to ensure that sufficient runway is available to stop the aircraft when landing on wet or icy runways. Review your landing performance charts or gross weights manual, but you will not find any information for anything other than wet conditions in an airline operation. The FAA, as previously mentioned, hasn't really considered anything other than wet runways due to the complexity of the problem and the present state of the art in determining braking coefficients in all cases. Here's where a good knowledge of the landing rules of thumb (see Chapter 15) comes in handy.

Know your exact landing weight and the correct approach and landing speeds. A stabilized, well-aligned approach on the numbers is the prerequisite to a good safe landing under adverse conditions.

Be well aligned on the roll-out before applying decelerating devices. Sometimes, in four-engine aircraft it may be desirable to use reverse on either the inboards or the outboards (depending on the aircraft and its characteristics and design features) where excess runway is available. Use reverse smoothly, taking precautions against high asymmetric reverse thrust that may cause directional control problems. Also use reverse thrust with caution when landing on a runway covered with loose snow, because it may raise a cloud of snow ahead of the plane, reducing visibility.

It is strongly recommended that, if possible, landings not be attempted on runways that have a covering of more than 1 inch of water, slush, or wet snow over an appreciable portion of the runway, especially in the portion where your speed will be greatest. (The 1 inch is a general figure; you must know the limitations of your particular aircraft.) Here again, it's difficult to actually know the runway conditions. Common sense and familiarity with the airport are helpful; you may get a tower report that is not official but an educated guess at best; an aircraft report of conditions may be relied upon; and any airline personnel on the field will have made periodic field condition inspections for their flights.

When landing on runways covered with slush, snow, or water, extra care must be taken to prevent damage to the wing flaps. This is one reason for depth limitations, and you might elect to retract the flaps immediately upon touchdown. On the other hand, if it is suspected that there is slush or ice on the flaps, which might damage them on retraction, raise them only to the takeoff position and taxi to the ramp for a visual inspection before fully retracting them. This inspection should include a check for damage to wing flaps and skin surfaces, ice formation in flight controls and their hinges and actuators, blockage of vent holes by ice or snow, and ice formation on wing surfaces and fillets. Then the whole process must start over again from the beginning if the flight is continuing and not terminating. It's not difficult—just use common sense, caution, and good judgment; plan ahead; and always inspect your aircraft yourself, particularly after having ice removed by ground personnel.

29

Thunderstorm Flight

A thunderstorm packs just about every weather hazard known to aviation into one vicious bundle. We all know what thunderstorms look like; much has been written about their mechanics and life cycles. They have been studied for many years; and while much has been learned, studies continue because much is still unknown. Knowledge and weather radar have modified our attitudes, but one rule continues to be firm; any storm recognizable as a thunderstorm should be considered hazardous until some measurements have proven it to be safe. That means safe for you and your aircraft. Almost any thunderstorm can spell disaster for the wrong combination of aircraft and pilot. Thunderstorms, and the multiplicity of hazards they produce, have been the primary cause of many accidents and incidents involving every type of aircraft from very small ones to large transports.

Since I'm not a meteorologist, I will restrict the following discussion of thunderstorm-related hazards to what I have learned by experience (some from others, but mostly my own) and cockpit observation. I'll reduce it to what the pilot needs to know.

Thunderstorm Hazards

TURBULENCE AND WIND SHEAR

Potentially hazardous turbulence is present in *all* thunderstorms, and a *severe* thunderstorm can *destroy* an aircraft. The strongest turbulence within the cloud occurs with shear between the updrafts and downdrafts. Outside the cloud, shear turbulence has been encountered several thousand feet above and 20 miles laterally from a severe storm. A low-level turbulent area is the shear zone associated with the gust front. Often a "roll cloud" on the leading edge of a storm marks the tops of the eddies in this shear and signifies an extremely turbulent zone. Gust fronts often move far ahead (up to 15 miles) of associated precipitation. Figure 29.1 shows a schematic cross section of a thunderstorm with areas outside the cloud where turbulence may be encountered.

It is almost impossible to hold a constant altitude in a thunderstorm, and maneuvering to do so produces greatly increased stress on the aircraft. It is understandable that the speed of the aircraft determines the rate of turbulence encounters and also their relative force. This is why you should slow to the recommended turbulence speed for your aircraft. Stresses are least if the aircraft is held in a constant attitude and allowed to "ride the waves." To date, even with radar and the new devices that show the location and presence of lightning, we have no *sure* way to pick the "soft spots" in a thunderstorm.

Thunderstorms are the primary and most hazardous producers of wind shear. They also produce tornadoes, which I'm sure you will agree should be avoided. I personally experienced a brief encounter with a tornadic tube while riding the jump seat of a DC-8 climbing out of Dulles southbound to Miami. A squall line of severe weather was moving east-northeast at 30 knots; the captain was avoiding the nearest cell by 15 miles to the west and climbing at 250 knots indicated airspeed. When we entered the lower clouds there was moderate turbulence, until suddenly at 4,000′ there was *violent* turbulence and the aircraft rolled rapidly the left to become *inverted*! The captain (an excellent pilot for whom I had flown copilot before upgrading to captain), knowing how to recover from unusual attitudes (Chapter 13), completed the roll to level flight, suffering an altitude loss down to 3,000′. This altitude loss was mostly due to the violent downdraft encountered, and the turbulence again be-

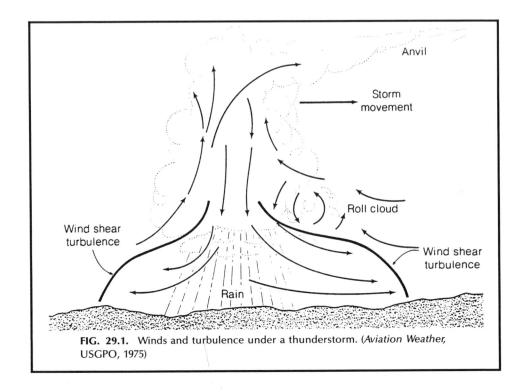

FIG. 29.1. Winds and turbulence under a thunderstorm. (*Aviation Weather*, USGPO, 1975)

came light to moderate. Feeling that no structural damage was done, we continued to Miami. The aircraft was inspected and the flight recorder showed an airspeed loss of 35 knots, −0.6 to +2.3 Gs, but no structural damage was found. Paul Kadlec, a recognized research meteorological scientist who was then the head of Eastern Air Lines weather department, felt that the aircraft had probably barely entered a tornadic tube with its left wing, thus causing the roll.

Another DC-8 was climbing at high altitude and the pilot could see the weather he wished to avoid, both on radar and visually. There was a large cell 10 miles to the right (again moving to the northeast as the aircraft climbed to the west) and a smaller cell to the left. There were lower clouds ahead connecting the two cells. The pilot hoped to clear their tops but entered them just below the peak. Suddenly the indicated airspeed dropped from 280 knots to *zero!* The aircraft had evidently entered the outer edge of a horizontally oriented tornadic tube rotating in such a manner as to put it into an instant tailwind environment equal to if not greater than the indicated airspeed. Violent turbulence was encountered, and then in seconds the plane was in the clear in a vertical dive! Still at climb power the airspeed increased beyond the redline almost instantly. The aircraft would not respond to the elevator, and the pilot reversed all four

engines to slow the speed. Maximum reverse thrust was required; one engine separated from its pylon, and another dangled beneath the wing and flamed out when the fuel line ruptured. The aircraft lost several thousand feet, but speed was slowed below buffet boundaries and a safe landing was made.

I could describe many other such incidents or accidents that were thunderstorm and tornado related; a Lockheed Electra encountering a tornadic tube in *clear* air and disintegrating and at least another half-dozen similar accidents, aircraft crashing on takeoff or landing in or near thunderstorms, and many other aircraft suffering structural damage in thunderstorms. Some either lost control or suffered structural damage to the degree that they crashed; others landed safely. Just realize that what has happened to others may happen to you unless proper avoidance operational techniques are used.

SQUALL LINES

A squall line, the most hazardous combination of thunderstorms, is a band of active storms. Often it develops on or ahead of a cold front in moist unstable air, but it may be too long to detour easily and too wide and severe to penetrate. It often contains steady-state thunderstorms and presents an extreme weather hazard to aircraft. It usually forms rapidly, generally reaching maximum intensity during the late afternoon and the first few hours after darkness. However, some exist through the hours of darkness and for long periods of time.

Squall lines, or very strong frontal activity, are often the subject of severe weather warnings. These will state that a line of severe weather exists from one geographic location to another, moving in a particular direction, and that moderate to severe turbulence will exist within a stated distance either side of it. In preflight planning with such a warning, if your route of flight will be in the area of such a weather line, you may estimate where you will encounter it or plan a route to avoid it if possible. Pay particular attention to its direction of movement. Tornadoes, if they exist—and they virtually always do aloft even though they may not reach the ground—will lie behind the line of weather opposite the direction of movement.

TORNADOES

The most violent or severe thunderstorms draw air into their cloud bases with great vigor. If the incoming air has any initial rotating motion, it forms an extremely concentrated vortex or rotating updraft from the surface well into the cloud. Meteorologists have estimated that wind in such a vortex can exceed 200 knots; pressure in the vortex is quite low. The strong winds gather dust and debris, and the low pressure generates a funnel-shaped cloud extending downward from the cumulonimbus base. If it does not reach the surface, it is a "funnel cloud"; if it touches the ground, it is a "tornado."

Tornadoes occur with both isolated and squall line thunderstorms. Reports or forecasts of tornadoes indicate that atmospheric conditions are favorable for this most violent form of turbulence. Since the vortex extends well into the cloud, any pilot caught flying on instruments in or near a severe thunderstorm could encounter a hidden vortex.

Families of tornadoes have been observed as appendages of the main cloud, extending several miles outward from the area of precipitation and lightning. Thus any cloud connected to a severe thunderstorm carries a threat of violence.

A severe thunderstorm is one whose vertical development penetrates the two primary causes of stopping vertical development in cumulus clouds; the tropopause and/or high-wind environment. The high-wind environment is what blows the anvil out in front of its direction of movement and is also the primary factor in the production of the tornadic tube.

Your knowledge of the existence of the severe storm is obtained, in most instances, by a severe weather warning or from air traffic control pointing out that a cell ahead of your direction of flight extends to an altitude above the level that you know is above the tropopause and also has penetrated high winds in its vertical development. You have a strong and severe thunderstorm ahead of you and it will produce tornadoes and tornadic tubes. Your problem is, therefore, knowing where the tornadic tubes will be and how you can avoid them.

My experience and study of the phenomenon has indicated that all severe thunderstorms do, in fact, create tornadic tubes that may or may not extend to the lower base of the clouds or be visible from the ground. They are born at an average altitude of 25,000–35,000' above ground level. The danger lies in a 45° quadrant behind them, opposite their direction of movement. Do not try to avoid the large cell of the severe thunderstorm by flying in that dangerous quadrant. If you must fly in that area, avoid the severe thunderstorm cell by at least 8 nautical miles at 35,000' and above and by 20–25 nautical miles at 12,000' and below.

ICING

Updrafts in a thunderstorm support abundant liquid; when carried above the freezing level, the water becomes supercooled. When temperature in the upward current cools to about −15°C, much of the remaining water vapor sublimates as ice crystals; above this level, at lower temperatures, the amount of supercooled water decreases.

Supercooled water freezes on impact with an aircraft. Clear icing can occur at any altitude above the freezing level; but at high levels, icing may be rime or mixed rime and clear. The abundance of supercooled water causes very rapid clear icing between 0° and −15°C, and encounters can be frequent in a cluster of cells. Thunderstorm icing can be extremely hazardous.

HAIL

Hail competes with turbulence as the greatest thunderstorm hazard to aircraft. Supercooled drops above the freezing level begin to freeze. Once a drop has frozen, other drops latch on and freeze to it, so the hailstone grows sometimes into a huge iceball. Large hail occurs with severe thunderstorms that have built to great heights. Eventually, the hailstones fall, possibly some distance from the storm core. Hail may be encountered in clear air several miles distant from the dark thunderstorm clouds; usually on the downwind side of the overhang.

As hailstones fall through air whose temperature is 0°C, they begin to melt and may reach the ground as either hail or rain. Therefore, rain at the surface does not mean the absence of hail aloft. You should anticipate possible hail with any thunderstorm, especially beneath the anvil of a large cumulonimbus. Hailstones larger than ½" in diameter can significantly damage an aircraft in a few seconds.

VISIBILITY AND LOW CEILING

Generally, visibility is at or near zero within a thunderstorm. Ceiling and visibility also may be restricted in precipitation and dust between the cloud base and the ground. The restrictions create the same problem as all ceiling and visibility restrictions; but the hazards are increased when associated with turbulence, hail, and lightning that make precision instrument flying virtually impossible.

EFFECT ON ALTIMETERS

Pressure usually falls rapidly with the approach of a thunderstorm then rises sharply with the onset of the first gust and arrival of the cold downdraft and heavy rain showers, falling back to normal as the storm moves on. The entire cycle of pressure change may occur in 15 minutes. If the pilot does not receive a corrected altimeter setting, the altimeter may be more than 100′ off.

Lightning

A lightning strike can puncture the skin of an aircraft and can damage communication and electronic navigation equipment. Lightning has been suspected of igniting fuel vapors, causing explosion; however, serious accidents due to lightning strikes are extremely rare. Nearby lightning can blind pilots rendering them temporarily unable to navigate by instrument or visual reference; it can also induce permanent errors in the magnetic compass. Lightning discharges, even distant ones, can disrupt radio communication low and medium frequencies. Though lightning intensity and frequency have no simple relationship to other storm parameters, severe storms, as a rule, have a high frequency of lightning.

Weather Radar

Weather radar detects droplets of precipitation size. Strength of the radar return (echo) depends on drop size and number. The greater the number and the larger the size, the stronger the echo. Drop size determines echo intensity to a much greater extent than drop number. Hailstones are covered with a film of water and are usually larger than raindrops. The

radar, therefore, sees them as huge water droplets, resulting in the strongest of all echoes.

Measuring such small targets, the radar must be in the contour (CTR) gain position—a preset gain value to correctly measure the drop size to give a return that will indicate a cell to be avoided. A normal gain is a lesser value and will indicate echoes or cells where they, in fact, do not exist, possibly being only very light rain. On aircraft equipped with color radar, the precipitation size and density is displayed in three discrete colors. Red indicates heavy precipitation; yellow, medium; and green, light. Cells indicated to be 75 miles and beyond (in older radars more than 30 miles) are areas of substantial precipitation.

GROUND WEATHER RADAR

Numerous methods have been used in an attempt to categorize the intensity of a thunderstorm. To standardize thunderstorm language between weather radar operators and pilots, a video integrator processor (VIP) is used to determine thunderstorm levels.

The National Weather Service radar operator is able to objectively determine storm intensity levels with VIP equipment and often can determine the height of the storm. These radar echo intensity levels are on a scale of one to six. If the maximum VIP levels are reported to be (Level 1) "weak" and (Level 2) "moderate," then light to moderate turbulence with lightning is possible. Level 3 is "strong," and severe turbulence with lightning is likely. Level 4 is "very strong," and again severe turbulence with lightning is likely. Level 5 is "intense" with severe turbulence, lightning, hail likely, and organized surface wind gusts. Level 6 is "extreme" with severe turbulence, lightning, large hail, extensive low-level wind gusts, and turbulence.

Thunderstorms may build and dissipate rapidly. Therefore, do not plan a course between echoes. Rapid changes occur and you may find yourself trapped.

The best use of ground radar information obtained from air traffic control is to isolate general areas and coverage of echoes. Radar used by control towers and air traffic control centers is set at a threshold of gain to display aircraft return targets; not precipitation echoes. You must avoid individual storms from inflight observations either by visual sighting or by airborne radar. It is better to avoid the whole thunderstorm area than to detour around individual storms unless they are scattered; the wider the better.

Airborne weather avoidance radar is, as its name implies, for avoiding severe weather, not for penetrating it. Whether to fly into an area of radar echoes depends on echo intensity, spaces between the echoes, and the capabilities of your aircraft. Remember that weather radar detects only precipitation drops; it does not detect turbulence. Therefore, the radar scope provides no assurance of avoiding turbulence. The radar scope also does not provide assurance of avoiding intense echoes; a clear area does not necessarily mean that you can fly between the storms and maintain visual sighting of them.

The Do's and Don'ts of Thunderstorm Flying

Above all remember this: *Never regard any thunderstorm lightly,* even when ground radar observers report the echoes are of light intensity. Avoiding thunderstorms is the best policy. Following are some do's and don'ts of thunderstorm avoidance:

1. *Don't* take off or land in the face of an approaching thunderstorm, or even one nearby. Sudden downdrafts or downbursts, horizontal wind changes from outflow, or a gust front of low-level turbulence can cause loss of control.

2. If possible, *avoid* flying under a thunderstorm even if you can see through to the other side. Turbulence and wind shear under the storm could be severe.

3. *Don't* fly without airborne radar into a cloud mass containing scattered embedded thunderstorms. Scattered thunderstorms not embedded usually can be visually circumnavigated.

4. *Don't* trust the visual appearance to be a reliable indicator of the turbulence inside a thunderstorm.

5. *Do* avoid by at least 20 miles any thunderstorm identified as severe or giving an intense radar echo. This is especially true in the tornado danger area behind the thunderstorm movement and under the anvil of a large cumulonimbus.

6. *Do* circumnavigate the entire area containing thunderstorms if it contains storms covering six tenths or more of the area.

7. *Do* remember that vivid and frequent lightning indicates the probability of a severe thunderstorm.

8. *Do* regard as extremely hazardous any thunderstorm with tops 35,000' or higher, whether the top is visually sighted or determined by radar.

Penetration of a Thunderstorm

If you cannot avoid penetrating a thunderstorm, do the following *before* entering the storm:

1. Tighten your seat belt, put on your shoulder harness, and secure all loose items.

2. Plan and hold your course to take you through the storm in minimum time.

3. To avoid the most critical icing, establish a penetration altitude below the freezing level or above the level of −15°C.

4. Verify that pitot heat is on and turn on engine anti-ice. Icing can be rapid at any altitude and can cause almost instantaneous power failure and/or loss of airspeed indication.

5. Establish power settings for the turbulence penetration airspeed recommended in your aircraft manual. Note the pitch attitude at that speed for a target attitude.

6. Turn up the cockpit lights to highest intensity to lessen temporary blindness from lightning.

7. If using automatic pilot, disengage altitude-hold and speed-hold modes. The automatic altitude and speed controls will increase maneuvers of the aircraft, thus increasing structural stress. The newer autopilots can do an excellent job if used properly, and their use is recommended in some aircraft.

8. If using airborne radar, tilt the antenna up and down occasionally. At high altitudes, well above freezing level, tilt it down enough to see the cell well below the freezing level. This will permit you to detect thunderstorm activity and echo cells at altitudes other than the one being flown. Remember, the radar depicts precipitation drops (rain) much better than frozen precipitation other than hail. Also, the "hooks, fingers, and scallops" of possible tornado presence are rarely seen on radar above 12,000'.

During the thunderstorm penetration:

1. *Do* keep your eyes on the instruments. Looking outside the cockpit can increase the danger of temporary blindness.

2. *Don't* change power settings; maintain settings for the recommended turbulence penetration airspeed. Expect and disregard indicated airspeed indications.

3. *Do* maintain a constant attitude; let the aircraft "ride the waves." Maneuvers in trying to maintain altitude significantly increase stress on the aircraft.

4. *Don't* turn back once you are in a thunderstorm. A straight course through the storm will most likely get you out of the hazards most quickly. In addition, turning maneuvers increase stress on the aircraft.

I clearly remember the first and the last thunderstorm I was involved with. So will you. Anyway, I hope the above helps you avoid them and wish you good luck in your next one.

30

Low-Level
Wind Shear

The purpose of this chapter is to provide guidance for recognizing the meteorological situations that produce the phenomenon known as "low-level wind shear." It describes both preflight and inflight procedures for predicting and detecting this phenomenon as well as suggesting pilot techniques that minimize its effects when inadvertently encountered on takeoff or landing. It is based on my own experience of over 50 years of flight and the knowledge acquired from comprehensive study of the subject. I used this material in teaching my 499 students during the 15 years I served as a check pilot and flight instructor for a major airline. Several of them have told me of their encounters with wind shear, but none of them knowingly entered an area of suspected potentially hazardous shear or had an accident resulting from encountering it inadvertently.

Wind shear is any change of wind speed and/or direction over a short distance in the atmosphere and along the flight path. This may be a downdraft or updraft (less than 12′ per second), a vertical wind called a "downburst" (12′ per second or greater), a horizontal change in direction and velocity, or a combination of both. Severe wind shear produces airspeed changes greater than 15 knots or vertical speed changes greater than 500′ per minute.

Under certain conditions, the atmosphere is capable of producing some very dramatic shears very close to the ground; for example, wind direction changes of 180° and speed changes of 50 knots or more within 200′ of the ground have been observed in very short distances. It has been said that wind cannot affect an airplane once it is flying except for influencing drift and groundspeed. However, studies have shown that this is not true if the wind changes faster than the aircraft mass can be accelerated or decelerated.

It may be said then that wind shear is a very important phenomenon that can affect an aircraft when it is most vulnerable, that is, near the ground after takeoff or on landing approach. This term, however, has been misunderstood at times and to some degree because it is used differently in aviation and in meteorology. The pilot needs to know how it may affect the aircraft from a flight operation viewpoint, that is, the aviation effects rather than meteorological analysis.

In aviation the effect of wind shear is experienced as the time variation of the winds encountered and their direction along the flight path. An aircraft may experience difficulty when it encounters a sudden change in the wind, both in direction and speed horizontally or vertically and combinations thereof, that is, its vectorial difference. The vectorial difference of the wind between two points on the flight path is the vector shear. This may be expressed mathematically:

$$\text{Vector wind shear} = \frac{\text{wind B} - \text{wind A}}{\text{flight time between A and B}}$$

Figure 30.1 shows the effect of downburst and outburst upon an aircraft during a final approach to landing. Of these, the most dangerous is the downburst, followed in hazard potential by tailwind and then crosswind bursts, encountered near the ground. Outburst is defined as being the strong outflow created when the downburst hits the ground and spreads out. Figure 30.2 is a pictorial expression of the vector shear, headwind shear, and crosswind shear likely to be encountered by an aircraft flying in or near a downburst. Noting these effects, it should be easy to picture the upper part of Figure 30.1 and realize that the same effects would also occur if encountered shortly after lift-off or at the departure end of the runway.

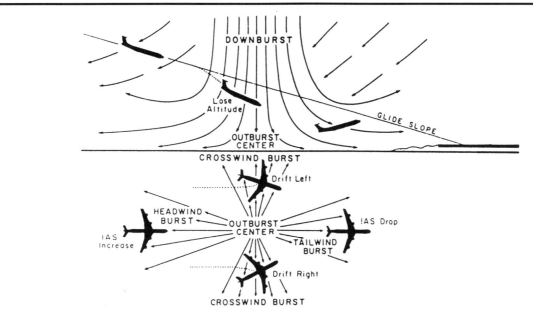

FIG. 30.1. Effects of a downburst and an outburst upon aircraft during a final approach. Of these the most dangerous is the downburst, tailwind burst, and crosswind burst encountered near the ground. Outburst is defined as being the strong outflow created when a downburst hits the ground and spreads out.

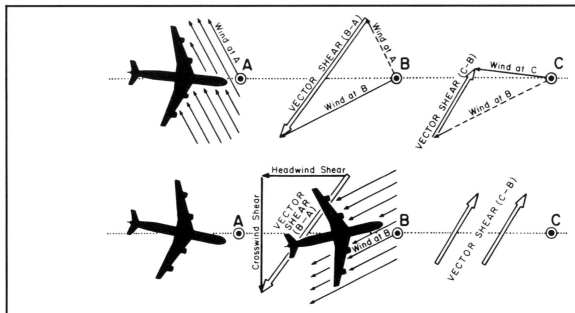

FIG. 30.2. Vector, headwind, and crosswind shear likely to be encountered by an aircraft flying in or near a downburst cell. When the aircraft flies further (A to B, B to C, etc.), the vector wind shear will vary successively.

Effects of Wind Shear

In discussing wind shear effects, I find (much to my amazement) that most pilots don't realize that their vertical speed indicator can be read in knots as well as feet per minute. A nautical mile being 6,080.27′ in length, vertical speed in feet per minute when converted to feet per hour and divided by 6,080.27 equals vertical speed in knots. For example, 700′ per minute × 60 = 42,000′ per hour ÷ 6080.27 = 6.91 knots. Simply round it off to 7 knots.

In most cases, accident reports or illustrations of wind shear indicate downdrafts and downbursts in feet per second. To determine the effect on the aircraft, they must be converted to feet per minute to determine how much effect or what specific effect they instantly have upon vertical speed. Simply stated, every 1′ per second equals 60′ per minute. For example, an aircraft descending at 700′ per minute on approach and encountering an 8.33′ per second downdraft will add 500′ per minute to its vertical speed. The rate of descent instantly becomes 1,200′ per minute rather than the original 700. The vertical speed in knots now becomes 72,000′ per hour or 11.84 (12) knots.

There is a practical application of the above. If you would take the time to use your calculator, you could determine the time to ground impact and/or recognition and recovery from various downbursts in relation to your normal descent rate in approach with an encounter at various heights above ground level. For example, an aircraft descending at 700′ per minute on approach (an average for most air carrier jets) and encountering a 12′ per second downdraft would instantly have an addition of 720′ per minute to the original 700 and would be descending at 1,420′ per minute × 60 = 85,200′ per hour ÷ 6080.27 = 14 knots (14.01253563 to be exact). If encountered at 200′ (0.03289 nautical miles) above the ground, there would be 8.5 seconds to recognize and recover from the downburst and avoid ground contact.

I stated that wind changes in excess of 50 knots or more have been observed within 200′ of the ground. A downburst of this velocity would be 84.44819′ per second, or 5,067′ per minute. Add that to 700′ per minute and you would be descending at 5,767′ per minute, 57 knots; to encounter such a shear at 200′ would give you 2.1 seconds to act.

Much higher velocities may be encountered. On August 1, 1983, the strongest downburst ever recorded was measured at Andrews Air Force Base, with winds exceeding 130 knots and a duration of almost 1 minute. That's a downburst of 219.57′ per second, or 13,173.92′ per minute! This occurred in heavy thunderstorm activity only 6 minutes after Air Force One landed with the president aboard.

Shear may affect an aircraft along any of its three axes. Shear causes action, pilots react, and the result-

ant correction keeps the aircraft on its desired course. If the force of the shear exceeds the capability of the aircraft to maintain its desired flight path, it would experience excessive deviations. Shear forces of such magnitude may be severe enough to induce physical sensations as entry cues to change the direction of flight.

CROSSWIND SHEAR

A sudden change in wind direction and/or speed, such as a "crosswind burst," may momentarily carry the aircraft sideways. This may be recognized by the pilot as a skid or slip. These sensations are not common when flying heavy aircraft and would be an alert to expect displacement from the flight path, from the localizer for example. Continuing an approach would therefore require sufficient altitude to regain the desired flight path.

HEADWIND OR TAILWIND SHEAR

A sudden change in the headwind or tailwind does not affect drift, and heading corrections are not required. Since acceleration and deceleration are common sensations in flying, these cues would possibly not alert a pilot who must recognize headwind or tailwind shear by *sudden* changes in indicated airspeed and vertical speed. Every aircraft has a speed associated with or related to power vs drag, a speed stability. Rapid changes of this type in the mass of air through which the aircraft is flying upset this relationship.

Headwind shear would cause indicated airspeed to momentarily *increase* until the aircraft has reestablished its speed stability. On an ILS approach, it would be necessary for the pilot to lower the nose and make a power reduction to remain on the glide slope; that would be the normal reaction. Since the indicated airspeed dissipates when the aircraft again achieves speed stability, additional power would again be required to recover airspeed to that desired. The ground speed would also decrease and time to the runway would increase. Rate of descent (vertical speed) would also be changed to a lesser value.

Tailwind shear will cause indicated airspeed to *decrease* rapidly and vertical speed to increase. Power must be applied simultaneously with raising the nose of the airplane in order to remain on the glide slope. When desired speed is recovered, the rate of descent will be greater than before and time to the runway decreased due to higher ground speed.

VERTICAL WIND SHEAR

The variation of the up- or downdraft along the flight path is "vertical wind shear," the variation of vertical winds along the flight path. When an aircraft flies into a strong downward current that does not vary horizontally, it will be blown toward the ground. Conversely, an updraft will cause the aircraft to gain altitude. It may be said, therefore, that an aircraft flying into a vertical wind shear will momentarily accelerate in the vertical plane and the vertical speed will change in direct proportion to the velocity of the vertical wind. Again, this may be recognized by the pilot as sensations similar to riding a fast elevator—either a sinking feeling for a downdraft or a heavier feeling for an updraft.

While updrafts can have a deleterious effect in low-altitude encounters during an approach to landing, they are generally considered safer than downdrafts; the aircraft is being blown away from the ground rather than toward it. In the downdraft the pilot would have to bring the nose up and increase power in order to maintain or regain the glide slope in approach or to prevent too much altitude loss shortly after takeoff. A strong- or long-duration downward vertical wind shear would require an unusually high body angle to create sufficient lift to maintain or recover altitude or glide slope. Often a very high body angle would be necessary to even stop descent and enable the aircraft to avoid ground contact or execute a missed approach.

In summary, it may be said that the four most important winds in shear and their effects are:

1. Downburst—The aircraft sinks abruptly.
2. Tailwind shear—Indicated airspeed drops suddenly and the aircraft sinks.
3. Crosswind shear—The aircraft drifts to right or left.
4. Headwind shear—Indicated airspeed increases suddenly and the aircraft gains altitude.

Meteorology of Wind Shear

The most prominent meteorological phenomena that cause low-level wind shear problems are thunderstorms and certain frontal systems at or near the airport.

THUNDERSTORMS

The winds around a thunderstorm are very complex (see Fig. 29.1). Wind shear can be found on all sides of a thunderstorm cell and in the vertical wind of the downdraft or downburst directly under the cell. There will often be several downburst cells in a thunderstorm, creating their own shear lines or gust fronts where their outflows meet. Though Figure 29.1 doesn't show it, as many as six downburst cells have been discovered by mesometeorological (study of weather in small areas) postaccident research.

The wind shift line or gust front associated with thunderstorms can precede the parent storm by 15 nautical miles or more. Consequently, if a thunderstorm is near an airport of intended takeoff or landing, low-level wind shear hazards not only may exist but most probably will.

FRONTS

A front separating two air masses is named by the relative temperature of the air moving in behind the frontal surface. When the two air masses of different temperature, dew point, etc., are brought into juxtaposition, a rather sharp boundary or discontinuity surface is formed between them. There is little tendency for the two air masses to mix. This discontinuity is called a "front" and given different names depending upon its direction of movement. It will always form with the colder, denser air underlying the warmer less dense air. Because the atmosphere is continually in motion, this front will not be a horizontal one but will slope upward from the surface in such a manner that the colder air will underly the warmer air in the form of a long flat wedge. This wedge, the frontal line aloft, rises 1 mile in 50 (1:50) for the fast-moving cold front to 1:150 for the slower moving front. The warm front aloft has a slope of 1:80–1:200.

The winds can be significantly different in the two air masses that meet to form a front. While the direction and velocities of the winds above and below a front may be accurately determined, existing procedures do not provide precise, current measurements of the height of the front above the airport or its near vicinity. However, there is a "rule of thumb" method for determining the height of the front and its associated wind shear with a reasonable degree of accuracy.

Rule of Thumb

The fast-moving cold front will rise approximately 106′ for every mile of movement or 122′ for every nautical mile; $5280 \div 50 = 105.6$, and $6080.27 \div 50 = 121.61$. The fast-moving warm front rises approximately 66′ per mile or 76′ per nautical mile; $5280 \div 80 = 66$, and $6080.27 \div 80 = 76$.

In practical application, round these figures off when departing or arriving at an airport with a front close by. Knowing its position in relation to the airport, you will find that you can determine the height of the front aloft fairly accurately, anticipate your penetration, and be prepared to cope with any shear problems you may encounter.

COLD FRONT

Low-level wind shear occurs during frontal passage and for a short period of time after the front passes the airport. A fast-moving front is a frontal movement of 30 miles per hour or more (26 knots), and it is the fast-moving front that moves the displaced air upward at a faster rate and also produces cumuliform clouds and thunderstorms from the surface front up to possibly 300 miles to the rear. It follows, therefore, that takeoff and landing during frontal passage, or with the front within 10 miles or less from the airport, virtually ensures low-level frontal wind shear in addition to any that may be present from associated thunderstorms.

WARM FRONT

With a warm front the most critical period is before the front passes the airport, but warm-front shear may exist below 5,000′ for up to 6 hours. It has been said that the problem usually ceases to exist after the front passes the airport. However, this is not always true and depends on your direction of flight in relation to the frontal position—particularly from behind the front from a "closed" low-pressure area and when the front has passed but is near the airport. With frontal passage a wind shift is noticed and there is an appreciable temperature rise. Warm fronts move at about half the speed of cold fronts. A warm front moving at 15 miles per hour is a fast-moving front. Also, data compiled on wind shear indicate that the amount of shear in warm fronts is much greater than that found in cold fronts (see Fig. 30.3).

It can be seen that any shear involved in the warm side will be in lower levels for a longer period of time than in the cold front. Figure 30.3 shows an aircraft penetrating the frontal zone from the cold air mass to the warm. Visualize the aircraft turned around and going in the opposite direction, from warm to cold. Turbulence may or may not exist in wind shear conditions in frontal penetration aloft; but if surface wind under the front is strong and gusty, there will be some turbulence associated with it.

To apply the rule of thumb to determine the height of the front, it is necessary that you have some idea of the position of the front for your arrival or departure, its distance from the airport, and speed of movement. This may be estimated from preflight weather analysis, from in-flight weather reports, or by asking air traffic control.

Localized Conditions

STRONG SURFACE WINDS

The combination of strong winds and small hills or large buildings that lie upwind of the approach or departure path can produce localized areas of shear.

Observing the local terrain and requesting pilot reports of conditions near the runway are the best means for anticipating wind shear from this source. This type of shear can be hazardous, particularly to light aircraft.

SEA BREEZE FRONTS

The presence of large bodies of water can create local airflows due to the differences in temperature between land and water. Changes in wind direction and velocity can occur in relatively short distances in the vicinity of airports situated near large lakes, bays, or oceans.

MOUNTAIN WAVES

These weather phenomena often create low-level wind shear at airports that lie downwind of the wave. Altocumulus standing lenticular clouds usually depict the presence of mountain waves, and they are clues that shear should be anticipated.

Detecting Wind Shear

Your aircraft may not be capable of safely penetrating all intensities of low-level wind shear. Pilots, therefore, should learn to detect, predict, and avoid severe wind shear conditions. Severe wind shear does not strike without warning; it can be detected by the following methods:

1. Analyze the weather before flight. If thunderstorms are observed at or near the airport, be alert for the possibility of wind shear in the departure or arrival area.

2. Check the surface weather charts for frontal activity; then check with the forecaster for its estimated proximity to the airport at the time of your departure or arrival, for the speed at which the front is moving, and for the surface temperature difference immediately across the front. A 10°F (5°C) or greater differential and/or a frontal speed of 25–30 knots (12–15 for a warm front) or more is an indication of the possible (virtually certain) existence of significant low-level wind shear.

3. Be aware of pilot reports (PIREPs) and be sure to obtain and read them for your route of flight. Part 1 of the *Airman's Information Manual* recommends (should require) that pilots report any wind shear to air traffic control. This report should be specific in terms and include the loss or gain of airspeed due to the shear, location, and the altitude(s) at which it was encountered. For example: "Atlanta departure, Eastern 301 (type and N number if a private aircraft) encountered wind shear; loss of 30 knots at 2,000 feet."

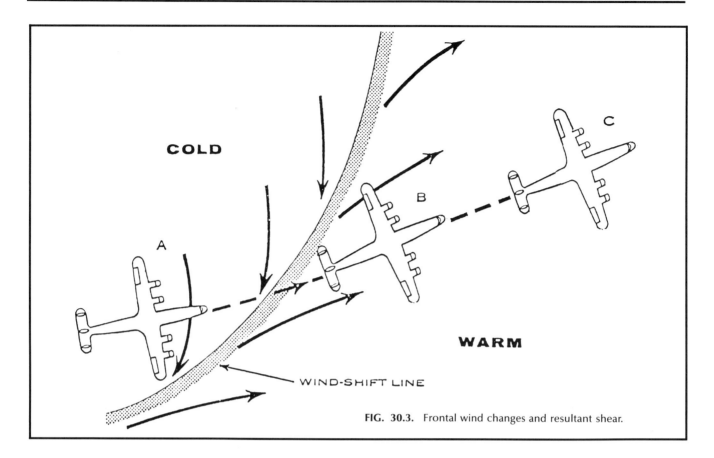

FIG. 30.3. Frontal wind changes and resultant shear.

This simple report is extremely important so that the pilot of the next aircraft can determine the safety of transiting the same area.

Generally speaking, I would recommend that reported shear causing losses in excess of 20 knots be avoided. Reported shears associated with thunderstorms should also be avoided due to the speed with which some storms move across the ground. The storm movement can cause an aircraft to encounter an airspeed increase that may appear harmless, the next aircraft can encounter a severe airspeed loss or a crosswind outburst, and the next aircraft may encounter the very center of a downburst cell.

4. Assume that wind shear is present when the following conditions exist in combination of any two or more:

    a. Extreme variations in wind direction in a relatively short time span.

    b. Evidence of a "gust front" such as blowing dust on the airport surface.

    c. Surface temperature in excess of 80°F (27°C).

    d. Dew point spread of 40°F (22°C).

    e. Virga (precipitation that falls from the bases of high-altitude clouds but evaporates before reaching the ground) is visible.

5. Examine the approach or takeoff area with the aircraft's radar to determine if thunderstorm cells are in the vicinity of the airport. A departure or approach *should not be flown* through or under a thunderstorm cell or even near it.

6. If frontal activity exists, note the surface wind direction and determine the location of the front in relation to the airport. If the aircraft will traverse the front, compare the surface wind to the direction and velocity of the winds above the front to determine potential wind shear during climbout or approach.

7. Use the airplane's instruments to detect wind shear. Pilots flying an aircraft equipped with an inertial navigation system, or any other similar system, should compare the winds at the initial approach altitude (1,500–2,000′ above ground level) with the reported runway surface winds to determine if there is a horizontal wind shear between the airplane and the runway.

Pilots flying airplanes equipped with ground speed readout capability should compare it with its indicated airspeed. Any rapid changes in the relationship of airspeed and ground speed is an indication of wind shear. Some operators have adopted the procedure of not allowing their aircraft to slow below a precomputed or minimum ground speed on approach. The minimum ground speed is obtained by subtracting the reported headwind component from the airspeed on the approach.

Pilots flying aircraft without navigation systems that read out ground speed, true airspeed, wind direction and velocity, etc., should closely monitor their airplane's performance when wind shear is suspected. When the rate of descent on an ILS approach differs from the nominal value for the aircraft, the pilot should be aware of a potential wind shear situation. Since the rate of descent on the glide slope is directly related to ground speed, a higher than expected descent rate would indicate a tailwind; conversely, a low rate of descent indicates a headwind. The power to hold the desired speed on the glide slope will be different from typical no-wind conditions. Less power than normal will be required to hold the glide slope at desired speed when a tailwind is present, and more power is needed for a strong headwind. Aircraft pitch attitude is also an important indicator. A pitch attitude that is higher than normal is a good indicator of a strong headwind and vice versa. By observing the aircraft's approach parameters—rate of descent, power, and pitch attitude—the pilot can acquire a "feel" for the wind encountered. Being aware of the wind-correction angle needed to stay on the localizer provides the pilot with an indication of the wind direction. Comparing wind direction in the initial phases of the approach with the reported surface winds provides an excellent first clue to the presence of wind shear before it is actually encountered.

I find, and again much to my amazement, that many pilots do not know how to determine their "nominal descent rate" (their expected descent rate) in approach. It's indicated on the lower left-hand corner of the ILS approach plate in relation to glide slope angle and ground speed. The typical glide slope angle is 3°, though some have a lesser slope, and you should always check this box in your review of the approach plate and briefing before executing the approach. Consider the reported wind and compute or estimate your ground speed on the glide slope in approach.

The nominal rates of descent on a 3° glide slope at various speeds are:

Ground speed (knots)	70	90	100	120	140	160
Descent rate (fpm)	377	485	539	647	755	862
Rate (rounded)	380	490	540	650	760	860

It's easily seen that a 10-knot change will affect rate of descent by about 50′ per minute. If you established a 140-knot approach speed, you would ex-

pect a rate of descent of about 750′ per minute in a calm condition. With a 10-knot headwind reported on the ground you would expect a 700′ per minute rate, but your vertical speed reads 600′ per minute. You might have had a 10-knot headwind component expected from reported wind, but you are in a 30-knot wind. Using this rule of thumb formula, you may determine the wind you are experiencing and therefore anticipate the wind change you will experience as you enter the environment of the reported surface wind. I'm sure that you can see that a tailwind would increase your rate of descent.

8. Utilize the low-level wind shear alert system (LLWSAS) at airports where it is available. The LLWSAS consists of five or six anemometers around the periphery of the airport that have their readouts automatically compared with the center field anemometer. If a wind vector difference of 15 knots or more exists between the center field anemometer and any peripheral anemometer, the tower informs the pilot of the wind from both directions. The pilot may then assess the potential for wind shear. An example of severe wind shear alert could be the following: "Center field wind is 230° at 7 knots and at the north north end of runway 35 is 180° at 40 knots." In this case, a pilot ready for takeoff on runway 35 would be taking off into an increasing tailwind condition that would result in significant loss of airspeed and, consequently, loss of indicated airspeed and altitude.

Airplane Performance in Wind Shear

The following provides a basis of understanding for the operational procedures recommended.

POWER COMPENSATION

Serious consequences may result on an approach when wind shear is encountered close to the ground when power has already been set for desired indicated airspeed and to compensate for wind. Figures 30.4 and 30.5 illustrate the situations when power is applied or reduced to compensate for the change in aircraft performance caused by wind shear. I always considered the throttles as flight controls in an approach, kept my hand on them in every approach even when using autothrottle, and required all my students and copilots to do the same.

HEADWIND

Consider that you are flying an approach at 140 knots indicated airspeed in a 20-knot headwind environment. Then without warning you encounter an instantaneous wind shear where the headwind shears

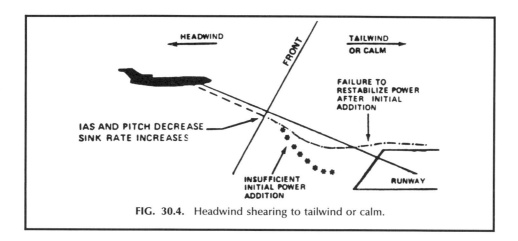

FIG. 30.4. Headwind shearing to tailwind or calm.

away completely. At that instant, several things will happen rapidly and simultaneously. The airspeed will drop from 140 to 120 knots, the nose will begin to pitch down, and the aircraft will begin to drop below the glide slope. The aircraft will then be both "low and slow" in a power-deficient state.

You will probably pull the nose up to an attitude even higher than it was before the shear to recover the glide slope. This will aggravate the airspeed situation until you advance the power levers (the reason you keep your hand on the throttles) and sufficient time elapses for the higher power setting to correct the power deficiency. If the airplane reaches the ground before the power deficiency is corrected, the landing (crash) will be short and very hard.

However, if there is sufficient time to regain the proper airspeed and glide slope before ground contact, then a "double reverse" problem arises. This is because the power is now too high for a stabilized approach in a calm or no-wind condition. So as soon as the power deficiency has been replenished, the power levers should be pulled back even further (slightly)

than they were before the shear and your indicated airspeed should be reduced to very near V_{ref}. The power required for a 3° glide slope in no wind is very much less than that for a 20-knot headwind. If you do not retard the throttles, your airplane will have an excess of power; you will be too high and too fast and may not be able to stop in the available runway length (see Fig. 30.5).

TAILWIND

When in an approach in a tailwind condition that shears into a calm wind or headwind, the reverse is true; the indicated airspeed and pitch will increase while vertical speed decreases, and the aircraft will "balloon" above the glide slope. Some recommend that power should be reduced to correct this on the premise that the approach may be high and fast with a danger of overshooting. I prefer to fly the glide slope with pitch control and recommend dropping the nose slightly to recapture the glide slope with little if any power reduction.

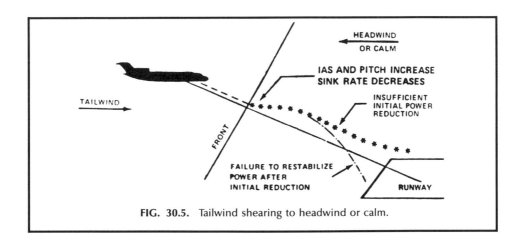

FIG. 30.5. Tailwind shearing to headwind or calm.

I have two reasons for my recommendation: first, if overshooting, a go-around is an easily accomplished and safe maneuver; second, if an initial power reduction is made and the aircraft is back on speed and glide slope, the "double reverse" comes into play once more. An appropriate power increase will be necessary to restabilize in the headwind environment. If this is not accomplished promptly, a high sink rate may develop and your landing again may be short and hard. I simply leave the power on, drop the nose, fly back to glide slope, and then wait for the double reverse before making power changes. Doesn't that make sense to you? This double reverse problem arises primarily in downdraft and frontal passage shears. Other shears may require a consistent correction throughout (see Fig. 30.4).

THUNDERSTORM

The classic thunderstorm "downburst cell" is illustrated in Figure 30.6. There is a strong downdraft (downburst) cell in the center of a thunderstorm cell, often accompanied by heavy rain in this vertical flow of air. As the vertical flow nears the ground, it turns 90° and gives rise to a strong horizontal wind flowing outward radially from the center. Point A in Figure 30.6 represents an aircraft that has not entered the cell's wind flow. The aircraft is on speed and on glide slope. At point B the aircraft encounters an increasing headwind as it enters the outflow from the cell. The airspeed increases and the aircraft climbs slightly above the glide path. Heavy rain may begin very shortly. At point C the situation changes very rapidly. If not fully aware of what is being encountered, the

pilot will react by trying to regain the glide slope by reducing the power and dropping the nose. Then in the very short time span between points C and D, the headwind ceases, a strong downdraft is entered, and a tailwind begins increasing. The power reduction has spooled the engines down, airspeed drops below V_{ref}, and the rate of descent becomes excessive. The pilot is in serious trouble and, depending upon the altitude above the ground, may not be able to recover.

You might notice that a missed approach started at point C or sooner would probably have been successful, since the aircraft is high and fast at this point. It may also be noted that the pilot of an aircraft equipped with ground speed readout would have recognized the telltale signs of a downburst cell shortly after point B—rapidly increasing airspeed with decreasing ground speed.

ANGLE OF ATTACK IN A DOWNDRAFT

Downbursts of falling air in a thunderstorm have gained considerable attention due to their role in wind shear accidents. When an aircraft flies into a downburst, the relative wind shifts so as to come from above the horizon, causing the wind to be vertically oriented to the aircraft to some degree up to and including straight down. This decreases the angle of attack, which in turn decreases lift, and the aircraft begins to sink rapidly. In order to regain the angle of attack necessary to support the weight of the aircraft, the pitch attitude must be significantly increased, often to an attitude that may seem uncomfortable to the pilot. But it is necessary because a normal pitch attitude will result in continued descent. The wing pro-

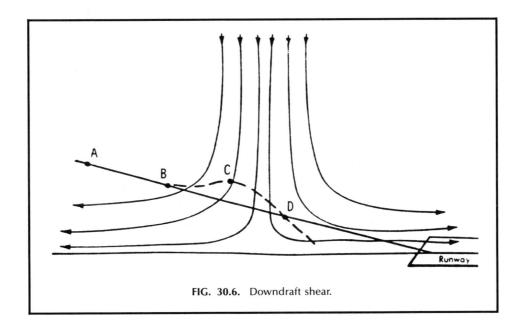

FIG. 30.6. Downdraft shear.

duces lift based on angle of attack, not pitch attitude. It is obvious that a vigorous application of power should accompany the pitch up.

The old "double reverse" then comes into action once more. Caution must be observed when the downdraft has been traversed. If the pilot does not lower the nose of the aircraft quickly as it leaves the downdraft, the angle of attack will become too large and may approach the stall angle; a stall is a function of angle of attack and lift demand on the wing, not airspeed. Recovering from a stall at low altitudes in wind shear conditions is also a dangerous situation.

CLIMB PERFORMANCE

In the takeoff or landing configurations, jet transports climb best at speeds near V_2 and V_{ref} depending upon configuration. Retracting gear and flaps will even improve climb performance if speed permits retraction. However, jet transport airplanes still have substantial climb performance without change of configuration (generally in excess of 700–1,000' per minute) at speeds to stall warning or stick shaker.

ENERGY TRADE

There are only two ways an aircraft can correct for wind shear; there can be an energy change created by changing angle of attack or a thrust change. Most pilots seem to use the thrust change first because they have little idea of how much an energy trade would benefit them. It is the primary reason an additional speed component is recommended for wind, gust, and anticipated wind shear. It gives you something extra to trade for altitude quickly without having to wait for additional speed to be generated by power application. The airspeed over V_{ref} in approach is like having extra money over your basic needs. It is an additional amount you may trade down to V_{ref} and still leave an additional amount before going nearly broke at stall warning or stick shaker. Therefore, a brief glimpse into the energy of flight seems warranted.

The energy of motion (kinetic energy) is equal to one-half mass times velocity squared ($\frac{1}{2} MV^2$), and kinetic energy is directly convertible to energy of vertical displacement, potential energy. Simply stated, airspeed can be traded for altitude and vice versa. Let's see how it works and how it affects the potential to produce the desired effect in vertical displacement when needed.

Mass is the weight of the airplane and V is its velocity or airspeed. An airplane weighing 140,000 pounds, with a V_{ref} of 130 knots and making an approach at 143 knots indicated airspeed, is 10% over V_{ref}. Its kinetic energy would be one-half 140,000 = 70,000 and 143^2 = 20,449, and 70,000 × 20,449 = 1,431,430,000. If the aircraft was at V_{ref}, the kinetic

energy would be 70,000 × 130^2, or 16,900 × 70,000 = 1,183,000,000. The actual kinetic energy at 143 knots, divided by the kinetic energy of V_{ref}, or 1,431,430,000 ÷ 1,183,000,000 = 1.21. It is important to note, as just illustrated, that a 10% increase in velocity (IAS) results in a 20% increase in kinetic energy to trade for potential energy by increasing pitch angle to reduce speed to a desired value. Carrying V_{ref} + 20 knots would create a potential energy of 70,000 × 150^2 or 1,575,000,000. Divide that by kinetic energy at V_{ref} (1,183,000,000) and you would find that the additional 7 knots gives you an additional 11%. You are now 33% over V_{ref}. That's like having $133 when you only need $100 for your basic needs. To go nearly broke at stick shaker (about 3% over actual stall or 103 knots in this case), you would be 46% over that speed and would have an additional $30 to spend before going nearly broke. That's 742,630,000 ÷ 1,183,000,000 = 63% of actual kinetic energy that could be traded in a severe wind shear encounter if necessary.

It may be said then that, independent of its mass, the capability of an airplane to trade airspeed for altitude increases as its airspeed increases. Conversely, a decrease in speed due to an increased angle of attack will increase climb rate and/or decrease descent rate.

TRADING ALTITUDE FOR SPEED

A pilot caught in a low-level wind shear at slower than normal airspeeds—even though at max power—might be tempted to lower the nose and regain speed by trading away altitude. This is trading potential energy for kinetic energy. However, flight research data (simulator and engineering performance analysis) shows that the penalty for doing this can be severe; a large sink rate is built up and a great deal of altitude is lost for a relatively small increase in airspeed. It is best to maintain the lower airspeed and rely on the airplane's climb performance at these lower speeds than to drop the nose and risk ground contact.

If you have a flight director that will attempt to maintain a given speed (such as V_2 + 10, etc.), it will call for trading altitude for speed if the airplane is below the proper speed. This has been observed in simulators during wind shear demonstrations where, following such a procedure, a flight director has resulted in flying into the ground in wind shear encounter. The pilot, not the flight director, should decide if trading altitude for speed is desirable.

TRADING SPEED FOR ALTITUDE

Conversely, a pilot caught in low-level wind shear may pull the nose up and trade speed for altitude, that is, trade kinetic energy for potential energy. If the speed is *above* V₂ or V_ref (as applicable), this trade may well be desirable. If at or below V₂ or V_ref, such a trade should be attempted only in extreme circumstances. In doing so, a pilot is achieving what may be a temporary increase in climb performance. After trading away all the airspeed felt to be safe, the pilot will then be left with a permanent decrease in climb performance. In addition, if ground contact is inevitable after the trade, there may be no airspeed margin with which to flare and soften the impact. However, in approach, wind shear simulations have shown that in many cases trading airspeed down to stick shaker prevented an accident, whereas maintaining V_ref resulted in ground contact.

ADDING SPEED FOR WIND SHEAR

The possibility of having to trade speed for altitude in wind shear makes it attractive to carry a great deal of extra speed in approach. However, on landing, if the speed is not used up in the shear and the airplane touches down at an excessive speed, the airplane may not be able to stop on the available runway, especially if the runway is wet. It has been generally agreed that if a speed margin in excess of 20 knots above V_ref appears to be required, the approach should not be attempted or continued.

DIFFICULTIES OF FLYING NEAR STICK SHAKER

In discussion of trading speed for altitude, I stated that in simulators wind shear accidents had been prevented by trading speed for altitude all the way down to stick shaker. But there are difficulties associated with flying at or near stick shaker that should be discussed.

1. The pilot often does not know stick shaker, that is, the speed at which it may be expected.

2. The stick-shaker mechanism may be miscalibrated or out of calibration, particularly in older airplanes.

3. The downdraft velocity may change (even while the airplane is in it), which often requires another change in pitch attitude.

4. It is hard to fly a precise airspeed in turbulence, which is often associated with wind shear.

5. Pilots have had little training in maintaining flight at or near stick shaker.

The stick-shaker or stall warning is just that, a warning of an approach to a stall, and occurs before the stall. It usually occurs 3–5% above the stall. That is to say, for example, if the aircraft was fully stalled at 100 knots, the stick-shaker warning would function at 103–105 knots.

But you need some method of determining in your own mind the speed at which you could expect the stick-shaker warning to occur, which you can read on your airspeed indicator rather than being entirely dependent on a mechanical device that may fail or be inaccurate, especially if wind shear may be present. That's simple enough. You basically operate in takeoff and landing with two numbers, V₂ and V_ref. V₂ is 1.20% over stall speed for the configuration and V_ref 1.30% over. If wind shear is a possibility, simply do a little arithmetic. Take the percentage over stall, divide the applicable number by that percentage, and add 3. For example, if you had a V₂ of 120 knots, divide that by 1.2 to determine stall speed (100) and add 3 knots. you could expect stick shaker at 103–105 knots indicated airspeed. If your weight is correct, you can make an energy trade of 15–17 knots. If you had a V_ref of 140 ÷ 1.3 = 107.69 (108) knots plus 3 equals 111 knots, you could trade 27–29 knots. This also gives you an idea of how much airspeed you can trade if you have additional speed. If percentages for V₂ and V_ref are different for your aircraft, simply use the correct ones.

General Procedures for Coping with Wind Shear

After the foregoing general discussion of wind shear, it is appropriate to discuss the two specific low-level wind shear situations where an aircraft is most vulnerable, takeoff and landing. Procedures and techniques will be briefly reviewed. This will include the recommendations initially made followed by common practice or by what I have been told by active airline pilots, instructors, and check pilots. It is "off the record" so to speak. It is not in any manual or flight training curriculum that I know of but is a commonsense application of the "numbers" associated with these two phases of flight. Strangely enough, it is exactly what I recommended and taught beginning in late 1974, even before simulator wind shear demonstrations, in-depth studies of the phenomena, bulletins on the subject, etc., became common.

The most important elements for the flight crew in coping with a wind shear environment are *awareness* of an impending wind shear encounter and the *decision to avoid* an encounter or to *immediately respond* if an encounter occurs.

TAKEOFF

If wind shear is expected on takeoff, the PIREPs and weather should be evaluated to determine if the phenomenon can be safely traversed within the capability of the aircraft. This is a judgment item on the part of the pilot and is based on many factors. Wind shear is not something to be avoided at all costs but rather to be assessed and avoided if severe.

An increasing headwind or decreasing tailwind will cause an increase in indicated airspeed. If the wind shear is great enough, the aircraft will initially pitch up due to increase in lift. Do not retrim the airplane. After encountering the shear, if and when the wind remains constant, ground speed will gradually decrease and indicated airspeed will return to its original value. This situation normally leads to increased aircraft performance, so it should not cause a problem if the pilot is aware of how this shear affects the airplane.

The worst situation occurs when the airplane encounters a rapidly increasing tailwind and/or downburst. Taking off into such circumstances will lead to decreased performance. The indicated airspeed will rapidly decrease and so will lift. The airplane will initially pitch down due to the decrease in lift in proportion to the airspeed loss. After encountering the shear, if you have enough altitude and the wind remains constant, ground speed will gradually increase and airspeed will return to its original value. Therefore, when the presence of severe wind shear is suspected on takeoff, the pilot should delay departure until conditions are more favorable.

If the suspected wind shear condition is judged to be safe for takeoff, the pilot should select the safest runway—considering length, wind direction, wind speed, and location of the storm and/or frontal areas— and make a maximum power takeoff using the minimum flaps allowable. After rotation the pilot should maintain lower than normal pitch attitude, a body angle that will result in acceleration to $V_2 + 25$. This speed and configuration should be held until at least 1,000' above ground level. Then (the FAA recommended when this bulletin was published in the 1970s) the normal noise abatement procedure should be flown.

If preflight planning shows that the airplane is limited by runway length or obstruction clearance is a problem, taking off into even a light wind shear using the $V_2 + 25$ procedure should not be attempted. This is because too much thrust is being used for acceleration, resulting in the V_2 flight path falling below the engine-out flight path at V_2. With an engine failure this would give insufficient clearance for an obstacle in close proximity to the departure end of the runway.

The preceding two paragraphs are the old recommendations, and I never used them. Long before they came out, I had adopted a different but somewhat similar operational technique to cope with wind shear.

Taking the last paragraph of the old recommendations first, I simply felt that taking off on a weight-limited runway with suspected wind shear was hazardous to my health. If there wasn't a longer more suitable runway available, I simply delayed departure until the conditions were safe.

I also never used the "drop the nose slightly and attain an additional 25 knots." It has been my experience that an airplane accelerates faster on the ground with a zero angle of attack than it does shortly after lift-off with gear in retraction; operation of the gear doors creates additional drag.

My recommendation, commonly used today by many, was (and is) to use the longest most suitable runway and determine the maximum weight for minimum flaps and use V speeds for that weight (set the speeds for actual weight), add at least 25 knots to V_R for rotation, and/or rotate in the last 2,000' of runway remaining, whichever seems most desirable considering the runway still in front of you. In some instances, staying on the ground until within the last 2,000' of runway may even cause some concern about tire speed.

An example of putting the above into practice could be planning a takeoff from Boston from runway 15R for a Boeing 727-225 with JT8D-7 engines. The runway is 10,090' in length, the wind is from 200° at 15 knots but has changed almost 90° in the past 10 minutes, and the temperature is 86°F. You will also be taking off over water that may be colder than ambient temperature. You have good reason to suspect that wind shear may be present in takeoff and departure. You will have a gross weight of 163,000 pounds and the certificate limit is 173,000 for taxi and 172,000 for takeoff. Your tires are in good condition and are speed rated at 225 miles per hour (195 knots).

Looking at your wind angle chart (Fig. 8.4), you determine you have a 10-knot headwind component and an 11° crosswind. Inspection of the takeoff gross weight charts shows that you could take off with 15° flaps up to 178,400 pounds, even with an obstructed ship channel, and would be climb limited at 180,000 pounds.

The V_1 at 163,000 is 135 and V_2 is 148; at 180,000 they would be 143 and 154 respectively. But if you used 5° flaps, you would have a V_1 of 154 and a V_2 of 164. However, you do not have a 5° gross weight chart for Boston. So I would set the 135 and 148 because those are the actual speeds for the weight and configuration but would accelerate *on the ground* to 160 knots before rotation. Then, runway and obstructions in the channel permitting, I would drop the nose during gear retraction to get at least another 25 knots if possible. That's a lot of excess speed and kinetic energy to trade if wind shear is encountered.

If severe tailwind shear or downburst is encountered on takeoff, the pilot should immediately confirm that maximum thrust is applied and trade the airspeed above actual V_2 for an increased rate of climb. Depending on gross weight, pitch attitudes of 15–22° are to be expected during this energy trade, particularly if a downburst is also present.

At this point, the pilot must quickly evaluate the airplane's performance in the shear and closely monitor airspeed and vertical velocity to ensure that an excessive rate of descent does not develop. If it becomes apparent that an unacceptable rate of descent cannot be prevented at V_2 speed or ground contact appears to be certain at the current descent rate, the pilot should gradually and smoothly increase pitch attitude to temporarily trade airspeed for climb capability to prevent further altitude loss. The trade should be terminated when stick-shaker speed occurs.

The airplane should be held in an attitude that will maintain an airspeed just above where stick-shaker speed was initially encountered. This would be accomplished by a slight reduction of back pressure, since, once again, retrimming is not recommended. A general rule, therefore, is to reduce pitch attitude very slightly when stick-shaker speed occurs. Further pitch reductions in the shear could result in a fast descent rate. As the airplane departs the shear or stabilizes in its new wind environment, the pilot should reduce pitch attitude and establish a normal climb.

APPROACH TO LANDING

Considerations involved in flying an approach and landing or go-around at an airport where wind shear is a factor are similar to those discussed for takeoff. When wind shear analysis of airplane performance indicates that a loss of airspeed will be experienced on an approach, the pilot should add to the V_{ref} as much speed as the loss expected up to a maximum of V_{ref} + 20 based on minimum allowable flaps in the approach. Here again, I recommend determining the maximum weight you could land on the runway in minimum flap configuration. Add the difference between V_{ref} for the actual weight and V_{ref} for maximum weight and then add 20. Set the numbers for actual weight but fly an indicated airspeed as if at maximum weight for the runway.

The pilot should fly a stabilized approach on the ILS localizer and glide slope, using the autopilot when available. In the shear, when airspeed loss is encountered, a prompt and vigorous application of thrust is essential, keeping in mind that if airspeed has previously been added for the approach, the thrust application should be aimed at preventing an airspeed loss below V_{ref}. An equally prompt and vigorous reduction of thrust is necessary after the shear has been traversed and normal target speed and glide path have been reestablished to prevent exceeding desired values. Early recognition of the need for thrust is essential. Along with the thrust addition is a need for nose-up rotation to minimize departure below the glide path. If the airplane is at or below 500′ above ground level and the approach becomes unstable, a go-around should be initiated immediately. Airspeed fluctuations, sink rate or vertical speed, and glide slope deviation should be assessed as a part of this decision.

Your chances of safely negotiating wind shear are better if you remain on instruments. Visual reference through a rain-splattered windshield and reduced visibility may interfere with cues that would indicate deviation from the desired flight path. At least one pilot should therefore maintain a constant instrument scan until a safe landing is ensured.

Some autothrottle systems may not effectively respond to airspeed changes in a shear. Accordingly, the thrust should be monitored closely (read that as keep your hand on the throttles). If autothrottles are used, be alert to override the autothrottles if response to increased thrust commands are too slow. Conversely, thrust levels should not be allowed to get too low during the late stages of an approach, since this will increase the time needed to accelerate the engines.

Should a go-around be required, the pilot should initiate a normal go-around procedure, evaluate the performance of the aircraft in the shear, and follow the procedures as outlined in Chapter 17.

CRITICAL STEPS

The following describes the critical steps in coping with low-level wind shear:

1. *Be prepared.* Use all the available forecasts and current weather information to anticipate wind shear. Also, make your own observations of thunderstorms, gust fronts, and telltale indicators of wind direction and velocity available.

2. *Giving and requesting PIREPs is essential.* Request them and report anything you encounter.

3. *Avoid known areas of severe shear.* When the weather, PIREPs, or your own observations indicate that severe wind shear is likely, delay your takeoff or approach.

4. *Know your aircraft.* Monitor the aircraft's power and flight parameters to detect the onset of a shear encounter. Know the performance limits of your particular aircraft so that they may be called upon in such an emergency situation.

5. *Act promptly.* Do not allow a high sink rate to develop when attempting to recapture a glide slope or to maintain a given airspeed. When it appears that a shear encounter will result in a substantial rate of descent, promptly apply full power and arrest the descent with a nose-up pitch attitude.

TAKEOFF

1. Use *maximum* takeoff thrust rather than reduced thrust.

2. Use the *longest* runway available with the least possibility of wind shear encounter. (Remember previous configuration and speed recommendations.)

3. *Do not* use the flight director. Use it as an attitude indicator only.

4. Check maximum weight for takeoff on the departure runway (wind-temperature adjusted weight

using zero wind and the climb limit). Find the takeoff speeds associated with the maximum takeoff weight determined and the performance-limited V_R for the weight. Compare actual gross weight V_R with the performance-limited V_R. Takeoff should *not* be attempted unless the performance-limited V_R *exceeds* the actual gross weight V_R by at least 10 knots. The selected *higher V speeds should be used for takeoff.*

5. Monitor airspeed closely during the takeoff roll. Be alert for any airspeed fluctuations, which are the first indication of wind shear. If wind shear is detected prior to rotation, the takeoff should be aborted.

6. At V_R, use the normal rotation rate toward the normal takeoff attitude.

7. Know the all-engine initial climb pitch attitude. Minimize reductions from the initial climb pitch attitude until terrain and obstruction clearance is ensured unless stick shaker (stall warning) activates.

8. If the airspeed should fall below the trim airspeed, unusual control column forces may be required to maintain the desired pitch attitude. Do not retrim. Stick-shaker or stall warning *must* be respected at all times.

9. On a weight-limited runway with wind shear totally unsuspected, if wind shear should occur or be encountered at or near V_R and the airspeed suddenly decreases, there may not be enough runway left to accelerate back to the normal V_R. Abort if possible. If there is insufficient runway left to stop, initiate a normal rotation of at least 1,000–2,000′ (depending on the aircraft and your knowledge of its performance capabilities) before the end of the runway even if the airspeed is low. You may expect that higher than normal attitudes may be required to lift off in the remaining runway. This is a critical situation. It is to be hoped that your correct use of avoidance analysis and procedures will ensure that the condition is only an isolated local occurrence and no further shear will be encountered.

10. Crew coordination and awareness are very important. Develop an awareness of normal values of airspeed, attitude, vertical speed, and airspeed buildup. Closely monitor vertical flight path instruments—vertical speed, altimeters, and airspeed. The pilot *not* flying should be particularly aware of vertical flight path instruments, especially rapid airspeed changes, and *call out* any deviation from normal or expected indications.

APPROACH AND LANDING

1. Select the minimum landing flap configuration consistent with field length.

2. Add an appropriate airspeed correction not to exceed 20 knots to V_{ref}. (Remember my recommendation of assuming you are at maximum weight for the runway and add 20 knots to the V_{ref} for that weight.)

3. Avoid large thrust reductions in regard to sudden airspeed increases, since these may be followed by airspeed decreases.

4. Cross-check flight director commands using vertical flight path instruments.

5. Crew coordination and awareness are always of prime importance, particularly at night or in marginal weather conditions. Closely monitor vertical flight path instruments such as vertical speed, altimeter, and glide slope displacement. Use of the autopilot and autothrottle during the approach may provide more monitor and recognition time. The pilot *not* flying should *call out* any deviation from normal or expected performance indicated.

Recovery from Wind Shear

TERRAIN AVOIDANCE

Recovery from wind shear at low levels is in reality a situation requiring a "terrain avoidance" maneuver. The flight technique, the manner in which terrain avoidance is accomplished, should be reviewed by pilots until it may be immediately accomplished by recall whenever the threat of ground contact exists.

Any of the following conditions are to be regarded as presenting a potential for imminent ground contact:

1. Activation of the PULL UP warning.

2. Below 500′ above ground level, inadvertent wind shear encounter or other situations resulting in marginal flight path control. In general, marginal flight path control may be recognized as uncontrolled changes from normal steady-state flight conditions in excess of the following: (a) 15 knots indicated airspeed (decrease), (b) 500′ per minute vertical speed (increase), (c) 5° pitch attitude (nose down), and (d) one dot displacement from glide slope (low).

CORRECTIVE ACTION

Disengage the autothrottle and aggressively position the thrust levers forward, commanding, "Takeoff power!" to ensure that maximum thrust is obtained. The pilot *not* flying should ensure maximum thrust.

Simultaneously disengage the autopilot and rotate smoothly at a normal rate toward an initial pitch attitude of 15°. Fly pitch attitude, not flight director commands. Pitch attitude in excess of 15° may be required to silence the PULL UP warning and/or avoid terrain contact. Stop rotation immediately if stick shaker, stall warning, or initial buffet of an impending stall occurs. In all cases, the pitch attitude that results in stick shaker is the upper pitch attitude limit. This may be even less than 15° in a severe wind shear encounter.

SIMULATOR EXPERIENCE

There may be those who disagree with some of this material, for example, using operational speeds for maximum takeoff and landing weights when the actual weight is lower or when gaining 25 knots over V_R on the ground and then, if runway distance and terrain and obstruction clearance permit, reducing pitch attitude after lift-off to gain another 25 knots. But I didn't arrive at my conclusions arbitrarily. Initially, I based them on my own experience, hangar flying sessions with others who had encountered unusual wind situations, extensive study of weather, and every bulletin published on the subject. Later, I included study of the two accidents that provided comprehensive data on the subject.

My findings and operational techniques for avoiding and coping with wind shear were a year in advance of data obtained from the two accidents that brought extensive research into the subject—Eastern's Flight 66, June 24, 1975, a landing accident; and Continental Flight 426, August 7, 1975, a takeoff accident. My conclusions were later confirmed when a statement was published stating, "In several wind shear accidents (both takeoff and approach), the National Transportation Safety Board has found that the full performance of the aircraft was not used following a severe wind shear encounter. Postaccident flight studies have shown that under similar circumstances, if flight techniques of an emergency nature had been used immediately, the aircraft could have remained airborne and the accident(s) averted."

I agree with that statement. I thoroughly studied the accident reports of all wind shear–related accidents, particularly the Eastern and Continental reports, and found that if the recommendations and procedures as described herein had been implemented by the pilots of those flights the accidents would not have occurred. In my 53 years of flight experience (38 when I first researched this subject), when encountering varying wind conditions and shear, I found that the procedures I recommend are both safe and effective when used judiciously.

I proved this when I flew the Flight 66 profile in the simulator. Boeing, Eastern (I wasn't one of this official group), the FAA, and the NTSB had flown 54 approaches with 18 crashes, 31 missed approaches, and "five approaches were flown successfully placing the simulator over the runway threshold in a position from which a landing could be attempted." Flying at the airspeeds that the flight used, I went around when my airspeed suddenly dropped 15 knots at 250'. Using the speeds and procedures I recommend, even though a go-around was indicated, I was able to land.

I also reviewed the Continental 426 accident report and found numerous definite indications of wind shear; temperature in excess of 80°F, thunderstorms present, gust fronts, pilot reports, tower report of "there have been reports of pretty stout up- and downdrafts at two to three hundred feet." The crew acknowledged the information and took off. Frankly, even then, I would have delayed my departure until more suitable conditions existed.

I programmed a simulator, based on data from the accident report, and did my own simulator research. I entered an instant tailwind component equivalent to the wind environment that Continental 426 encountered, to begin at 100' and end at 300' (the highest altitude reported by the pilots of the three previous departing flights when they encountered the shear). I set all the parameters: field elevation, 5,290'; temperature, 94°F; gross weight, 158,000; configuration, flaps 5°; V_R 143 and V_2 153; a surface wind with a tailwind component of 7 knots; crosswind component of 11 knots. They had an available runway of 11,500'; I used 10,000'.

They rotated 40 seconds after the beginning takeoff roll and lifted off in 5,260' from the threshold of runway 35L. They had 6,240' of runway still in front of them and heavy rain just as they rotated. They had an indicated airspeed of 158 knots and $V_2 + 5$ at 100' when they encountered the shear. The airspeed decreased to 116 knots, a 42-knot loss, in 5 seconds—an astounding 8.4 knots per second. The aircraft struck the ground on the right shoulder of the runway 390' short of the end.

Flying a normal takeoff and using the same procedure they used, I crashed the simulator. Using my procedures, accelerating to 168 knots for rotation, and then dropping the nose slightly to a position where I could see the end of the runway, I accelerated to 195 knots in 6 seconds and still had almost 2,000' of runway in front of me. I also lost 42 knots in about the same period of time, indicating normal V_{ref} of 153 knots; but 8.4 seconds is a long time when you are 42 knots over normal V_2. I flew out of the shear at V_1, which was 143 knots.

My job category at the time gave me the opportunity to "play" with the simulators when they were not in use. I experimented with these, replaying reported accidents and incidents to formulate my own conclusions and to develop and prove my theories of coping with wind shear, at least in simulators, which are obviously much safer than actual encounters in an aircraft.

All of the pilots to whom I taught these procedures, my students, and the many airline pilots I've had as copilots have found them effective. None of them have been involved in a wind shear accident. I hope those of you who read this material will also never have an accident as a result of a wind shear encounter.

Rain

Thus far I have discussed as thoroughly as possible the effects of wind and how it can affect an aircraft when it changes suddenly, causing "wind shear" and its effects. All of this is true and proven by engineering analysis and flight simulator study, as I stated. But I often wondered why another factor was not also considered in the studies—rain and its effects. Rain has weight and volume and falls at different rates to create forces that surely would affect aircraft performance upon sudden entry. In the accidents I studied and for which I flew simulator profiles to test my theories, wind alone was entered. Rain was involved in the two accident cases (Eastern 66 and Continental 426). The accident research, however, totally neglected this factor, which produces significant aerodynamic penalties and may have overestimated the effect of wind shear, particularly as a cause of thunderstorm-related accidents. I feel that rain-induced degradation of performance, associated with and above that induced by wind shear, may have made it highly improbable if not impossible for the pilot to have prevented the accident. Perhaps a look at rain and its effects is also pertinent.

I'm not an engineer, I'm just a country boy who has a predilection for observing, researching, and experimenting with flight procedures to prevent encounters with potentially hazardous situations and to learn how to cope with them if inadvertently encountered. That's why I advise that you especially remember never to take off or make approaches or land in, through, or even *near* thunderstorm activity. When you are near thunderstorms, you will encounter wind changes from their outflows. Flying into their cells, their rainfall, you will encounter a downdraft and/or downburst. If the rain is falling, the air is falling at the same rate, causing the wind shear we've discussed. I feel certain that the effect of rain is a hazard equal to or even greater than the wind shear alone.

When looking into some of the accident and incident reports, where the wind shear characteristic was reconstructed by analyzing and extrapolating flight recorder data, the findings cited wind shear as the cause when I believe that equally significant aerodynamic factors were also involved—those caused by rain. To the best of my knowledge, little if any consideration has been given to the effect of rainfall. Wind shear was determined to be the culprit in every case. Shear was involved, but I couldn't help but wonder if the rain could also have an effect. I think that rain is at least as significant as wind shear. In some of these accidents it could have been the total cause.

When you take a shower, you feel the pressure of the water on your body. If someone turns a water hose on you, you feel the pressure. If you flew an airplane through or into a waterfall, it too would feel the pressure and descend. Therefore, in addition to the wind effects we must also consider the aerodynamic effects of the rainfall.

IMPACT

I do not know the force of impact of rain colliding with the frontal surfaces of an aircraft, but it will induce at least a momentary moment and slow the indicated airspeed somewhat.

MOMENTUM LOSS

Both horizontal and vertical loss will occur from the impact of rain on aircraft. The volume of water in any given rain condition is uncertain, but I believe that a jet transport in the landing configuration, even when carrying additional speed to compensate for wind shear, will experience a significant horizontal momentum loss when encountering heavy rain that can bleed off airspeed. If this is true, some of the speed carried for shear will be lost and you may not have enough kinetic energy left to cope with the shear that is present.

WEIGHT INCREASE

It is obvious that a film of water will form on the aircraft and increase its weight. This should be negligible, perhaps 1–2%. However, the force of the falling rain, vertically oriented, will exert a downward moment in relation to its volume and rate.

AERODYNAMIC EFFECT

This film of water will roughen the airfoil and change the airflow pattern around the wing, fuselage, and all the aircraft surface. A roughened airfoil will move the center of lift, the transition point on the airfoil, toward the leading edge, thus increasing drag and decreasing the angle of attack at which maximum lift occurs. A light rain may increase drag 5%, moderate 20%, and heavy up to 30%. That will certainly decrease performance capability appreciably, leaving you less capability to cope with wind shear.

SUMMARY

I have every reason to believe that in the future the FAA will employ a multimillion dollar wind shear detection device. A doppler radar could measure and determine rainfall density and rate. There may even be the possibility of wind-sensing equipment, a sort of LLWAS, that works at higher altitudes. The vast majority of wind shear accidents have occurred in thunderstorms with heavy rain. What I am saying then is that we may not need all this elaborate and costly wind-sensing equipment if pilots will keep their aircraft out of heavy rain on takeoff or final approach.

I have always felt that pilots should strive to acquire knowledge of the environment in which they travel—the atmosphere and the effects of the meteorological phenomena within it—just as a good sailor studies the sea and the weather and its effects upon the vessel. I have tried to make clear to you that flight instruments can indicate wind changes causing shear and to emphasize how you can predict, recognize, and avoid wind shear. Heavy rain is a lot easier to see than wind. Airborne radar and the mark two eyeball can certainly spot heavy rain. Just do not take off into or make approaches into and through heavy rain. It's that simple. You don't want your aircraft embedded in a heavy rainfall cell, experiencing its effects and at the same time also experiencing the effects of wind shear.

I always added extra speed even if making an approach or taking off into light or moderate rain, even if the surface wind was light. Even if you can see through the rain shower on approach and the runway is in sight, add another 10 knots to your wind and gust factor. If rainfall is moderate, add 20 knots; if heavy, don't attempt to land.

Note: Research on rain and its effects is being conducted by Dr. James Luers at the University of Dayton, Dayton, Ohio, using a wind tunnel. The effects noted so far are very near those described by me, but the research is far from complete; much remains to be learned.

INDEX